Progress in Systems and Control Theory
Volume 19

Series Editor
Christopher I. Byrnes, Washington University

Optimal Design and Control

Proceedings of the Workshop on Optimal Design and Control Blacksburg, Virginia April 8-9, 1994

Jeffrey Borggaard, John Burkardt,
Max Gunzburger, and Janet Peterson
Editors

Springer Science+Business Media, LLC

Editors:

Jeffrey Borggaard
John Burkardt
Max Gunzburger
Janet Peterson
Air Force Center for Optimal Design and Control
Interdisciplinary Center for Applied Mathematics
Virginia Tech
Blacksburg, Virginia 24061-0531

Library of Congress Cataloging In-Publication Data

Workshop on Optimal Design and Control (1994 : Blacksburg, VA.)
Optimal design and control : proceedings of the workshop on Optimal Design and Control, Blacksburg, Virginia, April 8-9, 1994 / Jeffrey Borggaard...[et al.], editors.
p. cm. -- (Progress in systems and control theory ; v. 19)
Includes bibliographical references.
ISBN 978-1-4612-6916-8 ISBN 978-1-4612-0839-6 (eBook)
DOI 10.1007/978-1-4612-0839-6
1. Automatic control--Congresses. 2. Mathematical optimization--Congresses. I. Borggaard, Jeffrey, 1964- . II. Title.
III. Series.
TJ212.W675 1994 95-10230
629.8--dc20 CIP

Printed on acid-free paper

Originally published by Birkhäuser Boston in 1995
Softcover reprint of the hardcover 1st edition 1995

ISBN 978-1-4612-6916-8

Typeset by the editors

CONTENTS

CONTRIBUTORS

Richard J. Adams – Flight Dynamics Directorate, Wright Laboratory, Wright-Patterson Air Force Base OH 45433-7531, USA

Natalia Alexandrov – Institute for Computer Applications in Science and Engineering, Mail Stop 132C, NASA Langley Research Center, Hampton VA 23681-0001, USA

Eyal Arian – The Weizmann Institute of Science, Rehovot 76100, Israel and Institute for Computer Applications in Science and Engineering, Mail Stop 132C, NASA Langley Research Center, Hampton VA 23681, USA

Jasbir S. Arora – Optimal Design Laboratory, Civil and Environmental Engineering, Mechanical Engineering, The University of Iowa, Iowa City IA 52242, USA

Siva S. Banda – Flight Dynamics Directorate, Wright Laboratory, Wright-Patterson Air Force Base OH 45433-7531, USA

Jordan M. Berg – Flight Dynamics Directorate, Wright Laboratory, Wright-Patterson Air Force Base OH 45433-7531, USA

John T. Betts – Mathematics and Engineering Analysis, Research and Technology Division, Boeing Computer Services, P.O. Box 24346, MS 7L-21, Seattle WA 98124-0346, USA

Jeffrey Borggaard – Air Force Center for Optimal Design and Control, Interdisciplinary Center for Applied Mathematics, Virginia Tech, Blacksburg VA 24061-0531, USA

Kathryn E. Brenan – The Aerospace Corporation, 2350 E. El Segundo Blvd., El Segundo CA 90245, USA

John Burkardt – Air Force Center for Optimal Design and Control, Interdisciplinary Center for Applied Mathematics, Virginia Tech, Blacksburg VA 24061-0531, USA

John Burns – Air Force Center for Optimal Design and Control, Interdisciplinary Center for Applied Mathematics, Virginia Tech, Blacksburg VA 24061-0531, USA

Eugene Cliff – Air Force Center for Optimal Design and Control, Interdisciplinary Center for Applied Mathematics, Virginia Tech, Blacksburg VA 24061-0531, USA

Michel C. Delfour – Centre de Recherches Mathématiques et Département de Mathématiques et de Statistique, Université de Montréal, C.P. 6128 Succ. Centreville, Montréal, Québec, H3C 3J7, Canada

J. E. Dennis, Jr. – Department of Computational and Applied Mathematics, Rice University, P. O. Box 1892, Houston TX 77251-1892, USA

Richard H. Fabiano – Department of Mathematics, Texas A&M University, College Station TX 77843, USA

Manfred Fruth – Institute of Scientific Computing, Technische Universität Dresden D-0-8027 Dresden, Germany

P. Gilmore – National High Magnetic Field Laboratory, Florida State University, Tallahassee FL 32306-3016, USA

Andreas Griewank – Institute of Scientific Computing, Technische Universität Dresden D-0-8027 Dresden, Germany and Mathematics and Computer Science Division, Argonne National Laboratory, Argonne IL 60439-4844, USA

Max Gunzburger – Air Force Center for Optimal Design and Control, Interdisciplinary Center for Applied Mathematics, Virginia Tech, Blacksburg VA 24061-0531, USA

Wayne P. Hallman – The Aerospace Corporation, 2350 E. El Segundo Blvd., El Segundo CA 90245, USA

Jaroslav Haslinger – KFK MFF UK, Ke Karlova 5, Faculty of Mathematics and Physics, Charles University, 12000 Prague, The Czech Republic

H. Heidemüller – GIWEP, Tristanstr. 12, D-45473 Mülheim (Ruhr), Germany

W. Heyman – Department of Applied Mathematics, University of Virginia, Charlottesville VA 22903, USA

Dale A. Holtz – Optimal Design Laboratory, Civil and Environmental Engineering, Mechanical Engineering, The University of Iowa, Iowa City IA 52242, USA

Kazufumi Ito – Center for Research in Scientific Computation, North Carolina State University, Raleigh NC 27695-8205, USA

H. Jäger – Universität Trier, FB IV – Mathematik, D-54286 Trier, Germany

C. T. Kelley – Department of Mathematics and Center for Research in Scientific Computation, North Carolina State University, Raleigh NC 27695-8205

Hongchul Kim – Interdisciplinary Center for Applied Mathematics, Virginia Tech, Blacksburg VA 24061-0531, USA

H. Klammer – GIWEP, Tristanstr. 12, D-45473 Mülheim (Ruhr), Germany

Narendra N. Kota – Optimal Design Laboratory, Civil and Environmental Engineering, Mechanical Engineering, University of Iowa, Iowa City IA 52242, USA

I. Lasiecka – Department of Applied Mathematics, University of Virginia, Charlottesville VA 22903, USA

Hyung-Chun Lee – Air Force Center for Optimal Design and Control, Interdisciplinary Center for Applied Mathematics, Virginia Tech, Blacksburg VA 24061-0531, USA

James C. Malas III – Materials Directorate, Wright Laboratory, Wright-Patterson Air Force Base OH 45433-7531, USA

F. Maschler – GIWEP, Tristanstr. 12, D-45473 Mülheim (Ruhr), Germany

C. T. Miller – Department of Environmental Sciences and Engineering, 105 Rosenau Hall, University of North Carolina, Chapel Hill NC 27599-7400, USA

Perry Newman – Mail Stop 159, NASA Langley Research Center, Hampton VA 23681-0001, USA

Janet Peterson – Air Force Center for Optimal Design and Control, Interdisciplinary Center for Applied Mathematics, Virginia Tech, Blacksburg VA 24061-0531, USA

David L. Russell – Department of Mathematics, Virginia Tech, Blacksburg VA 24061-0123, USA

E. Sachs – Universität Trier, FB IV – Mathematik, D-54286 Trier, Germany

Jeffrey S. Scroggs – Center for Research in Scientific Computation, North Carolina State University, Raleigh NC 27695-8205, USA

Ajit Shenoy – Air Force Center for Optimal Design and Control, Interdisciplinary Center for Applied Mathematics, Virginia Tech, Blacksburg VA 24061-0531, USA

Shlomo Ta'asan – The Weizmann Institute of Science, Rehovot 76100, Israel and Institute for Computer Applications in Science and Engineering, Mail Stop 132C, NASA Langley Research Center, Hampton VA 23681, USA

Hien T. Tran – Center for Research in Scientific Computation, Box 8205, North Carolina State University, Raleigh NC 27695-8205, USA

G. A. Williams – Department of Environmental Sciences and Engineering, 105 Rosenau Hall, University of North Carolina, Chapel Hill NC 27599-7400, USA

G. Woelk – Rhein.-Westf. Technische Hochschule Aachen, Lehrgebiet Industrieofenbau, Kopernikusstr. 16, D-52056 Aachen, Germany

Xianon Wu – Air Force Center for Optimal Design and Control, Interdisciplinary Center for Applied Mathematics, Virginia Tech, Blacksburg VA 24061-0531, USA

Jean-Paul Zolésio – Institut Non Linéaire de Nice, CNRS, 1361 Route des Lucioles, 06904 Sophia Antipolis Cédex, France

PREFACE

This volume is the proceedings of the Workshop on Optimal Design and Control that was held in Blacksburg, Virginia, April 8-9, 1994. The workshop was sponsored by the Air Force Office of Scientific Research through the Air Force Center for Optimal Design and Control (CODAC) at Virginia Tech.

The workshop was a gathering of engineers and mathematicians actively involved in innovative research in control and optimization, with emphasis placed on problems governed by partial differential equations. The interdisciplinary nature of the workshop and the wide range of subdisciplines represented by the participants enabled an exchange of valuable information and also led to significant discussions about multidisciplinary optimization issues. One of the goals of the workshop was to include laboratory, industrial, and academic researchers so that analyses, algorithms, implementations, and applications could all be well-represented in the talks; this interdisciplinary nature is reflected in these proceedings.

An overriding impression that can be gleaned from the papers in this volume is the complexity of problems addressed by not only those authors engaged in applications, but also by those engaged in algorithmic development and even mathematical analyses. Thus, in many instances, systematic approaches using fully nonlinear constraint equations are routinely used to solve control and optimization problems, in some cases replacing ad-hoc or empirically based procedures.

Although many different applications (e.g., metal forging, heat transfer, contact problems, structures, and acoustics) are addressed here, flow control plays a central role throughout these proceedings. Optimal control of incompressible viscous flows and compressible flows, including flows with shock waves, are considered by many of the authors.

Many algorithmic issues are addressed in these proceedings. These include sensitivity analyses, novel optimization methods especially tailored for specific applications, multilevel methods, and programming techniques. At times, different methods for similar problems are presented so that one can compare the different approaches. For example, in sensitivity based optimal control algorithms, the sensitivity equations approach and the automatic differentiation approach are discussed. Another set of contrasting approaches are the differential equation sensitivity view and the discrete equation sensitivity (including grid sensitivity) approach of different authors. Other algorithmic issues considered include sparse least squares methods for aerospace design and implicit filtering.

Another recurring theme throughout these proceedings is shape control, i.e., the use of the shape of the boundary to effect control or to achieve an optimal configuration. Two papers specifically address problems in this area; one is on contact shape optimization and another addresses issues related to the existence and computation of optimal domains.

The overall impression one is left with is that there has been significant progress in the analysis and computation of control and optimization problems involving complex systems, and that these results are being applied to interesting and important applications. On the other hand, there are many issues that have not been completely resolved. These include the reduction of the cost of obtaining optimal states and controls and a more complete understanding of shape optimization for nonlinear, three-dimensional systems. Thus, there is much interesting work still to be done!

We would like to acknowledge the efforts of the Organizing Committee for the Workshop on Optimal Design and Control (Bernard Grossman, John Burns, Eugene Cliff, Terry Herdman, and Janet Peterson, all of Virginia Tech) and Melissa Chase for putting together the interesting and informative workshop that led to these proceedings. We also acknowledge the support of the Air Force Office of Scientific Research and, in particular, Charles Holland and Marc Jacobs of that office, for supporting the Workshop under AFOSR URI Grant Number F49620-93-1-0280.

Jeffrey Borggaard, John Burkardt
Max Gunzburger, and Janet Peterson
Blacksburg, 1995

Multilevel Algorithms for Nonlinear Optmization*

Natalia Alexandrov
ICASE
Mail Stop 132C
NASA Langley Research Center
Hampton VA 23681-0001, USA

J. E. Dennis, Jr.
Department of Computational and Applied Mathematics
Rice University
P. O. Box 1892
Houston TX 77251-1892, USA

Multidisciplinary design optimization (MDO) gives rise to nonlinear optimization problems characterized by a large number of constraints that naturally occur in blocks. We propose a class of multilevel optimization methods motivated by the structure and number of constraints and by the expense of the derivative computations for MDO. The algorithms are an extension to the nonlinear programming problem of the successful class of local Brown-Brent algorithms for nonlinear equations. Our extensions allow the user to partition constraints into arbitrary blocks to fit the application, and they separately process each block and the objective function, restricted to certain subspaces. The methods use trust regions as a globalization startegy, and they have been shown to be globally convergent under reasonable assumptions. The multilevel algorithms can be applied to all classes of MDO formulations. Multilevel algorithms for solving nonlinear systems of equations are a special case of the multilevel optimization methods. In this case, they can be viewed as a trust-region globalization of the Brown-Brent class.

1. Introduction

This work is concerned with a class of methods, called multilevel optimization algorithms, for solving the nonlinear equality constrained optimization problem, i.e.,

Problem EQC:

$$\begin{aligned} &\text{minimize } f(x) \\ &\text{subject to } C(x) = 0, \end{aligned}$$

where $f : \mathbb{R}^n \to \mathbb{R}$ and $C : \mathbb{R}^n \to \mathbb{R}^m$, $m \leq n$, are at least twice continuously differentiable.

The proposed class of algorithms can be used to solve any general nonlinear equality constrained optimization problem, but its development has been motivated

*Research was supported by AFOSR-89-0363, AFOSR F49620-92-J0203, DOE DE-FG05-86ER25017, NSF/CRPC CCR 9120008. Also supported in part by the National Aeronautic and Space Administration under NASA Contract No. NAS1-19480 while the first author was in residence at the Institute for Computer Applications in Science and Engineering (ICASE), NASA Langley Research Center, Hampton, VA 23681-0001.

by the engineering design problems that give rise to large-scale optimization formulations with constraints occuring naturally in blocks. In particular, in the multidisciplinary design optimization (MDO) environment, the sheer number of constraints, the structure of the problems, and the expense of the derivative computations necessitate the development of flexible algorithms that allow the user to partition the problem into a set of smaller problems.

While there is a number of nonlinear optimization methods that attack large problems by decomposing them into several smaller ones, these methods require the problems to have a special structure, for example, separability and convexity.

In particular, in engineering, decomposition and multilevel optimization have been used to solve large problems for some time. See [22] and [31] for a survey. The process of decomposition and multilevel formulation generally depends on identifying groups of variables and constraints that influence each other only weakly. The problem is then decomposed into such weakly coupled subproblems in various possible formulations, some hierarchic, some nonhierarchic. Recent developments in formulations can be found in [5] and [11]. Some of the approaches in [5] have been proven to be successful for many problems. In order to be more widely applicable, it requires the development of theoretical foundations.

We propose a class of multilevel optimization methods (see [2], [3]) for solving the nonlinear equality constrained optimization problem characterized by the following features:

- The constraints of the problem can be partitioned into blocks in any manner suitable to an application, or in any arbitrary manner at all. The analysis of the methods assumes certain standard smoothness and boundedness properties, but no other assumptions are made on the structure of the problem. There is no need to identify the weakly coupled groups of variables and constraints, although that may be helpful in practice. If all constraints and variables are strongly coupled, the partitioning can be done according to any other criterion useful to a particular application, for example, just the size of constraint blocks. The algorithm then solves progressively smaller dimensional subproblems to arrive at the trial step.

- The multilevel methods belong to the class of out-of-core methods. To the authors' knowledge, the multilevel algorithms are the only algorithms for general nonlinear optimization problems that require only a currently processed part of the constraints to be held in memory. Thus, theoretically, there is no limit to the size of the problem the methods can handle.

- The trial steps computed by the algorithm are required to satisfy very mild conditions, both theoretically and computationally. In fact, the substeps comprising the trial step can be computed in the subproblems using different optimization algorithms. The substeps are only required to satisfy a mild decrease condition for the subproblems and a reasonable boundedness condition — both satisfied in practice by most methods of interest. This feature is of great practical significance because in applications like MDO various constraint blocks may originate from different disciplines and may require different approaches to solving the subproblems.

- The class uses trust regions as a globalization strategy. The algorithms are proven to converge under reasonable assumptions.
- The algorithms together with their convergence theory provide a foundation for developing the algorithms and analyses of the general multilevel optimization formulations.

The proposed multilevel class of algorithms differs from the conventional algorithms in that its major iteration involves computing an approximate solution of not one model over a single restricted region, but of a sweep of models, each approximately minimized over its own restricted region. Each model approximates a block of constraints and, finally, the objective function, restricted to certain subspaces. Each model is computed at a different point. The case of a single block of constraints is included.

In the next section we introduce the foundations on which the proposed class of algorithms rests. Section 3 is devoted to the description of the class. Section 4 briefly describes current theoretical results. Section 5 concludes with a summary and discussion of current and future research.

2. Preliminaries

The proposed class of algorithms may be viewed as an extension of several areas of research. In this section we describe the existing algorithms and analysis schemes which serve as a foundation for the multilevel optimization methods.

2.1. The local Brown-Brent class of methods

Theoretical origins of this research lie in the method for solving nonlinear systems of equations, $F(x) = 0, F : \mathbb{R}^n \rightarrow \mathbb{R}$, introduced by Brown in [7], [8], [9]. In [6], Brent viewed Brown's method from a different perspective, which allowed Brent to propose a class of methods, among which Brown's original method was a special case. Gay [20] and Martinez [25], [26] provided further modifications and generalizations of the methods.

The following statement of the general Brown-Brent algorithm was condensed from the descriptions in Gay [20] and Dennis [12]. In these works the algorithm is described in terms of one-dimensional blocks.

Denote the components of $F(x)$ by $F_1(x), \ldots, F_n(x)$.

Algorithm 1. Local Brown-Brent algorithm for nonlinear systems.

Let x_c be the current approximation to the solution.

Outer Loop: Do until convergence:

$y_0 = x_c$

$H_0 = \mathbb{R}^n$

Inner Loop: Do $k = 1, n$

1. Form the linearization, L_k about y_{k-1} of F_k restricted to $\bigcap_{i=0}^{k-1} H_i$. $L_k = 0$ defines H_k, an $(n-k)$-dimensional hyperplane in $\mathbb{R}^n$.
2. Move from $y_{k-1} \in \bigcap_{i=0}^{k-1} H_i$ to $y_k \in \bigcap_{i=0}^{k} H_i$.

End Inner Loop

$x_c = y_n$

End Outer Loop

The point y_n of intersection of all the hyperplanes is the point where all the linearizations vanish. The way in which the steps 1-2 of the inner loop are actually done determines the particular kind of Brown-Brent method. In Brent's method, $s_k = y_k - y_{k-1}$ is the shortest ℓ_2 norm step from y_{k-1} to H_k. In Brown's method, s_k is the shortest ℓ_2 norm step from y_{k-1} to H_k parallel the k-th coordinate axis.

When applied to a linear system of equations, i.e., when $F(x) = Ax - b$, Brown's method is equivalent to Gaussian elimination with pivoting about the maximum row element of the reduced matrix [7], while Brent's method is equivalent to factoring A into a product of a lower triangular matrix and an orthogonal matrix [6]. It can be shown, based on [33], that there exists a Brown-Brent analog for any matrix decomposition in the linear case.

Brown [7], [9], Brown and Dennis [10], Brent [6], and Gay [20] established local quadratic convergence of variants of the algorithm, both for analytic and finite difference derivatives. To the authors' knowledge, there had been no theoretically supported global extensions of Brown-Brent methods until [2].

2.2. Trust-region methods

Consider the following unconstrained minimization problem.

Problem UNC:

$$\begin{aligned}&\text{minimize } f(x)\\ &x \in \mathbb{R}^n,\end{aligned}$$

where $f : \mathbb{R}^n \to \mathbb{R}$ is continuously differential. Given x_c, the current approximation to the solution, a trust-region algorithm for solving the problem finds a trial step by solving the following trust-region subproblem approximately:

$$\begin{aligned}&\text{minimize } f(x_c) + \nabla f(x_c)^T s + \frac{1}{2} s^T H_c s\\ &\text{subject to } \|s\| \le \delta_c\,,\end{aligned} \tag{1}$$

where $f, \delta_c \in \mathbb{R}$, $\nabla f, s \in \mathbb{R}^n$, $H_c = H_c^T \in \mathbb{R}^{n\times n}$ is the Hessian of f or an approximation to it, $\delta_c > 0$ is the trust-region radius, and $\|\cdot\|$ denotes the ℓ_2 norm.

The idea is to accept the trial step when the quadratic model adequately predicts the behavior of the function, and to recompute the step in a smaller region if it does not.

The trust-region approach to the problem of solving systems of nonlinear equations is just a special case of the approach to the problem above; namely, for nonlinear equations, the objective function $f(x)$ is taken to be $\| F(x)\|_2^2$.

Detailed treatment of the trust-region approach to unconstrained optimization and nonlinear equations can be found in Dennis and Schnabel [14], Sorensen [32], Moré [27], Moré and Sorensen [28], Powell [29], and Shultz, Schnabel, and Byrd [30].

For the equality constrained optimization problem, the successive quadratic programming (SQP) algorithm is used commonly. Its step is found by computing a minimum of the quadratic model of the Lagrangian at the current point, subject to linearized constraints. A trust-region algorithm based on SQP adds the trust-region constraint to the subproblem and additional constraints designed to ensure that the trust-region constraint and the linearized constraints are consistent.

2.2.1. Merit functions

In order to evaluate a trial step, trust-region algorithms use merit functions, which are functions related to the problem in such a way that the improvement in the merit function signifies progress toward the solution of the problem.

For unconstrained minimization, a natural choice for a merit function is the objective function itself. Let

$$\phi(s) = f(x_c) + \nabla f(x_c)^T s + \frac{1}{2} s^T H_c s \tag{2}$$

denote the quadratic model of the merit function. We define two related functions.

The actual reduction is defined as

$$ared_c(s_c) = f(x_c) - f(x_c + s_c)\,, \tag{3}$$

and **the predicted reduction** is defined as

$$\begin{aligned} pred_c(s_c) &= \phi(0) - \phi(s_c) \\ &= -\nabla f(x_c)^T (s_c) - \frac{1}{2} s_c^T H_c s_c\,, \end{aligned} \tag{4}$$

so that the predicted reduction in the merit function is an approximation to the actual reduction in the merit function.

The standard way to evaluate the trial step in trust-region methods is to consider the ratio of the actual reduction to the predicted reduction. A value lower than a small predetermined value causes the step to be rejected. Otherwise the step is accepted.

For nonlinear systems of equations, the norm of the residuals serves as a merit function. For the constrained optimization, the merit function is some expression that involves both the objective function and the constraints.

We shall see that conventional merit functions prove to be inadequate for multilevel algorithms.

2.2.2. Fraction of Cauchy decrease

To assure global convergence of a trust-region algorithm for problem UNC, the trial step is required to satisfy a **fraction of Cauchy decrease** condition. This mild condition means that the trial step, s_c, must predict at least a fraction of the decrease predicted by the Cauchy step, which is the steepest descent step for the model within the trust region. We must have for some fixed $\kappa > 0$

$$\phi(s_c) - \phi(0) \leq \kappa[\phi(s_c^{CP}) - \phi(0)], \tag{5}$$

where

$$s_c^{CP} = -\alpha_c^{CP} \nabla f(x_c) \text{ with}$$

$$\alpha_c^{CP} = \begin{cases} \frac{\|\nabla f(x_c)\|^2}{\nabla f(x_c)^T H_c \nabla f(x_c)} & \text{if } \frac{\|\nabla f(x_c)\|^3}{\nabla f(x_c)^T H_c \nabla f(x_c)} \leq \delta_c \\ \frac{\delta_c}{\|\nabla f(x_c)\|} & \text{otherwise.} \end{cases}$$

See Dennis and Schnabel [14], pp. 139—141, for details on the Cauchy point.

The fraction of Cauchy decrease property implies a weaker condition which has a more convenient form and is frequently used as a technical lemma in the global convergence proofs.

Lemma 1. *Let s_c satisfy (5). Then*

$$\phi(0) - \phi(s_c) \geq \frac{\kappa}{2} \|\nabla f(x_c)\| \min \left\{ \frac{\|\nabla f(x_c)\|}{\|H_c\|}, \delta_c \right\}. \tag{6}$$

References: Powell [29]; Moré [27].

Either (5) or (6) is necessary to establish global convergence theoretically.

2.2.3. Global convergence results

Powell's global convergence theorem [29] for any unconstrained minimization trust region algorithm serves as a prototype for most trust-region related convergence results.

Theorem 2. *Let $f : \mathbb{R}^n \to \mathbb{R}$ be continuously differentiable and bounded below on the level set $\{x \in \mathbb{R}^n \mid f(x) \leq f(x_0)\}$. Assume that $\{H_i\}$ are uniformly bounded above. Let $\{x_i\}$ be the sequence of iterates generated by a trust-region algorithm that satisfies (5) or (6). Then*

$$\liminf_{i \to \infty} \|\nabla f(x_i)\| = 0.$$

Detailed treatment of the unconstrained minimization theory and practice can be found in Moré [27], Moré and Sorensen [28], Sorensen [32], and Shultz, Schnabel and Byrd [30].

2.2.4. Tangent-space methods for constrained optimization

The multilevel methods proposed here may be viewed as a generalization of an approach to nonlinear programming known as the null space or generalized elimination approach (see Fletcher [19]).

Different authors refer to different methods as "null space methods", but the general idea of a null space method for equality constrained minimization is to reduce the dimension of the problem by first taking the step intended to solve the constraint equations, and then to minimize the model of the function restricted to the null space of the linearized constraints. The resulting minimization problem is of a lower dimension than the original one.

A well-known local method of this type is the GRG (Generalized Reduced Gradient) algorithm. Details of GRG and other null space methods can be found in Lasdon [23], Fletcher [19], Avriel [4], and Gill, Murray and Wright [21].

A class of global trust-region algorithms that use the same general principle of reducing the problem's dimension is known as the class of tangent space methods. The tangent space approach was introduced to avoid the possibility of inconsistency of the constrained trust-region subproblem.

Recent work on these methods can be found in Maciel [24] and Dennis, El-Alem and Maciel [13]. The main feature of the class is that the trial step is computed as a sum of two substeps, the first of which is made toward the linearized constraints in the direction orthogonal to the null space of the constraint Jacobian, while the second is made to minimize the model of the Lagrangian in the null space of the linearized constraints. The function and derivative information is computed at a single point x_c.

The multilevel methods proposed here generalize the tangent space methods in the sense that their trial steps are sums of not two substeps but of as many substeps as there are constraint blocks together with a substep on the model of the objective function with the model information computed at the points resulting from taking the substeps one-by-one.

3. Multilevel algorithms for nonlinear optimization

In this section we present the class of multilevel optimization algorithms for the nonlinear equality constrained minimization problem. Since the time of its introduction in [2], the class has undergone changes. In [2], the globalization and extension to constrained optimization only of local Brent's method was proposed. Recent developments (see [3]) have extended the results to provide globalization and extension to constrained optimization of the entire local Brown-Brent class.

3.1. Notation

Due to arbitrary blocking of the constraints, the notation becomes cumbersome. To ease the reading effort, we omit the subscripts and superscripts where possible. Here is an explanation of the notation conventions.

Unless specified otherwise, all norms are ℓ_2 norms.

From here on, we assume that the equality constraints of problem EQC are partitioned into M blocks of arbitrary size and composition. Let the constraints of the first block be numbered from $n_1 = 1$ to $n_2 - 1$; the constraints of the second block—from n_2 to $n_3 - 1$; and so on, until the constraints of the last block are numbered from n_{M-1} to $n_M = m$.

The algorithms will be formally considered to have an outer loop, in which we make the decision about the acceptability of the step, and the inner loop, in which we solve a sequence of minimization subproblems. The sum of the substeps produced as solutions of these subproblems yields the total trial step. The outer loop counter is i; the inner loop counter is k. Thus k corresponds to the block number of constraints. If the subscript k is used with a constant, that constant refers to the properties of the k-th block of constraints, independent of the iterates. Note that the term "inner loop" is formal. The purpose of the inner loop is to compute a basis for the null space of the Jacobian of our constraint system, but step-by-step, using information at different points, instead of the simultaneous computation of, say, the Newton's method.

We denote the sequence of points generated by the outer loop of the algorithm by $\{x_i\}$ when we consider the iterates as members of the sequence in convergence analysis. Otherwise, we use x_c, x_-, and x_+ to denote the current, the previous, and the next iterates, respectively.

We denote the sequence of points generated within each inner loop by y_k, $k = 0, \ldots, M+1$, when we need not consider the outer loop iteration number. Most of the time we shall be discussing entities within a single iteration. Otherwise, we use subscripts and superscripts. For example, y_k^i or y_k^c denote the inner k–th iterate within the i–th outer loop. Note that $y_0 = x_c$ and $y_{M+1} = x_+$.

The substep produced by solving the k–th subproblem of the inner loop is denoted by s_k, $k = 1, \ldots, M+1$. The sum $s_1 + \ldots + s_{M+1}$ yields the total trial step $\hat{s}_c$. Again, we use subscripts, e.g., $\hat{s}_i$, to denote the total step as a part of the sequence of steps produced by the algorithm.

We denote the radius of the trust region for subproblem k, centered at y_{k-1}, by δ_k, $k = 1, \ldots, M+1$. The radius of the total trust region, centered at $x_c = y_0$ is $\hat{\delta}_c$ or $\hat{\delta}_i$.

We donote the projector onto the intersection of null spaces of $\nabla C_1(x), \ldots, \nabla C_k(x)$ by P_k.

Again, when we omit superscripts, we refer to the objects within a single outer loop. For example, $C_k(y_{k-1})$ refers to $C_k(y_{k-1}^i)$ or $C_k(y_{k-1}^c)$.

Additional notation will be introduced as needed.

3.2. General description

The general glass of multilevel algorithms can be described in the following way. The constraint system of the problem is partitioned into M arbitrary blocks. In practice, this block decomposition is obvious in most cases. At the current approximation to a solution of problem EQC, x_c, we set $y_0 = x_c$. The trial step is computed as follows.

We find an approximate minimizer, s_1, of the quadratic Gauss-Newton model about y_0 of the first block of constraints in the trust region of radius δ_1. The step is required to satisfy a fraction of Cauchy decrease condition for this model and a mild boundedness condition disussed in the next subsection. The step is taken to yield the point $y_1 = y_0 + s_1$.

We then find an approximate minimizer of the quadratic model of the second block of constraints, restricted to the null space of the Jacobian of the first block. This model is built using **the information at the new point**. It is important to emphasize that all the function and derivative information for the second block is computed at the new point y_1. The next step, s_2, bounded by its own trust-region, is obtained to satisfy a fraction of Cauchy decrease condition for this restricted model of the second block. The step is taken to yield the point y_2.

The process of computing steps that satisfy sufficient predicted decrease for the restricted models of progressively smaller dimensions continues. Again, **the model for each block is built by using the function and derivative information at the most recently computed point**. The final step on the constraints, s_M, is obtained to produce sufficient predicted decrease in the quadratic model, at y_{M-1}, of the last block of constraints, restricted to the intersection of the null spaces of the Jacobians of all previous blocks.

When all the constraint blocks have been processed, $n - m$ degrees of freedom still remain. The remaining variables are used in building a model of the objective function, so that the final substep, s_{M+1}, is obtained to produce sufficient predicteddecrease in the quadratic model at y_M of the objective function, restricted to the intersection of the null spaces of the Jacobians of all constraint blocks. The final step is taken to yield the next major iterate, i.e., the next approximation to a solution of problem EQC. Thus, the total trial step from x_c to x_+ is the sum of the substeps in the inner sweep, i.e., $\hat{s}_c = s_1 + \ldots + s_{M+1}$.

Unless the convergence criterion is met, the total trial step is evaluated, and the algorithm returns to process again the first block of constraints in a trust region determined by the success or failure of the trial step.

3.2.1. Computing the substeps

During the constraint elimination stage, the substeps solve the following subproblems:

$$\text{minimize } \tfrac{1}{2}\|C_k(y_{k-1}) + \nabla C_k(y_{k-1})^T s\|_2^2$$

subject to $\nabla C_j(y_{j-1})^T s = 0$, $j = 1, \ldots, k-1$,
and possibly an additional constraint
on the step direction,
and $\|s\|_2 \leq \delta_k$,

for $k = 1, \ldots, M$. (Note that for $k = 1$ there is no null space constraint.) Then the objective function subproblem is:

minimize $f(y_M) + \nabla f(y_M)^T s + \frac{1}{2} s^T H(y_M) s$
subject to $\nabla C_j(y_{j-1})^T s = 0$, $j = 1, \ldots, M$,
and possibly an additional constraint
on the step direction,
and $\|s\|_2 \leq \delta_{M+1}$.

If there is no additional constraint on the direction of the step, the subproblems produce a trust-region generalizaton of the local Brent step. A constraint requiring that the step be parallel to some coordinate hyperplane would be a generalization of the local Brown step. In practice, there is no explicit constraint for the generalization of the Brown step; rather it is computed implicitly (see [3]).

Let Q_{k-1} be a matrix the columns of which form a basis for the intersection of the null spaces of $\nabla C_1(y_0), \ldots, \nabla C_{k-1}(y_{k-2})$. A change of variables, $v = Q_{k-1}s$, converts the constrained subproblems to unconstrained ones.

For relatively small problems, the null space bases can be computed by using the QR decomposition to find the basis for null space of $\nabla C_1(y_0)$, and then by updating the decomposition for subsequent subproblems to find a basis for the null space intersections. For larger problems, the QR decomposition becomes prohibitively expensive. In that case, reduced basis projectors can be used. More details about the null space basis computations can be found, for example, in [3] and [21].

There are various methods for solving large-scale trust region subproblems. We are holding much hope for the method recently developed by D. Sorensen of Rice University.

However, once the subproblems with null space constraints are converted into unconstrained trust-region subproblems, the steps may be chosen in any manner, as long as they satisfy two mild conditions.

1. As mentioned earlier, if there are no additional constraints on the subproblem k, its solution, a Levenberg-Marquard step for the reduced problem, produces a generalization of the Brent step. That is, the substep is orthogonal to the linearized constraint hyperplane, for all blocks numbered $k+1, \ldots, M$. However, we do not require that the substeps be orthogonal. We require that each substep satisfies

$$\|s_k\| \leq \Lambda_k \|C_k(y_{k-1})\| \tag{7}$$

for some positive constant Λ_k that depends only on the properties of that particular constraint block but is independent of the iteration. Other conditions are possible to assure global convergence. However, this condition, first formalized in [13], is reasonable in that it is enforced automatically by any algorithms of interest for computing linearly feasible points. For instance, this is easily shown for the extensions of both Brown and Brent steps (see [3]).

2. We also require for each substep to satisfy a fraction of Cauchy decrease condition for the particular subproblem that substep solves. This is also a very mild condition—it is satisfied by all reasonable methods. Note that we do not place any conditions on the total trial step—only on the substeps.

   It is easy to show that if s_k^{B-B} is an unconstrained Brown or Brent substep (or any substep out of the local Brown-Brent class), we can claim the following: *If* $\|s_k^{B-B}\| \leq \delta_k$, *then let* $s_k = s_k^{B-B}$. *Otherwise let*

$$s_k = \frac{\delta_k * s_k^{B-B}}{\|s_k^{B-B}\|}. \tag{8}$$

   Then s_k *satisfies the fraction of Cauchy decrease condition on subproblem* k. The proof is given in [3].

   Thus, we see that simply truncating the unconstrained Brown or Brent substep to the size of the trust region will produce sufficient predicted decrease in the models of the constraint blocks.

3.2.2. The merit function and its model

Merit functions used to evaluate the progress of single-block trust-region algorithms consist of some combination of the objective function and the constraints. One common merit function is the ℓ_2 penalty function $f(x) + \rho\|C(x)\|_2^2$, where ρ is the penalty parameter.

In the process of the multilevel algorithm development, it has become apparent that conventional merit functions are inadequate for measuring progress of the multilevel methods, because a conventional merit function does not take into account the order in which minimization proceeds.

The difficulty can be summarized as follows:

- The result of the k-th minimization subproblem predicts decrease for the k-th component from point y_{k-1} to point y_k. It predicts no change for all previous blocks. However, there is no prediction at all about how $s_1 + \ldots + s_k$ changes and likely increases the norms of the blocks numbered $k+1, \ldots, M$. Neither does any substep, except s_{M+1} predict the behavior of the objective function.

This observation brought us to the conclusion that the merit function must take into account the multilevel structure of the scheme. Consider the following modified ℓ_2 penalty function:

$$\begin{aligned}\bar{\mathcal{P}}(x;\rho_1,\ldots,\rho_M) &= f(x)+\rho_M(\|C_M(x)\|^2\\ &\rho_{M-1}(\|C_{M-1}(x)\|^2+\rho_{M-2}(\|C_{M-2}(x)\|^2+\ldots\\ &+\rho_2(\|C_2(x)\|^2+\rho_1\|C_1(x)\|^2))))\\ &= f(x)+\sum_{k=1}^{M}(\prod_{j=k}^{M}\rho_j)\|C_k(x)\|^2,\end{aligned}$$

where $\rho_k \geq 1$, $k = 1, \ldots, M$. where $\rho_k \geq 1$, $k = 1, \ldots, M$. The initial choice $\rho_k = 1$ is arbitrary and scale-dependent. The only requirement is that $\rho_k \geq 1$. For theoretical purposes, the problem is assumed to be well-scaled.

The new merit function penalizes for the possible predicted increase in the constraint blocks $k, \ldots, M$, or in the objective function that may have occured during inner loop iterations $1, \ldots, k-1$.

At $y_{M+1} = x_+ = x_c + \hat{s}_c$, we model each $\|C_k(x_+)\|^2$ by $\|C_k(y_{k-1})+\nabla C_k(y_{k-1})s_k\|^2$, and so we model the merit function at x_+ by

$$\begin{aligned}\mathcal{M}_c(s_1,\ldots,s_{M+1};\rho_1^c,\ldots,\rho_M^c) &= f(y_M)+\nabla f(y_M)^T s_{M+1}\\ &+\frac{1}{2}s_{M+1}^T H(y_M)s_{M+1}+\|C_M(y_{M-1})+\nabla C_M(y_{M-1})^T s_M\|^2\\ &+\rho_{M-1}^c(\|C_{M-1}(y_{M-2})+\nabla C_{M-1}(y_{M-2})^T s_{M-1}\|^2\\ &+\rho_{M-2}^c(\|C_{M-2}(y_{M-3})+\nabla C_{M-2}(y_{M-3})^T s_{M-2}\|^2+\ldots\\ &+\rho_2^c(\|C_2(y_1)+\nabla C_2(y_1)^T s_2\|^2+\rho_1^c\|C_1(y_0)+\nabla C_1(y_0)^T s_1\|^2)))\\ &= f(y_M)+\nabla f(y_M)^T s_{M+1}+\frac{1}{2}s_{M+1}^T H(y_M)s_{M+1}\\ &+\sum_{k=1}^{M}(\prod_{j=k}^{M}\rho_j)\|C_k(y_{k-1})+\nabla C_k(y_{k-1})^T s_k\|^2.\end{aligned}$$

We define the actual reduction as the difference between the merit function values at x_c and x_+, and we define the predicted reduction as the difference between the value of the merit function at x_c and the value of the model at x_+.

3.2.3. Updating the penalty parameters

This penalty parameter updating scheme for multilevel methods generalizes the scheme proposed in El-Alem [15], [16]. It ensures that our merit function has an essential property, namely, that unless an iterate is optimal, the predicted reduction should always be positive. We use the following procedure:

Algorithm 2. Penalty parameter updating algorithm. *(Done on completion of each inner sweep of minimization problems.)*

Denote the set $\{s_1, \ldots, s_k\}$ by S_k and denote the set $\{\rho_1, \ldots, \rho_k\}$ by ρ_k.

At the beginning of a multilevel algorithm, set $\rho_1^- = \ldots = \rho_M^- = 1$ and choose $\beta \in (0, 1)$.

1. Compute $Cpred_1(s_1) = \|C_1(y_0)\|^2 - \|C_1(y_0) + \nabla C_1(y_0)^T s_1\|^2$.
2. **Do** $k = 1, M$

   Update ρ_k.

   Compute

   $$Cpred_{k+1}(S_{k+1};\ \rho_{k-1}^c, \rho_k^-) = [\|C_{k+1}(y_0)\|^2 - \|C_{k+1}(y_k) + \nabla C_{k+1}(y_k)^T s_{k+1}\|^2 + \rho_k^- Cpred_k(S_k;\ \rho_{k-1}^c).$$

   if $Cpred_{k+1}(S_{k+1};\ \rho_{k-1}^c, \rho_k^-) \geq \frac{\rho_k^-}{2} Cpred_k(S_k;\ \rho_{k-1}^c)$ **then**

   $\rho_k^c = \rho_k^-$.

   $Cpred_{k+1}(S_{k+1};\ \rho_{k-1}^c, \rho_k^c) = Cpred_{k+1}(S_{k+1};\ \rho_{k-1}^c, \rho_k^-)$.

   else

   $\rho_k^c = \bar{\rho}_k + \beta$,

   where $\bar{\rho}_k = \frac{2[\|C_{k+1}(y_k)+\nabla C_{k+1}(y_k)^T s_{k+1}\|^2 - \|C_{k+1}(y_0)\|^2]}{Cpred_k(S_k;\ \rho_{k-1}^c)}$.

   Compute $Cpred_{k+1}(S_{k+1};\ \rho_{k-1}^c, \rho_k^c)$.

   end if

   end Do
3. Update ρ_M.

   Compute

   $$pred(S_M;\ \rho_{M-2}^c, \rho_{M-1}^-) = [f(y_0) - \phi_M(s_M)] + \rho_M^- Cpred_M(S_M;\ \rho_{M-1}^c).$$

   if $pred(S_M;\ \rho_{M-2}^c, \rho_{M-1}^-) \geq \frac{\rho_M^-}{2} Cpred_M(S_M;\ \rho_{M-1}^c)$ **then**

   $\rho_M^c = \rho_M^-$.

   $pred(S_M;\ \rho_{M-2}^c, \rho_{M-1}^c) = pred(S_M;\ \rho_{M-2}^c, \rho_{M-1}^-)$.

   else

   $\rho_M^c = \bar{\rho}_M + \beta$,

   where $\bar{\rho}_M = \frac{2[\phi_M(s_M) - f(y_0)]}{Cpred_M(S_M;\ \rho_{M-1}^c)}$.

   Compute $pred(S_M;\ \rho_{M-2}^c, \rho_{M-1}^c)$.

   end if

End

Note that without updating the penalty parameters we can be assured of the positive predicted reduction from x_c only for the first block of constraints, i.e., only $Cpred_1(s_1)$ is definitely positive without additional considerations. To ensure that $Cpred_2(s_1, s_2;\ \rho_1)$ is positive, we may have to increase ρ_1. Now that

$Cpred_2(s_1, s_2; \rho_1)$ is positive, we use it to ensure that the next partial predicted reduction is positive, and so on. So, for each each substep s_k, the predicted reduction accumulated by the step $s_1 + \ldots + s_k$ is at least a fraction of the predicted decrease accumulated by the step $s_1 + \ldots + s_{k-1}$.

Thus the predicted reduction of the first block is the most heavily penalized one.

It should be emphasized that the step computation is completely independent of the penalty parameter computation.

3.2.4. Step evaluation and trust-region radii updating

Although there are various schemes of evaluating the trial step and updating the trust region radii, for the sake of simplicity in discussion, we adopt the following strategy:

- The total trial step is evaluated outside the inner loop.
- All individual trust region radii are equal and are updated simultaneously by the same factor.

Other strategies for practical implementations are discussed in [1]. We would like to emphasize that the simultaneous expansion or contraction of the trust region radii is not a technical requirement.

The algorithm for evaluating the step and updating the trust region radii follows.

Algorithm 3. Step evaluation / trust region update.

Given $\delta_k > 0, k = 1, \ldots, M$ (or $k = 1, \ldots, M + 1$ for optimization), $\delta_{max} > 0, \delta_{min} > 0, 0 < \eta_1 < \eta_2 < 1, \alpha_1 \in (0, 1], \alpha_2 > 1, x_c \in \mathbb{R}^n$, $ared$, $pred$,

Compute $r = \frac{ared}{pred}$.
if $r < \eta_1$ **then** (step not accepted)
  $\delta_k = \alpha_1 * \delta_k$.
else if $r \geq \eta_2$ **then** (step accepted)
  $\delta_k = \min\{\delta_{max}, \max\{\delta_{min}, \alpha_2 * \delta_k\}\}$.
  $x_c = x_+$.
else (step accepted)
  $\delta_k = \max\{\delta_{min}, \delta_k\}$.
  $x_c = x_+$.
end if

We note that if the step is not accepted, the trust region radii are decreased withoutany safeguard. However, if the step is accepted, the next trust region radius is set to be no smaller than a predetermined positive value δ_{min}. This strategy is extremely important in the global convergence theory. It ensures that the trust region radius is bounded away from zero and hence that the penalty parameters are bounded from above. This technique was introduced in [18].

3.2.5. The stopping criteria

We use the first order necessary conditions for problem EQC to terminate the algorithm and require that

$$\begin{aligned} \|C_1(y_0)\| &\leq \epsilon_{tol} \\ \|C_2(y_1)\| &\leq \epsilon_{tol} \\ &\dots \\ \|C_M(y_{M-1})\| &\leq \epsilon_{tol} \\ \|P_M^T \nabla f(y_M)\| &\leq \epsilon_{tol} \end{aligned} \tag{9}$$

hold simultaneously.

Since

$$\|s_k\| = O(\|C_k(y_{k-1})\|),$$

if $\|C_k(y_{k-1})\|$ is small, $\|s_k\|$ will be small and the inner loop iterates y_k will be close to each other, and in the limit we can show ([2], [3]) that at least a subsequence of the generated sequence of the outer loop iterates will converge to a stationary point of problem EQC.

The tolerance parameters ϵ_{tol} need not be the same, but for convenience, they are taken to be the same throughout the convergence analysis.

The reason for requiring such a stopping criterion is theoretical and practical. The conventional test for the entire norm of the constraint residual being close to 0 does not differentiate between the individual $\|C_k(y_{k-1})\|$. It is essential for the convergence proof to determine how close to feasibility an iterate must be in order for the penalty parameters not to be increased. This is a measure of feasibility versus optimality. The conventional stopping criterion allows only the total feasibility to be measured and thus to determine when ρ_M does not have to be increased. But even if ρ_M is not increased, $\rho_1, \dots, \rho_{M-1}$ may have to be increased because of the relative sizes of the component block norms. The conventional criterion does not allow us to measure **relative feasibility** of one block of constraints with respect to the others.

In practice, we do not wish to evaluate the residuals at the same point just for the sake of the stopping criterion.

Other stopping criteria are possible, but the one above is the most natural one.

3.2.6. The statement of the algorithm

The formal description of the algorithm follows.

Let the constraints be partitioned into M blocks.

Algorithm 4. Multilevel algorithm for equality constrained optimization.
Given $\delta_k > 0, k = 1, \dots, M, \delta_{max} > 0, \delta_{min} > 0, 0 < \eta_1 < \eta_2 < 1, \alpha_1 \in (0, 1], \alpha_2 > 1, x_c \in \mathbb{R}^n$.

Outer Loop: Do until convergence:
$y_0 = x_c$.
Compute the trial step.
Inner Loop: Do $k = 1, M$
If y_{k-1} is not feasible then
Compute s_k that satisfies a fraction of Cauchy decrease condition on $\frac{1}{2}\|C_k(y_{k-1}) + \nabla C_k(y_{k-1})s\|_2^2$ restricted to the intersection of the null spaces of $\nabla C_j(y_{j-1})^T s = 0, j = 1, \dots, k-1$, and $\|s_k\| \le \Lambda_k \|C_k(y_{k-1})$ (satisfied automatically).
$y_k = y_{k-1} + s_k$.
End if
End Inner Loop
Compute s_{M+1} to satisfy the fraction of Cauchy decrease condition on the subproblem: minimize $\phi_M(s_M)$ restricted to the intersection of the null spaces of $J_j(y_{j-1})s = 0, j = 1, \dots, M$, and $\|s\|_2 \le \delta_M$.
$y_{M+1} = y_M + s_{M+1}$.
$x_+ = y_{M+1}$.
The trial step is: $\hat{s}_c = s_1 + \dots + s_{M+1}$.
Update the penalty parameters
Evaluate the step and update the trust region radius
If the step is accepted, set $x_c = x_+$.
End Outer Loop

We should note that there is an option to eliminate only a subset of constraints via the described procedure. In this case, the rest of the constraints and the objective function would be restricted to the intersection of the null spaces of the Jacobians of the processed constraints, and the resulting reduced optimization problem would be solved by a chosen method. The discussion of this approach is left for later work.

4. Global convergence results

In this section we give a summary of the global convergence theory for multilevel algorithms.

4.1. Basic ingredients of a global convergence proof

Our proof contains the general ingredients of a global convergence analysis for a trust-region method. The first three are requires for a typical analysis of an unconstrained minimization algorithm.

1. The trial step must be shown to satisfy a sufficient predicted decrease condition, usually the FCD condition. Our algorithm assumes that the substeps satisfy the FCD condition on the subproblems. It remains for us to show that the total step from x_c to x_+ satisfies a suitable decrease condition.

2. The difference between the actual and predicted reduction must be bounded above by at least a constant multiple of the square of the total step norm plus multiples of higher powers of the step norm. This is easily shown multilevel algorithms.

3. The algorithm must be shown to be well-defined, i.e., we must prove that the ratio of the actual reduction to predicted reduction can be made greater than a given $\eta_1 \in (0, 1)$ after a finite number of trial step computations. Given 2, it is easy to show that as the trust region radius approaches zero, the ratio of the actual reduction to predicted reduction approaches one. For the algorithm to be well-defined we must show that the ratio of the predicted to actual reduction approaches one faster than the trust region radius goes to zero. This is easily established for our algorithm.

   An algorithm for constrained optimization that uses penalty parameters in its merit function requires the fourth ingredient.

4. The penalty parameter in the merit function must be shown to be bounded. The technique is to prove that the product of the penalty parameter and the trust region radius is bounded by a constant independent of the iterates. The sequence of the trust region radii is then shown to be bounded away from zero. Here a crucial role is played by the trust region updating technique introduced in [18]: after a successful iteration and before starting the next iteration, the trust region radius is set to be no smaller than a pre-defined value. This way of updating allows us to prove that the sequence of penalty parameters is bounded from above.

   The method for updating the penalty parameters ensures that the sequence of penalty parameters is nondecreasing *, which, together with its boundedness, allows us to conclude that the penalty parameter sequence converges and, moreover, remains constant after a finite number of increases. This fact is used in the global convergence theorem.

*The global convergence theory for algorithms with nonmonotone penalty parameters has been investigated by Mahmoud El-Alem [17].

4.2. Assumptions

We make the following assumptions on the problem and the sequence of steps and iterates:

- f, C are at least twice continuously differentiable.
- The gradient of the constraints has full rank. This is a strong assumption, but it is a standard practice to require it for the sake of convergence proofs. Practical experience suggests that the breakdown of this assumption does not necessarily diminish the efficacy of our algorithm. Not assuming full rank would allows us to prove a slightly weaker convergence result.
- $f(x)$, $\nabla f(x)$, $\nabla^2 f(x)$, H_M, $C(x)$, $\nabla C(x)$, $\nabla C_k(x)$, $\nabla^2 C_j(x)$, $j = 1, \ldots, m$, $\{[P_{k-1}^T \nabla C_k(x)]^T [P_{k-1}^T \nabla C_k(x)]\}^{-1}$, $k = 1, \ldots, M$, are all uniformly bounded in norm for all x in the domain of interest.

Since we require that the Hessian of the objective function be only bounded, we can even take it to be 0. Of course, such an approximation would lower the effectiveness of the algorithm.

4.3. Summary of the proof

In this subsection we provide an overview of steps in the convergence proof. The details can be found in [3].

- We show that under our assumptions, the norm of any intermediate sum of the substeps is bounded by a costant times the norm of the total trial step.
- Several technical results provide workable expressions of the FCD (fraction of Cauchy decrease) condition similar to the one used for unconstrained optimization.
- A standard result provides and upper bound on the error between actual reduction and predicted reduction.
- By virtue of the penalty parameter updating scheme, the multilevel algorithms have the property that if an iterate is feasible, the penalty parameters are not increased. We show that if the iterates are sufficiently close to feasibility, the penalty parameters are not increased either. This result is crucial to the proof of convergence, giving a sufficient condition for the penalty parameters not to be increased.
- Next we establish an upper bound on the product of the penalty parameters with the trust region radii. This result allows us to conclude that the radii are bounded below if the penalty parameters increase. The penalty parameter sequences are shown to be nondecreasing, which, together with their

boundedness from above, allows us to conclude that the penalty parameters tend to a limit, and, moreover, stay constant after a finite number of outer iterations. The limit is shown to exist, but its explicit expression is not known.

- We have shown that the total trust region radius is bounded away from zero if any of the penalty parameters are increased. Now we show that radius is always bounded away from zero. The trust region updating strategy ensures that is is bounded from above.

- The next result guarantees that the algorithm is well defined, i.e., that after a finite number of outer loop iterations an acceptable step $\hat{s}_c$ with

$$\frac{ared}{pred} \geq \eta_1$$

  will be found.

- In the global convergence result, we show that if the objective function is bounded below, then the sequence of iterates generated by a multilevel algorithm has a subsequence convergent to a stationary point of the equality constrained minimization problem.

- As a corollary, we can now conclude that the multilevel algorithm for nonlinear equations is also globally convergent.

5. Discussion and concluding remarks

We have described a broad new class of multilevel algorithms for solving the nonlinear equations problem and the equality constrained optimization problem. The class can be considered as a globalization and an extension of the local class of algorithms of Brown and Brent for solving nonlinear systems of equations.

The main practical appeal of the multilevel algorithms is that in the case of equality constrained optimization, they allow the user to partition the constraint system arbitrarily, to fit the application, and to process the blocks of constraints separately. In their finite-difference derivative form, they require fewer function evaluations than the Newton's method.

The multilevel class is characterized by requiring very mild conditions to be imposed on the trial steps. All reasonable algorithms satisfy these conditions automatically.

We have established global convergence theory for the entire class. The theory implies convergence of the nonlinear equations solver, which, to the author's knowledge, is the first theoretically supported method for globalizing Brown-Brent methods. The global convergence theory was made possible by the introduction of the new merit function that takes into account the order of the constraint processing. The nested penalty parameters are updated by an extension of the scheme proposed by El-Alem [15].

The algorithms are expected to be applicable to the problem of the multidisciplinary design optimization and to serve as a foundation for the study of the general multilevel optimization problem.

We would like to mention one more application. The design of complex engineering systems is by nature a multicriteria optimization problem. The design projects are distinguished by very large numbers of variables, constraints, and expensive analyses. To solve the problem, it is necessary to break it into disciplines, each of which produces its own optimal design. The discipline designs are then incorporated into a total design. The multilevel methods proposed here would allow researchers to integrate constraints obtained from different sources.

To solve the multicriteria optimization problem, it is necessary to decide when an iterate is optimal. One of the approaches to optimality is the statement of the multicriteria problem as a multilevel optimization problem, i.e., the problem of minimizing a function on a feasible set, which is an optimal set for another function, and so on. In such an approach, the user places priorities on the optimization problems that are to be solved sequentially. We believe that the multilevel algorithms proposed here will serve as a beginning for a detailed study of the general multilevel optimization problem.

Directions of research in progress include local convergence rates, implementation, extensive testing on applications, incorporation of bound and inequality constraints, and extensions to general nonlinear bilevel and multilevel optimization.

References

[1] N. Alexandrov; On implementation of multilevel algorithms for nonlinear equations and equality constrained optimization, in progress.

[2] N. Alexandrov; *Multilevel Algorithms for Nonlinear Equations and Equality Constrained Optimization*, PhD Thesis, Rice, Houston, 1993. (Also: Technical Report TR93-20, Department of Computational and Applied Mathematics, Rice, Houston, 1993.)

[3] N. Alexandrov and J. Dennis; A class of general trust-region multilevel algorithms for systems of nonlinear equations: global convergence theory, in preparation.

[4] M. Avriel, *Nonlinear Programming*, Prentice-Hall, Englewood Cliffs, 1976.

[5] R. Balling and J. Sobieszczanski-Sobieski; Optimization of coupled systems optimization: a critical overview of approaches, AIAA Paper No. 94-4330, 1994. (Also: to appear in *Proceedings of the 5th AIAA/USAF/NASA/ISSMO Symposium on Multidisciplanary Analysis and Optimization,* Panama City, 1994.)

[6] R. Brent; Some efficient algorithms for solving systems of nonlinear equations, *SIAM J. Numer. Anal.* **10**, 1973, 327–344.

[7] K. Brown; *A Quadratically Convergent Method for Solving Simultaneous Nonlinear Equations*, PhD Thesis, Purdue, West Lafayette, 1966.

[8] K. Brown; Algorithm 316. Solution of simultaneous nonlinear equations, *Com. ACM* **10**, 1967, 728–729.

[9] K. Brown; A quadratically convergent Newton-like method based on Gaussian elimination, *SIAM J. Numer. Anal.* **6**, 1969, 560–569.

[10] K. Brown and J. Dennis; On the Second Order Convergence of Brown's derivative-free method for solving simultaneous nonlinear equations, Research Report No. 71-7, Department of Computer Science, Yale, New Haven, 1971.

[11] E. Cramer, J. Dennis, P. Frank, R. Lewis, and G. Shubin; Problem formulation for multidisciplinary optimization, Technical Report TR93-34, Department of Computational and Applied Mathematics, Rice, Houston, 1993.

[12] J. Dennis; Non-linear least squares and equations, in *The State-of-the-Art in Numerical Analysis*, 1976.

[13] J. Dennis, M. El-Alem, and M. Maciel; A global convergence theory for general trust-region-based algorithms for equality constrained optimization; Technical Report TR92-28, Department of Computational and Applied Mathematics, Rice, Houston, 1992.

[14] J. Dennis and R. Schnabel; *Numerical Methods for Unconstrained Optimization and Nonlinear Equations*, Prentice-Hall, Englewood Cliffs, 1983.

[15] M. El-Alem; *A Global Convergence Theory for a Class of Trust Region Algorithms for Constrained Optimization*, PhD Thesis, Rice, Houston, 1988.

[16] M. El-Alem; A global convergence theory for the Celis-Dennis-Tapia trust region algorithm for constrained optimization, *SIAM J. Numer. Anal.* **28**, 1991, 266–290.

[17] M. El-Alem; Robust trust-region algorithm with non-monotonic penalty parameter scheme for constrained optimization, Technical Report TR92-30, Department of Computational and Applied Mathematics, Rice, Houston, 1992.

[18] M. El Hallabi and R. Tapia; A global convergence theory for arbitrary norm trust-region methods for nonlinear equations, Technical Report TR87-25, Department of Computational and Applied Mathematics, Rice, Houston, 1987.

[19] R. Fletcher; *Practical Methods of Optimization*, Wiley, 1989.

[20] D. Gay; *Brown's Method and Some Generalizations, with Applications to Minimization Problems*, PhD Thesis, Cornell, Ithaca, 1975.

[21] P. Gill, W. Murray, and M. Wright; *Practical Optimization*, Academic, 1981.

[22] R. Haftka and Z. Gürdal; *Elements of Structural Optimization*, Kluwer, 1993.

[23] L. Lasdon and A. Waren; Generalized reduced gradient software for linearly and nonlinearly constrained problems, in *Design and Implemetation of Optimization Software* (Ed. by H. Greenberg), Sijthoff and Noordhoff, Netherlands, 1978, 335—362.

[24] M. Maciel; *A Global Convergence Theory for a General Class of Trust Region Algorithms for Equality Constrained Optimization*, PhD Thesis, Rice, Houston, 1993.

[25] J. Martínez; Generalization of the methods of Brent and Brown for solving nonlinear simultaneous equations, *SIAM J. Numer. Anal.* **1**, 1979, 434–448.

[26] J. Martínez; Solving nonlinear simultaneous equations with a generalization of Brent's method, *BIT* **20**, 1980, 501–510.

[27] J. Moré; Recent developments in algorithms and software for trust region methods, in *Mathematical Programming: The State of the Art* (Ed. by A. Bachem, M. Grötschel, and B. Korte), Springer, 1983.

[28] J. Moré and D. Sorensen; Computing a trust region step, *SIAM J. Sci. Stat. Comput.* **4**, 1983, 553–572.

[29] M. Powell; Convergence properties of a class of minimization algorithms, in *Nonlinear Programming* 2, (Ed. by O. Mangasarian, R. Meyer, and S. Robinson), Academic, 1975.

[30] G. Shultz, R. Schnabel, and R. Byrd; A family of trust-region-based algorithms for unconstrained minimization with strong global convergence pProperties, *SIAM J. Numer. Anal.* **22**, 1985, 47–67.

[31] J. Sobieszczanski–Sobieski, B. James, and Au. Dovi; Structural optimization by multilevel decomposition, AIAA Paper No. 83-0832-CP, 1993. (Also: in *Proc. AIAA/ASME/ASCE/AHS 24th Structures, Structural Dynamics and Materials Conference*, Lake Tahoe, 1983.)

[32] D. Sorensen; Newton's method with a model trust region modification, *SIAM J. Numer. Anal.* **19**, 1982, 409–426.

[33] G. Stewart, Conjugate direction methods for solving systems of linear equations, *Numer. Math.* **21**, 1973, 285–297.

Shape Optimization in One Shot*

Eyal Arian and Shlomo Ta'asan
The Weizmann Institute of Science
Rehovot 76100, Israel

Institute for Computer Applications in Science and Engineering
NASA Langley Research Center
Hampton VA 23681, USA

The multigrid one shot method for optimal shape design problems, governed by elliptic systems, is introduced for the infinite dimensional design space. In this case the design variable is a function whose discrete representation involves increasing number of variables with grid refinement. The minimization algorithm uses Lagrange multipliers to calculate sensitivity gradients. A gradient descent algorithm is accelerated by a set of coarse grids. It optimizes for different scales in the representation of the boundary position on different discretization levels. Numerical results have been obtained for an optimal shape design of a two dimensional geometry with a finite element discretization. The computer time for solving the optimization problem, up to the discretization error, was of the same order of magnitude as solving the constraint PDE just a few times.

1. Introduction

Optimal shape design (OSD) problems involve the performance improvement of physical systems by modification of geometrical configurations. An example is the design of an airfoil wing which will provide maximum lift over drag under given flow conditions. Recently an increased effort is being made in the numerical solution of OSD problems, especially for aerodynamic systems. An overview of the literature on OSD can be found in [10], [15], [16], [19].

Mathematically, the problem is to minimize a cost function under certain constraints, which are a set of PDEs called *state equations*. The cost function is defined to measure the performance of the physical system. A standard solution process involves an iterative algorithm where each iteration is composed of two steps. First, the shape variables are updated with the sensitivity gradients, which are gradients of the cost function with respect to the shape variables. Then the state variables are updated by solving the constraint equation with the new values of the shape variables. The repeated solution of the constraint PDE makes this computation extremely time consuming and in some cases not practical.

A crucial step in the iterative solution is to determine the sensitivity gradients of the cost function with respect to the design variables, i.e., the position of the boundary. The first formula for the sensitivity of a PDE with respect to the shape of the domain is found in Hadamard [8] (1910). Since then several methods were

*This research was supported by the National Aeronautics and Space Administration under NASA Contract No. NAS1-19480, and for the first author also by the Nathan and Emily Blum Scholarship for Ph.D. studies at the Weizmann Institute of Science.

developed to calculate the sensitivity gradients; among them is the *adjoint method* [7], [11], [15]. In the adjoint formulation, a Lagrangian is defined together with Lagrange multipliers functions, which are also called *costate variables*. Costate and design equations are derived from the variation of the Lagrangian, and together with the state equation form the necessary conditions for a minimum. The sensitivity gradients are the design equation residuals calculated with the solutions of the state and costate PDEs.

Several works have been done to improve the performance of the existing algorithms using multiscale ideas, some using an hierarchical approach [5], [14], and some using multigrid methods [2], [11], [12], [17], [18]. Multigrid (MG) methods were originally developed to accelerate the convergence rate of the numerical solutions of PDEs [3], [9], [20]. Such methods were first applied to aerodynamic shape design by A. Jameson in 1988 [11]. A multigrid solver was used there to accelerate the convergence of the solution of the state and costate equations, thus reducing the complexity of each optimization iteration to $O(N)$ operations, where N is the total number of grid points. However, the number of iterations depends on the problem and might be very large for some problems.

The adjoint and MG methods were combined to solve an optimal design problem, in "one shot", for the finite dimensional design by S. Ta'asan (1991) [17]. In the finite dimensional approach the design variables are represented as a finite sum of some preassigned base functions. The main idea in [17] is to represent the state, the costate and the design equations on coarser grid with the full approximation scheme (FAS)[3]. It is shown in [17] that in general the use of Lagrange multipliers is essential to achieve acceleration of the fine grid solution process by coarser grids. The algorithm optimizes the design variables on coarse grids, thus, eliminates the repeated solution of fine grid equations per optimization step. The one shot algorithm was applied successfully to the small disturbance approximation of an aerodynamic wing design problem in a subsonic flow by S. Ta'asan, G. Kuruvila and M. D. Salas (1992) [18]. Recently this work was extended to a shape design problem where the grid is updated as the shape changes [12] (1994). The performance of the algorithm in [12] and [18] is a substantial improvement in terms of computational cost. However, the performance of the finite dimensional one shot algorithm depends on on the choice of base functions and on the level on which the different design variables were optimized.

In [2] (1994) the multigrid one shot algorithm was extended to the infinite dimensional design space. The performance of the algorithm in [2] was independent of grid size, and the minimum was reached in a cost of solving the analysis problem just a few times. An analysis, which is based on the reduction of the problem to the boundary, was performed both to predict the convergence rate of a two grid algorithm and to determine the minimization step.

A different approach of multilevel shape optimization was recently proposed by Beux and Dervieux [5] (1993). A descent direction was computed with alternatively different levels of parametrizations for describing smoothly a two dimensional nozzle shape in a subsonic Euler flow. Later Marco and Beux [14] (1993) used the hirarchical approach in [5] combined with ideas from [17]. Results in [14] are reported as improvement of those in [5], but still depend on the number of levels and are restricted to coarse grids. Although the complexity of the hirarchical algorithm is more than $O(N)$, as in multigrid algorithms, it has an advantage that it is easier to apply.

In this paper we apply the infinite dimensional multigrid one shot method [2] to optimal shape design of a two dimensional nozzle geometry with elliptic constraint. In the discrete space we define the design variables to be the locations of the grid points which define the discrete boundary, thus the number of design variables increases with the grid refinement. The sensitivity gradients which are used in the minimization algorithm are computed with the adjoint method (Lagrange multipliers) using the necessary conditions for a minimum. The solution of the fine grid problem is accelerated by a set of coarse grids, where on each grid two processes take place: the relaxation, which smooths the errors of the constraint equation variables, the Lagrange multipliers variables as well as the errors in the boundary position and the coarse grid correction which eliminates the smooth errors in these variables. The coarse grid sensitivity gradients are computed using the full approximation scheme (FAS) of the necessary conditions. In that way a reduction in the computational time is gained by reducing the number of optimization steps required to reach the minimum. Due to the elliptic characteristic of the problems it is possible to perform the optimization step in a local region neighboring the boundary, reducing the computational cost of each optimization step to only $O(\sqrt{N})$ operations.

2. Problem definition and necessary conditions for a minimum

Consider the minimization problem defined by:

$$\min_{\Gamma(x)} \int_{\Gamma(x)} \left(\frac{\partial \phi}{\partial y} - f^*(x)\right)^2 dx \tag{1}$$

with $f^*(x)$ a given function and where ϕ satisfies the state equation:

$$\begin{cases} \Delta\phi = f & \text{on } 0 < x < 1\ ;\ \Gamma(x) < y < 1 \\ \phi = g(x) & \text{on } y = 1 \\ \phi = \phi_0 & \text{on } y = \Gamma(x)\,. \end{cases} \tag{2}$$

The location of the boundary defines the design variable, i.e., the function $\Gamma = \Gamma(x)$. The domain of solution is denoted by $\Omega = \left\{(x, y)\middle| 0 \le x \le 1\ ;\ 0 \le y \le 1\right\}$.

2.1. Derivation of the necessary conditions

We apply the adjoint method to the optimal boundary design problem (1),(2). The variable space is enlarged by adding Lagrange multiplier functions or costate variables denoted by λ. A Lagrangian is defined to be the sum of the functional and a linear term in the costate variables which vanishes as the constraint equation is satisfied;

$$E(\phi,\lambda,u)=\int_{\Gamma(x)}\Big(\frac{\partial\phi}{\partial y}-f^*(x)\Big)^2dx+\int_{\Omega}\lambda\Delta\phi dxdy\,. \tag{3}$$

A perturbation of the Lagrangian with respect to all the variables independently, i.e., state, costate and design, result in a variation of the Lagrangian. That is,

$$\begin{aligned}\phi &\leftarrow \phi+\varepsilon\tilde{\phi}\\ \lambda &\leftarrow \lambda+\varepsilon\tilde{\lambda} && (4)\\ \Gamma(x) &\leftarrow \Gamma(x)+\varepsilon\tilde{\alpha}(x) && (5)\end{aligned}$$

with $\tilde{\phi},\tilde{\lambda}\in L_2(\Omega)$ and $\tilde{\alpha}(x)$ is a real function in the same class as $\Gamma(x)$. The variation of the Lagrange function, δE, in the first order approximation in ε, is given in the following form:

$$\begin{aligned}\delta E(\phi,\lambda,\Gamma)_{\varepsilon\tilde{\alpha}}=\varepsilon\Big[\int_{\Gamma}2(\frac{\partial\phi}{\partial y}-f^*)(\frac{\partial\tilde{\phi}}{\partial y}+\tilde{\alpha}\frac{\partial^2\phi}{\partial y^2})dx\\ +\int_{\Omega}\tilde{\phi}\Delta\lambda dxdy+\int_{\Gamma}(\lambda\frac{\partial\tilde{\phi}}{\partial n}-\tilde{\phi}\frac{\partial\lambda}{\partial n})d\sigma\Big]+O(\varepsilon^2)\,.\end{aligned} \tag{6}$$

The relation between the integration elements on the boundary and on the x axis is given by $dx=d\sigma\cos\theta$. The y derivative of $\tilde{\phi}$ is decomposed into a normal and tangential derivatives,

$$\frac{\partial\tilde{\phi}}{\partial y}=\sin\theta\frac{\partial\tilde{\phi}}{\partial\sigma}+\cos\theta\frac{\partial\tilde{\phi}}{\partial n} \tag{7}$$

where $\theta=\theta(x)$ is the angle between the boundary, $\Gamma(x)$, and the x axis. Substituting (7) in (6) gives

$$\begin{aligned}\delta E_{\varepsilon\tilde{\alpha}}=\varepsilon\int_{\Omega}\tilde{\phi}\Delta\lambda dxdy\\ +\varepsilon\int_{\Gamma}\Big[2\cos\theta(\frac{\partial\phi}{\partial y}-f^*)\Big(\sin\theta\frac{\partial\tilde{\phi}}{\partial\sigma}+\cos\theta\frac{\partial\tilde{\phi}}{\partial n}+\tilde{\alpha}\frac{\partial^2\phi}{\partial y^2}\Big)\\ +\lambda\frac{\partial\tilde{\phi}}{\partial n}-\tilde{\phi}\frac{\partial\lambda}{\partial n}\Big]d\sigma+O(\varepsilon^2)\,.\end{aligned} \tag{8}$$

Integrating $\frac{\partial\tilde{\phi}}{\partial\sigma}$ by parts, and substituting $\tilde{\phi} = -\tilde{\alpha}\frac{\partial\phi}{\partial y}$, gives

$$\begin{aligned}\delta E_{\varepsilon\tilde{\alpha}} = \varepsilon \int_{\Omega} \tilde{\phi}\Delta\lambda dxdy \\ +\varepsilon \int_{\Gamma} \frac{\partial\tilde{\phi}}{\partial n}\Big[\lambda + 2\cos^2\theta(\frac{\partial\phi}{\partial y} - f^*)\Big]d\sigma \\ +\varepsilon \int_{\Gamma} \tilde{\alpha}\frac{\partial\phi}{\partial y}\Big[\frac{\partial}{\partial\sigma}\Big(\sin(2\theta)(\frac{\partial\phi}{\partial y} - f^*)\Big)\Big]d\sigma \\ +\varepsilon \int_{\Gamma} \tilde{\alpha}\Big[2\frac{\partial^2\phi}{\partial y^2}\cos\theta(\frac{\partial\phi}{\partial y} - f^*) + \frac{\partial\phi}{\partial y}\frac{\partial\lambda}{\partial n}\Big]d\sigma + O(\varepsilon^2)\,.\end{aligned} \tag{9}$$

At a minimum the first order terms vanish, resulting in the necessary condition for a minimum which include the state equation (2) and the costate and the design equations:

The costate equation

$$\begin{cases} \Delta\lambda = 0 & \text{on } 0 < x < 1\;;\; \Gamma(x) < y < 1 \\ \lambda = 0 & \text{on } y = 1 \\ \lambda + 2\cos^2\theta(\frac{\partial\phi}{\partial y} - f^*) = 0 & \text{on } y = \Gamma(x)\,. \end{cases} \tag{10}$$

The design equation

The design equation is simplified by using the costate boundary condition yielding

$$\mathcal{A}(\phi, \lambda, \Gamma) = 0 \quad on \quad y = \Gamma(x) \tag{11}$$

where

$$\mathcal{A}(\phi, \lambda, \Gamma) = \frac{\partial\phi}{\partial y}\frac{\partial}{\partial\sigma}\Big(\lambda\tan\theta\Big) + \frac{\lambda}{\cos\theta}\frac{\partial^2\phi}{\partial y^2} - \frac{\partial\phi}{\partial y}\frac{\partial\lambda}{\partial n}\,. \tag{12}$$

If the state and costate equations are satisfied then the variation in the Lagrangian is given by

$$\delta E_{\varepsilon\tilde{\alpha}} = \varepsilon \int_{\Gamma} \tilde{\alpha}\mathcal{A}(\phi(\Gamma), \lambda(\Gamma), \Gamma)d\sigma + O(\varepsilon^2)\,. \tag{13}$$

Therefore the choice $\tilde{\alpha} = -\varepsilon\mathcal{A}$ reduces the functional. This can be used to derive a gradient descent algorithm for the OSD problems.

3. The small disturbance approximation

In order to analyze the numerical solution of OSD problems we reduce it to an optimal boundary control problem on a fixed domain although in the numerical computation the domain is not fixed. The resulting control problem can be analyzed by techniques presented in [2]. This analysis help to design an efficient numerical method as well as to predict its performance.

Consider a solution ϕ_0 of the state equation (2) in a domain Ω_0. Let Γ_0 be part of the boundary $\Gamma_0 \subset \partial\Omega$, and consider a perturbation of the boundary position Γ_0 with (4). The perturbed boundary,$\Gamma_{\varepsilon\tilde{\alpha}}$, defines a domain, $\Omega_{\varepsilon\tilde{\alpha}}$, with a state and costate solutions $\phi_{\varepsilon\tilde{\alpha}}$ and $\lambda_{\varepsilon\tilde{\alpha}}$ respectively. Solutions are extended analytically to a neighborhood containing $\Omega \bigcup \Omega_{\varepsilon\tilde{\alpha}}$. The extentions will be denoted as the original functions. The following are relations between quantities on the perturbed domain, $\Omega_{\varepsilon\tilde{\alpha}}$, and those on the original one, Ω.

$$\Delta\phi_{\varepsilon\tilde{\alpha}}\Big|_{\Omega_{\varepsilon\tilde{\alpha}}} = \Delta\phi\Big|_{\Omega_0} + \varepsilon\Delta\tilde{\phi}\Big|_{\Omega_0} + O(\varepsilon^2) \tag{14}$$

$$\phi_{\varepsilon\tilde{\alpha}}\Big|_{\Gamma_{\varepsilon\tilde{\alpha}}} = \phi\Big|_{\Gamma_0} + \varepsilon\tilde{\phi}\Big|_{\Gamma_0} + \varepsilon\tilde{\alpha}\frac{\partial\phi}{\partial y}\Big|_{\Gamma_0} + O(\varepsilon^2) \tag{15}$$

where

$$\tilde{\phi} = \lim_{\varepsilon\to 0}\frac{\phi_{\varepsilon\tilde{\alpha}} - \phi}{\varepsilon}.$$

The second order terms in (14) and (15) can be neglected for sufficiently small ε, depending on $\tilde{\alpha}$, i.e., for $\varepsilon \leq \varepsilon(\tilde{\alpha})$.

3.1. The small disturbance minimization problem

Relations (14) and (15) are used to reduce the OSD problems to a minimization problem on a fixed domain with some unknown boundary data, i.e. an optimal boundary control (OBC) problem. In this problem Ω and ϕ fixed, while $u = \tilde{\alpha}$ and $\varphi = \tilde{\phi}$ are the design and the state variables respectively. The OBC problem is

$$\min_{u(x)} \int_{\Gamma(x)} \left(\frac{\partial\varphi}{\partial y} + u\frac{\partial^2\phi}{\partial y^2} + \frac{\partial\phi}{\partial y} - f^*\right)^2 dx \tag{16}$$

where φ satisfies

$$\begin{cases} \Delta\varphi = 0 & \text{on } 0 < x < 1\ ;\ \Gamma(x) < y < 1 \\ \varphi = 0 & \text{on } y = 1 \\ \varphi = -u\dfrac{\partial\phi}{\partial y} & \text{on } y = \Gamma(x)\,. \end{cases} \tag{17}$$

These type of problems were treated in [2].

3.2. Derivation of the optimality conditions

The small disturbance Lagrangian is given by

$$\begin{aligned} E \;=\; & \int_{\Gamma(x)} \left(\frac{\partial\varphi}{\partial y} + u\frac{\partial^2\phi}{\partial y^2} + \frac{\partial\phi}{\partial y} - f^*\right)^2 dx \\ & + \int_{\Omega} \varphi\Delta\kappa\, dx dy + \int_{\Gamma(x)} \left(\kappa\frac{\partial\phi}{\partial n} - \varphi\frac{\partial\kappa}{\partial n}\right) d\sigma \end{aligned} \tag{18}$$

where κ is a Lagrange multiplier. The necessary conditions for a minimum of the small disturbance problem are the state equation (17) together with the costate and design equations.

Costate equation

$$\begin{cases} \Delta\kappa = 0 & \text{on } 0 < x < 1\,;\ \Gamma(x) < y < 1 \\ \kappa = 0 & \text{on } y = 1 \\ \kappa + 2\cos^2\theta\Big(\frac{\partial\varphi}{\partial y} + u\frac{\partial^2\phi}{\partial y^2} + \frac{\partial\phi}{\partial y} - f^*\Big) & \text{on } y = \Gamma(x)\,. \end{cases}$$

Design equation

$$\mathcal{A}(\kappa, \Gamma) = 0 \quad on \;\; y = \Gamma(x). \tag{19}$$

where

$$\mathcal{A}(\kappa, \Gamma) = -\frac{\partial\phi}{\partial y}\frac{\partial}{\partial\sigma}\Big(\kappa\tan\theta\Big) - \frac{\kappa}{\cos\theta}\frac{\partial^2\phi}{\partial y^2} + \frac{\partial\phi}{\partial y}\frac{\partial\kappa}{\partial n}\,. \tag{20}$$

The analysis of numerical processes for the OSD will be done using the small disturbance OBC problem.

3.3. Fourier analysis

A useful analysis for understanding different issues regarding the numerical solution of the OSD via the OBC is done by considering high frequency changes in the design variable u. This is done by freezing the coefficients, and performing the analysis presented in [2]. Freezing the coefficient is justified as long as the change in the control variable are highly oscillatory compared to changes in the coefficients appearing in the OBC problem. Applying this analysis to the gradient descent algorithm we obtained relations between the changes in the control and changes in the design equation residuals. Let $\bar{\varphi}$, $\bar{\kappa}$ and $\bar{u}$ stand for the errors in the variables φ, κ, u respectively. Examining $\bar{u}$ of the form

$$\bar{u} = \hat{u}e^{i\omega x} \tag{21}$$

we obtain the solutions of the state and costate errors,

$$\begin{aligned} \bar{\varphi}(x, y) &= -\frac{\partial\phi}{\partial y}(x_0)\hat{u}e^{i\omega x}e^{-|\omega|y} \\ \bar{\kappa}(x, y) &= -2\cos^2\theta_0\Big(\frac{\partial^2\phi}{\partial y^2}(x_0) + |\omega|\frac{\partial\phi}{\partial y}(x_0)\Big)\hat{u}e^{i\omega x}e^{-|\omega|y}\,. \end{aligned} \tag{22}$$

The Fourier transform of the design equation residuals is given by

$$\hat{A}(\omega) = \hat{T}(\omega)\hat{u}(\omega) \tag{23}$$

where

$$\hat{T}(\omega) = \Big(\frac{\partial^2\phi}{\partial y^2}(x_0) + |\omega|\frac{\partial\phi}{\partial y}(x_0)\Big)\cdot \\ \Big[\sin 2\theta_0\Big(\frac{\partial\phi}{\partial y}(x_0)\cos\theta_0 i\omega + \sin\theta_0|\omega|\Big) \\ +2\cos\theta_0\frac{\partial^2\phi}{\partial y^2}(x_0) + 2\cos^2\theta_0\frac{\partial\phi}{\partial y}(x_0)|\omega|\Big]. \tag{24}$$

For large ω there exist a positive constant C duch that

$$|\hat{T}(\omega)| \geq C|\omega|^2 \tag{25}$$

thus high frequency errors in the shape are amplified in the residuals of the design equation. Note that for $\theta_0 = 0$ the symbol $\hat{T}(\omega)$ may vanish if

$$\frac{\partial\phi}{\partial y}(x_0)\frac{\partial^2\phi}{\partial y^2}(x_0) < 0$$

and difficulties are expected in the convergence for such cases.

4. Discretization

In discretizing the OSD problem one can distinguish several approaches. One approach is to use a fixed grid in the interior of the domain and local grid patches near the boundary [4]. Another approach, which is much easier from programming point of view, is to use a body fitted grid. The grid is refined towards the design boundary. Two common ways to discretize the equations are finite volume cell centered grid [12] and finite elements vertex grid [5], [14], [15].

Finite element discretization

We used a finite element discretization and developed the costate and design equations in the discrete space. The equations are treated in their weak formulation:

$$\int_\Omega \nabla W \cdot \nabla\phi \, dxdy = \int_\Omega W f dxdy \tag{26}$$

for every $W \in H_0^1(\Omega)$, where ϕ satisfies the boundary conditions. $H_0^1(\Omega)$ is defined by

$$H_0^1(\Omega) = \{w \in H^1(\Omega) : w\Big|_{\partial\Omega} = 0\} \tag{27}$$

and $H^1(\Omega)$ is the Sobolev space of order 1 which is defined on the domain Ω. The domain was divided into triangles. The base functions $W_{ij}(x, y)$ were chosen to be linear elements on triangular meshes, and ϕ is given by

$$\phi(x, y) = \sum_{ij}\phi_{ij}W_{ij}(x, y). \tag{28}$$

The costate and design equations were derived by perturbing the boundary grid points in the y direction and following a similar derivation to that presented in the previous sections.

The discrete minimization problem is given by

$$\min_{\Gamma^h} \sum_{i=1}^{N-1} \left(\frac{\partial \phi^h}{\partial y}_i - f_i^{*h} \right)^2 \tag{29}$$

subject to

$$\begin{cases} \Delta^h \phi^h = f^h & \text{on } 0 < x < 1 \, ; \; \Gamma(x) < y < 1 \\ \phi^h = g^h(x) & \text{on } y = 1 \\ \phi^h = \phi_0^h & \text{on } y = \Gamma(x) \end{cases} \tag{30}$$

The discrete costate and design equations are of the form

$$\begin{cases} \Delta^h \lambda^h = 0 & \text{on } 0 < x < 1 \, ; \; \Gamma(x) < y < 1 \\ \lambda^h = 0 & \text{on } y = 1 \\ G^h(\phi^h, \lambda^h, \Gamma^h) = 0 & \text{on } y = \Gamma(x) \end{cases} \tag{31}$$

$$\mathcal{A}^h(\phi^h, \lambda^h, \Gamma^h) = 0 \text{ on } y = \Gamma(x)$$

for appropriate $G^h(\phi^h, \lambda^h, \Gamma^h)$ and $\mathcal{A}^h(\phi^h, \lambda^h, \Gamma^h)$.

5. A multigrid one shot minimization method

The gradient descent algorithm is preconditioned by applying it on a sequence of nested grids where each coarse grid accelerates the convergence rate of its finer grid. On each grid two processes are employed: relaxation and coarse grid correction (of all the variables, including the design variables). On coarse grids the state, the costate and the design equations are restricted from the finer grid with the full approximation scheme (FAS) [3].

5.1. Relaxation

On each level a relaxation is performed on the state, costate and design variables. The state and costate equations (which are elliptic PDEs) are relaxed by a Gauss-Seidel scheme. Updating the boundary position is done in the following way. A small change is introduced to the boundary position with the design equation residuals,

$$\Gamma^h \leftarrow \Gamma^h + \delta \mathcal{A}^h(\Gamma^h) \, . \tag{32}$$

Then its effect on the state and costate solutions is computed approximately. The amplitude of the actual change in the boundary's position, β, is then chosen so as to minimize the residuals of the design equation:

$$\min_{\beta} \|\mathcal{A}^h(\phi + \beta\tilde{\phi}(\delta), \lambda + \beta\tilde{\lambda}(\delta), \Gamma + \beta\tilde{\alpha})\|_{L_2}^2 \tag{33}$$

where $\tilde{\phi}^h(\delta)$ and $\tilde{\lambda}^h(\delta)$ are the changes in the solution of the state and costate equations as a result of a perturbation in the shape variables by (32).

Let $\mathcal{A}'$ and $\mathcal{A}''$ represent the first and second derivatives of $\mathcal{A}^h(\delta)$ at $\delta = 0$, then the solution of (33) is given approximately by

$$\beta = -\frac{(\mathcal{A}^h, \mathcal{A}')}{2(\mathcal{A}'', \mathcal{A}^h) + (\mathcal{A}', \mathcal{A}')}. \tag{34}$$

However, the quantity $\mathcal{A}^h$ is vanishing at the minimum while $\mathcal{A}'$ is not. Therefore close enough to the minimum we can neglect the second derivative term in the denominator and use

$$\beta \approx -\frac{(\mathcal{A}^h, \mathcal{A}')}{(\mathcal{A}', \mathcal{A}')}. \tag{35}$$

In the discrete level we take the first order approximation for $\mathcal{A}'$, i.e.,

$$\mathcal{A}'^h = \frac{\mathcal{A}^h(\delta_0) - \mathcal{A}^h(0)}{\delta_0}. \tag{36}$$

The line search algorithm for the minimization step size, denoted by $\mathcal{S}^h$, is given below.
<u>Line search</u>

1. Perturb the boundary position variables with a small trial amplitude δ_0.

2. Compute the effect of the perturbation by solving the state and costate equations for $\tilde{\phi}(\delta_0)$ and $\tilde{\lambda}(\delta_0)$ in a narrow strip neighboring the boundary.

3. Compute the change in the residuals of the design equation with $\mathcal{A}'^h = \frac{\mathcal{A}^h(\delta_0) - \mathcal{A}^h(0)}{\delta_0}$.

4. Set $\beta = -\frac{\langle \mathcal{A}^h, \mathcal{A}'^h \rangle}{\langle \mathcal{A}'^h, \mathcal{A}'^h \rangle}$.

Let us assume that the residuals $\mathcal{A}^h(\Gamma^h)$ are dominated by high frequency Fourier modes. By the small disturbance approximation this is equivalent to performing a high frequency change in the boundary data (see Section 3). In elliptic systems a perturbation of the boundary condition with a Fourier mode $e^{i\omega\xi}$ has an exponential decaying effect on the interior solution, as discussed in [2]. Therefore, both in the trial perturbation stage (32) and after the actual boundary position update, the state and costate solutions and the grid points position relative to the boundary, should be updated only in a narrow strip neighboring the boundary. Therefore the computational cost of each optimization step is only in the order of $O(\sqrt{N})$.

Summarizing, the relaxation sweep for the optimization problem, $\mathcal{R}^h$, is given below.

Relaxation

1. `Perform one relaxation of the state equation for` ϕ^h.
2. `Perform one relaxation of the costate equation for` λ^h.
3. `Compute the amplitude of the perturbation,` β`, by performing the line search algorithm,` $\mathcal{S}^h$.
4. `Update the boundary position:` $\Gamma^h \leftarrow \Gamma^h + \beta \mathcal{A}^h(\Gamma^h)$.
5. `Update the grid to fit the new boundary.`
6. `Perform a few local relaxations of the state and costate equations in a narrow strip near the boundary.`

5.2. The one shot minimization algorithm

The one shot multigrid algorithm is done in the framework of full multigrid (FMG) scheme. In this framework the starting approximation of the solution on each fine grid is obtained by solving the problem on a coarser grid and then interpolating the solution. On the coarsest grid the problem is solved in a single grid algorithm. On finer grids the FMG scheme uses a V cycle multigrid algorithm in order to solve the problem. The V cycle is composed of a recursive applications of a relaxation and coarse grid correction. The problem is solved in one application of an FMG solver.

FMG cycle

The following is a n-FMG(ν_1, ν_2) cycle to solve the problem with M grids. The coarsest mesh size is denoted by h_c.

1. `Start with the coarsest grid,` $(h = h_c)$`, and solve the problem with a standard minimization algorithm.`

2. Interpolate the solution to a finer grid, rescale $h \to \frac{h}{2}$.

3. Perform n times $V^h(\nu_1, \nu_2)$ cycles.

4. If the finest grid is reached then stop, else go to 2.

Vcycle

The following is a $V^h(\nu_1, \nu_2)$ cycle to accelerate the convergence of a fine grid with mesh size h. The parameters ν_1 and ν_2 are integers (typically $\nu_1 = 2$ and $\nu_2 = 1$).

1. Perform ν_1 relaxation sweeps, $\mathcal{R}^h$.

2. Restrict the state, the costate and the design equations to the coarse grid with $H = 2h$,

   State

$$\begin{aligned} \Delta^H \phi^H &= I_h^H f_\phi^h + \Delta^H \tilde{I}_h^H \phi^h - I_h^H \Delta^h \phi^h \text{ on } \Omega^H \\ \phi^H &= I_h^H \phi_0^h \text{ on } \Gamma^H . \end{aligned}$$

   Costate

$$\begin{aligned} \Delta^H \lambda^H &= I_h^H f_\lambda^h + \Delta^H \tilde{I}_h^H \lambda^h - I_h^H \Delta^h \lambda^h \text{ on } \Omega^H \\ G^H(\phi^H, \lambda^H, \Gamma^H) &= I_h^H g_\lambda^h + G^H(I_h^H \phi^h, I_h^H \lambda^h, I_h^H \Gamma^h) \\ &\quad - I_h^H G^h(\phi^h, \lambda^h, \Gamma^h) \text{ on } \Gamma^H \end{aligned}$$

   Design

$$\begin{aligned} A^H(\phi^H, \lambda^H, \Gamma^H) &= I_h^H f_\Gamma^h + A^H(I_h^H \phi^h, I_h^H \lambda^h, I_h^H \Gamma^h) \\ &\quad - I_h^H A^h(\phi^h, \lambda^h, \Gamma^h) \text{ on } \Gamma^H \end{aligned}$$

   Rescale $h \to 2h$.

3. If the coarsest level is not reached go to 1.

4. Solve the problem with a standard minimization algorithm.

5. `Interpolate the coarse grid correction to the finer grid:`

$$\begin{aligned} \phi^h &\leftarrow \phi^h + I_{2h}^h(\phi^{2h} - I_h^{2h}\phi^h) \text{ on } \Omega^h \\ \lambda^h &\leftarrow \lambda^h + I_{2h}^h(\lambda^{2h} - I_h^{2h}\lambda^h) \text{ on } \Omega^h \\ \Gamma^h &\leftarrow \Gamma^h + \bar{I}_{2h}^h(\Gamma^{2h} - \bar{I}_h^{2h}\Gamma^h) \text{ on } \Gamma^h . \end{aligned}$$

`Rescale` $h \to \frac{h}{2}$.

6. `Perform` ν_2 `relaxation sweeps,` $\mathcal{R}^h$.

7. `If the finest grid is reached then stop, else go to 5.`

A similar algorithm, which spends more time on coarse grids, is the $W(\nu_1, \nu_2)$ cycle (see [4]). The computational cost of both the V and the W cycles is $O(N)$ operations.

6. Numerical results

The numerical experiments were done with a Gaussian target function (see problem definition (1)):

$$f^*(x) = ae^{-30(x-\frac{1}{2})^2}. \tag{37}$$

This target function contains all Fourier frequencies and the parameter a controls the curvature of the target shapes. The starting approximation for the shape, on the coarsest grid, was a flat surface: $\Gamma(x) = 0$.

Convergence performance

Figures 1A and 1B show the convergence rate for a range of mesh sizes showing a mesh independent convergence rate. In Figure 1A the target function, f^* was taken with the parameter $a = 0.1$, (see (37)), while in Figure 1B it was $a = 0.2$ (meaning that the curvature of the optimal shape geometry in the later case was much higher). The results show a mesh size independent convergence rate for both cases. We see that as the geometry becomes more curved the convergence rate deteriorated.

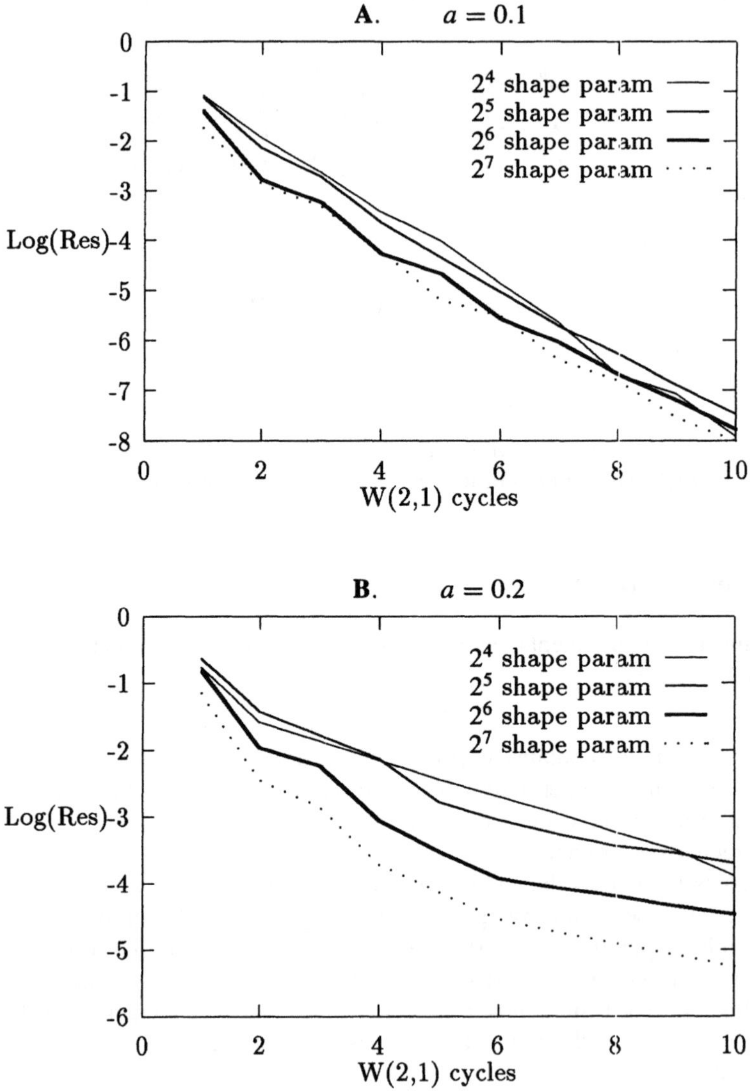

Figure 1. Convergence rates of the optimal shape design algorithm, with a finite element discretization, for mesh sizes in the range $2^4 \times 2^4$ to $2^7 \times 2^7$ grid points, on a logarithmic scale. The number of design parameter on the finest grid is 128.

Discretization error

Figure 2 shows the optimal shape solution, $\Gamma^h(x)$, for the different levels in the FMG algorithm. Level 5, for example, corresponds to the optimal shape solution on a grid which is composed of $2^5 \times 2^5$ points. We can see that the shapes approach the continuum solution (target shape) in an $O(h)$ behavior. This is confirmed by Figure 3A where the error in the boundary's position, $\|\Gamma - \Gamma^*\|_{L_2}$, is plotted as a function of the MG-W cycle, for $a = 0.2$, where the target shape $\Gamma^*(x)$ is known. It is seen that after $3 - 4$ cycles the discretization error is reached. Therefore, it is unnecessary to perform 10 MG cycles since solutions beyond the discretization error are of no interest. Numerical experiments show that a 4FMG-W cycles were enough to reach the discretization error on all levels. Figure 3B shows the discretization error as a function of the mesh size, h, in a logarithmic scale. The dependence of the error on the mesh size satisfies

$$\|\Gamma - \Gamma^*\|_{L_2} \propto O(h). \tag{38}$$

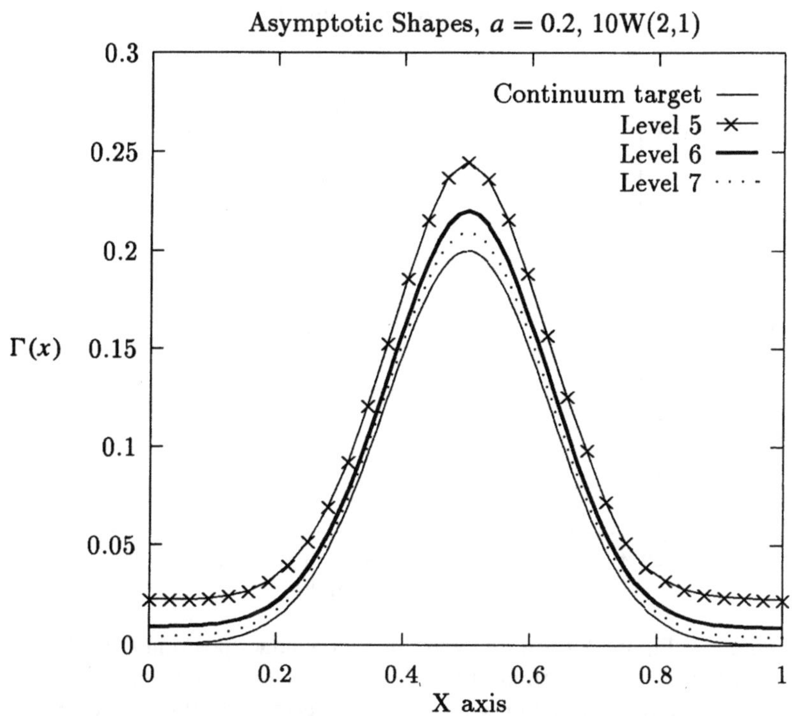

Figure 2. Asymptotic shapes for grids with number of design variables in the range 2^5 to 2^7.

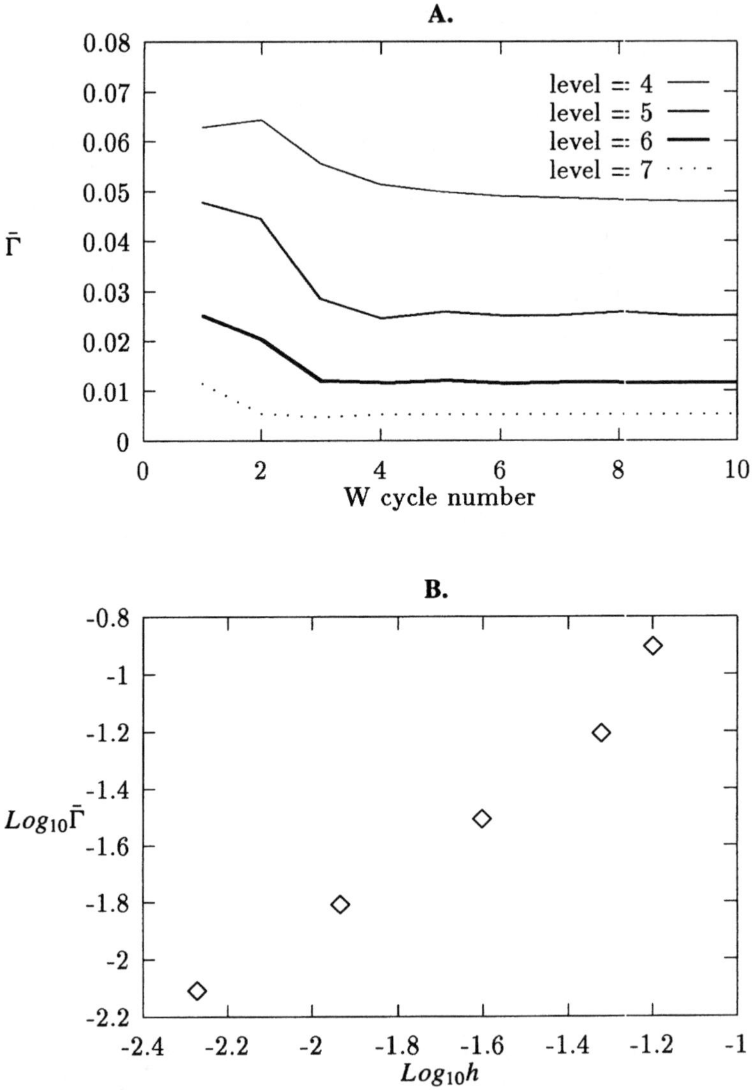

Figure 3. Figure A shows the error in the boundary position $\bar{\Gamma} = \Gamma - \Gamma^*$ where Γ^* is the optimal shape. The error is given in the maximum norm for a target function with $a = 0.2$, performing W(2,1) cycles. It shows that a cycle of the form 4FMG W(2,1) is enough to reach the discretization error. Figure B shows the asymptotic shape error, in L_2 norm. It shows that the discretization error is of the order of the meshsize: $\|\bar{\Gamma}\|_{L_2} \propto O(h)$.

7. Conclusion

Numerical results have been obtained for an optimal shape design of a two dimensional OSD problem with a finite element discretization. A cost function was defined on the design boundary and measured the distance (in L^2 norm) of the normal derivative, of the constraint PDE solution, from a desired given function. The state, costate and design variables, were smoothed at each relaxation sweep, which was accelerated using coarse grids, on all levels leading to a very efficient converging algorithm. With this "one shot" full multigrid algorithm a mesh size independent of convergence rate has been achieved for meshes ranging between $2^2 \times 2^2$ to $2^7 \times 2^7$ grid points. The computer time for solving the optimization problem, up to the discretization error, for 128 design parameters, was of the same order of magnitude as solving just a few times the constraint PDE.

References

[1] S. Agmon; *Lectures on Elliptic Boundary Value Problems*, D. Van Nostrand, 1965.

[2] E. Arian and S. Ta'asan; Multigrid one shot methods for optimal design problems: infinite dimensional control, ICASE Report No. 94-52, Hampton, 1994.

[3] A. Brandt; Multilevel Adaptive Solutions to Boundary Value Problems, Math. Comp. 31, pp 333-390 (1977).

[4] A. Brandt; *Multigrid Techniques: 1984 Guide, with Applications to Fluid Dynamics*, 1984.

[5] F. Beux and A. Dervieux; A hierarchical approach for shape optimization, INRIA Rapports de Recherche N. 1868, 1993.

[6] D. Chenais; On the existence of a solution in a domain identification problem, *J. Math. Anal. Appl.* **52**, 1975, 189–289. Shape optimization in shell theory, *Eng. Opt.* **11**, 1987, 289–303.

[7] R. Fletcher; *Practical Methods of Optimization*, Wiley, 1981.

[8] J. Hadamard; *Lecon Sur le Calcul des Variation* Gauthier-Villars, paris, 1910.

[9] W. Hackbush; *Multi-Grid Methods and Applications*, Springer, Berlin, 1985.

[10] J. Haslinger and P. Neittaanmϕaki; *Finite Element Approximation for OSD: Theory and Application*, Wiley, 1988.

[11] A. Jameson; Aerodynamic design via control theory, *J. Scient. Comp.* **3**, 1988, 233–260.

[12] G. Kuruvila, S. Ta'asan, and M. Salas; Airfoil optimization by the one-shot method, *Optimum Design Methods in Aerodynamics*, AGARD-FDP-VKI Special Course, 1994.

[13] M. Lighthill; A new method of two dimensional aerodynamic design. R & M 1111, Aeronautical Research Council, 1945.

[14] N. Marco and F. Beux; Multilevel optimization: application to shape optimum design with a one-shot method, INRIA Rapport de Recherche N. 2068, 1993.

[15] O. Pironneau; *Optimal Shape Design for Elliptic Systems*, Springer, Berlin, 1983.

[16] J. Sokolowski and J. Zolesio; *Introduction to Shape Optimization*, Springer, Berlin, 1992.

[17] S. Ta'asan; "One shot" methods for optimal control of distributed parameter systems I: finite dimensional control, ICASE Report No. 91-2, Hampton, 1991.

[18] S. Ta'asan, G. Kuruvila, and M. Salas; Aerodynamic design and optimization in one shot, AIAA Paper No. 92-0025, 1992.

[19] von Karman Institute; *Optimum Design Methods in Aerodynamics*, AGARD-FDP-VKI Special Course, April 25-29, (1994)

[20] P. Wesseling; *An Introduction to Multigrid Methods*, Wiley, 1991.

Computational Aspects of Differential Dynamic Programming and Nonlinear Programming Techniques for Optimal Control*

Jasbir S. Arora, Dale A. Holtz, and Narendra N. Kota
Optimal Design Laboratory
Civil and Environmental Engineering
Mechanical Engineering
The University of Iowa
Iowa City IA 52242, USA

In this paper, differential dynamic programming (DDP) and nonlinear programming (NLP) techniques are applied to problems of optimal control of nonlinear heating ventilation and air conditioning systems, and structural/mechanical systems. The two approaches are described and their computational aspects are discussed. It is shown that implementation details of the methods can have considerable impact on the performance of the methods. Example problems are solved to compare performance of the methods.

1. Introduction

The optimal control problem (OCP) considered in the paper is to minimize a cost functional

$$J = g(\mathbf{x}(T), T) + \int_0^T F(\mathbf{x}(t), \mathbf{u}(t), t)\, dt \tag{1}$$

subject to the state equations

$$\dot{\mathbf{x}} = \mathbf{f}(\mathbf{x}, \mathbf{u}, t)\,; \qquad \mathbf{x}(0) = \mathbf{x}_0 \tag{2}$$

and also the constraints

$$\phi_i(\mathbf{x}, \mathbf{u}, t, T) \leq 0\,; \quad i = 1, \ldots, p \tag{3}$$

which may represent functional, dynamic response or final time constraints. The vectors $\mathbf{x}$ and $\mathbf{u}$ are referred to as the state and control variables, respectively, and are defined as

$$\mathbf{x}(t) = [x_1(t)\ x_2(t) \cdots x_n(t)]^T\,; \quad \mathbf{u}(t) = [u_1(t)\ u_2(t) \cdots u_m(t)]^T\,. \tag{4}$$

For the purposes of this paper the functions g, F and $\mathbf{f}$ are assumed to be twice continuously differentiable with respect to its arguments and $\mathbf{x}(t)$ and $\mathbf{u}(t)$ are assumed to be continuous functions over the time interval $[0, T]$. However, less restrictive continuity requirements may be acceptable for some solution procedures.

*Financial support provided by the Iowa Energy Center and NASA Langley Research Center for this research is gratefully acknowledged. Also the authors would like to thank Dr. John M. House for providing NLP results for HVAC system.

Several computational procedures have been explored in the literature to solve this OCP. Many methods discretize the problem from the beginning and use nonlinear programming methods to solve the resulting finite dimensional optimization problem. Using this approach both the control and state variables may be treated as unknowns of the optimization problem ([5]). Or, with the use of design sensitivity analysis principles, only the control variables may be treated as unknowns of the optimization problem ([17]). Other procedures stay in the continuum domain, derive the computational methods, and introduce discretizations for only numerical calculations. Again approaches using both control and state variables ([10], [14]), or only the control variables ([4]) as unknowns have been explored. Still other methods based on the principle of optimality have been studied ([12], [13]).

In this paper, computational aspects of two methods to solve the OCP are explored – the nonlinear programming (NLP) method described by Tseng and Arora [17] and the differential dynamic programming (DDP) method described by Lin and Arora [12].

The partial derivative of a scalar function with respect to a vector variable is defined as a column vector

$$F_{\mathbf{x}} = \left[\frac{\partial F}{\partial x_i}\right]_{n\times 1} . \tag{5}$$

For the partial derivative matrix of a vector function with respect to a vector variable, the transpose of the vector function is taken, and then each column is treated as partial derivative of scalar function with respect to the vector variable as in Eq. (5):

$$\mathbf{f_u} = \frac{\partial \mathbf{f}^T}{\partial \mathbf{u}} = \left[\frac{\partial f_i}{\partial u_j}\right]_{m\times n} . \tag{6}$$

Total derivatives are expressed using the notation $\frac{d}{dt}(\cdot)$. This notation is used throughout the paper.

In Section 2, the NLP approach is summarized. Section 3 describes the DDP approach. Section 4 presents applications of the two techniques to optimal control of heating, ventilation and air conditioning systems. Section 5 presents structural and mechanical system example problems, and finally, concluding remarks are given in Section 6.

2. Nonlinear programming

In this section, the NLP method described by Tseng and Arora [17] is introduced. This method treats only the control variables as unknown; the state variables are treated as implicit functions of the control variables.

Using this method, the control variables are discretized with a finite dimensional approximation using a set of basis functions as

$$u_i(t) \approx \sum_j \phi_{ij}(t) b_{ij}, \quad i = 1, \ldots, m \tag{7}$$

where ϕ_{ij} are time dependent basis functions and b_{ij} are the unknown variables of the optimization problem, called the design variables. Let $\mathbf{b}$ represent an N dimensional vector of all the design variables. Many different types of basis functions can be used but piecewise polynomial splines are the most popular. The resulting finite dimensional optimization problem can be solved using NLP techniques which have been studied extensively and many robust and efficient codes are available. This is sometimes called the direct shooting or state variable elimination technique.

Unlike some other methods, the state variables are not treated as design variables; instead they are treated as dependent variables that change as a function of the design variables. Therefore, to determine the total derivatives of a cost or constraint function that contains state variables, design sensitivity analysis (DSA) techniques must be used ([17]). Two methods are widely used to determine the total derivatives. The first is the direct differentiation method (DDM). This method determines the sensitivity of a state variable to a change in the design variables by differentiating the state equation (2) as

$$\frac{d\dot{\mathbf{x}}}{d\mathbf{b}} = \mathbf{f}_{\mathbf{x}}^T \frac{d\mathbf{x}}{d\mathbf{b}} + \mathbf{f}_{\mathbf{b}}^T ; \quad \frac{d\mathbf{x}}{d\mathbf{b}}(0) = \mathbf{0} . \tag{8}$$

This sensitivity can be used to determine the total derivative of any functional, for instance, the cost functional as

$$\frac{dJ}{d\mathbf{b}} = \frac{d\mathbf{x}^T}{d\mathbf{b}} g_{\mathbf{x}}|_{t=T} + \int_0^T \left(F_{\mathbf{b}} + \frac{d\mathbf{x}^T}{d\mathbf{b}} F_{\mathbf{x}} \right) dt . \tag{9}$$

Another method, called the adjoint variable method (AVM), uses an adjoint variable to directly determine the sensitivity of a functional with respect to the design variables. For the cost functional, this derivative is given as

$$\frac{dJ}{d\mathbf{b}} = \int_0^T (F_{\mathbf{b}} + \mathbf{f}_{\mathbf{b}} \lambda) \, dt \tag{10}$$

where the adjoint variable is determined by

$$-\dot{\lambda} = F_{\mathbf{x}} + \mathbf{f}_{\mathbf{x}} \lambda ; \quad \lambda(T) = g_{\mathbf{x}}(\mathbf{x}(T), T) . \tag{11}$$

If piecewise polynomial splines are used as the basis functions, the design variables will effect only a small interval of time and the partial derivatives are non-zero only for this interval which greatly simplifies the above equations. To determine which method to use, the number of design variables and the number of active constraints must be compared. If the number of active constraints becomes larger than the number of variables, the DDM becomes more efficient, otherwise the AVM should be used.

Algorithm 1 *The general NLP algorithm is*

Step 1 *Choose a discretization scheme for the control variables and an initial estimate* $\mathbf{b}^{(0)}$ *for design variables. Also, determine* $\mathbf{x}(t)$ *and evaluate cost and constraint functions. Set* $k = 0$

Step 2 *Determine total derivatives using DDM or AVM.*

Step 3 *Stop if convergence criteria are satisfied.*

Step 4 *Determine a search direction* $\mathbf{d}^{(k)}$.

Step 5 *Determine a positive step size* α_k *which reduces a descent function* Φ *(i.e.,* $\Phi(\mathbf{b}^{(k)} + \alpha_k \mathbf{d}^{(k)}) < \Phi(\mathbf{b}^{(k)})$

Step 6 *Set* $\mathbf{b}^{(k+1)} = \mathbf{b}^{(k)} + \alpha_k \mathbf{d}^{(k)}$ *and* $k = k + 1$. *Go to Step 2.*

While this is a general algorithm for any NLP method, many variations for each step are available ([1]) which may greatly affect the performance of the algorithm. Some of these variations are discussed in Sections 4 and 5.

3. Differential dynamic programming

In this section, the basic idea of differential dynamic programming (DDP) is described and an algorithm for its implementation for a continuous unconstrained OCP is presented. The DDP method, originally proposed by Mayne ([13]), is a successive approximation technique for determining optimal control of nonlinear systems. It is derived from Taylor's expansion of the Hamilton-Jacobi-Bellman (HJB) equation. A method based on second order Taylor's expansion is described here, but methods based on first order Taylor's expansion are also possible. The method starts with an initial approximation for the control vector u and updates it at each iteration until the optimum solution is reached. The HJB equation is given as follows:

$$J_t^*(\mathbf{x}(t), t) + \min_{\mathbf{u}} H(\mathbf{x}(t), \mathbf{u}(t), J_{\mathbf{x}}^*, t) = 0 \tag{12}$$

where superscript "$*$" indicates optimal value, $J^*(\mathbf{x}(t), t)$ is the optimal cost for the interval $[t, T]$ with the initial condition $\mathbf{x}(t)$ and H is the Hamiltonian defined as

$$H(\mathbf{x}(t), \mathbf{u}(t), J_{\mathbf{x}}^*, t) = F(\mathbf{x}(t), \mathbf{u}(t), t) + \left[J_{\mathbf{x}}^*(\mathbf{x}(t), t)\right]^T \mathbf{f}(\mathbf{x}(t), \mathbf{u}(t), t). \tag{13}$$

Let the desired control and corresponding state variables be expressed as $\mathbf{u} = \tilde{\mathbf{u}} + \delta\mathbf{u}$ and $\mathbf{x} = \tilde{\mathbf{x}} + \delta\mathbf{x}$ where $\tilde{\mathbf{u}}$ and $\tilde{\mathbf{x}}$ are the known nominal values which satisfy the state equations but do not, in general, satisfy the HJB equation. The variations $\delta\mathbf{u}$ and $\delta\mathbf{x}$ are chosen now to satisfy HJB equation. Substituting these values into Eq. (12) yields

$$-J_t(\tilde{\mathbf{x}} + \delta\mathbf{x}, t) = \min_{\delta\mathbf{u}} H(\tilde{\mathbf{x}} + \delta\mathbf{x}, \tilde{\mathbf{u}} + \delta\mathbf{u}, J_{\mathbf{x}}, t) \tag{14}$$

where the superscript $*$ is dropped for simplicity. A second order Taylor's expansion is taken for J_t and H about the nominal values and the minimum of H with respect to $\delta\mathbf{u}$ is used to determine $\delta\mathbf{u}$ as a function of $\delta\mathbf{x}$ as

$$\delta\mathbf{u} = \alpha + \beta\delta\mathbf{x} \tag{15}$$

where α and β are given as

$$\alpha = -H_{\mathbf{uu}}^{-1} H_{\mathbf{u}}\, ; \qquad \beta = -H_{\mathbf{uu}}^{-1}\left(H_{\mathbf{ux}} + J_{\mathbf{xx}} f_{\mathbf{u}}^{T}\right)^{T} . \tag{16}$$

For this relationship to be valid $H_{\mathbf{uu}}$ must be positive definite. Substituting $\delta\mathbf{u}$ into the Taylor's expansion of Eq. (14) and equating the coefficients of like terms of $\delta\mathbf{x}$ yields the following set of differential equations:

$$-\dot{a} = -\frac{1}{2} H_{\mathbf{u}}^{T} H_{\mathbf{uu}}^{-1} H_{\mathbf{u}} \tag{17}$$

$$-\dot{J}_{\mathbf{x}} = H_{\mathbf{x}} - \left(H_{\mathbf{ux}} + J_{\mathbf{xx}} f_{\mathbf{u}}^{T}\right) H_{\mathbf{uu}}^{-1} H_{\mathbf{u}} \tag{18}$$

$$\begin{aligned} -\dot{J}_{\mathbf{xx}} = {} & H_{\mathbf{xx}} + f_{\mathbf{x}} J_{\mathbf{xx}} + J_{\mathbf{xx}} f_{\mathbf{x}}^{T} \\ & - \left(H_{\mathbf{ux}} + J_{\mathbf{xx}} f_{\mathbf{u}}^{T}\right) H_{\mathbf{uu}}^{-1} \left(H_{\mathbf{ux}} + J_{\mathbf{xx}} f_{\mathbf{u}}^{T}\right)^{T} \end{aligned} \tag{19}$$

with the terminal conditions

$$a(T) = 0\,, \qquad J_{\mathbf{x}}(T) = g_{\mathbf{x}}(\mathbf{x}(T), T)\,, \qquad J_{\mathbf{xx}}(T) = g_{\mathbf{xx}}(\mathbf{x}(T), T)\,. \tag{20}$$

The variable a in Eq. (17) is the difference between the cost function evaluated at the nominal and the predicted values, assuming second order approximation at the updated values of $\mathbf{u}$ and $\mathbf{x}$. Note that Eq. (19) is a differential Riccati equation and can become unbounded in some circumstances.

For OCP's that involve state and control constraints, the augmented Lagrangian method can be used to transform the problem into a series of unconstrained OCP's which can then be solved with the method described above. For more details on the use of the augmented Lagrangian method, see [11].

Algorithm 2 *The following steps summarize the unconstrained DDP algorithm:*

Step 1 *Choose $\tilde{\mathbf{u}}$. Determine $\tilde{\mathbf{x}}$ and evaluate the cost function.*

Step 2 *Integrate Eqs. (12) and store vector α and matrix β at each grid point. If integration fails, go to Step 3, otherwise go to Step 4.*

Step 3 *Let $\beta = 0$ and $\alpha = -H_{\mathbf{uu}}^{-1} H_{\mathbf{u}}$ where the adjoint variable λ from Eq. (11) is used in place of $J_{\mathbf{x}}$ in Eq. (13)*

Step 4 *Determine a step size η in the control law*

$$\mathbf{u} = \tilde{\mathbf{u}} + \eta\alpha + \beta\delta\mathbf{x} \tag{21}$$

which decreases the cost function. Note that $\delta\mathbf{x}(0) = 0$ since initial conditions are specified.

Step 5 *Check a stopping criterion; if it is not satisfied, go to Step 2.*

Several stopping criteria can be used. The first one is to check the actual reduction in the cost function from iteration to iteration, and when the reduction becomes small, the algorithm is stopped. Another criterion is to use the predicted decrease of the cost function which is similar to the gradient norm (i.e., $a(0) \leq \varepsilon$). Step 3 is included in the algorithm to improve its robustness. Without this step the method may fail for some problems where an accurate initial guess is not known. Jacobson and Mayne ([9]) suggest other modifications to improve robustness of the method.

Note that unlike NLP algorithm of Section 2 no particular discretization scheme is assumed for derivation purposes; however, to numerically integrate the differential equations and store the required function values, some discretization scheme must be introduced for computer implementation.

4. Application to HVAC systems

General purpose programs are developed to solve OCP for HVAC systems using the DDP and NLP algorithms of Sections 2 and 3. All programs are written in FORTRAN 77 and compiled on an HP 9000 Series 715/75 workstation. Some details of the implementation are:

1. The user supplies only the expressions in subroutines that describe the system functions $\mathbf{f}$, F, g and partial derivatives of these functions with respect to $\mathbf{x}$ and $\mathbf{u}$.

2. In order to approximate functions (such as $\mathbf{x}$, $\mathbf{u}$, α, etc.), values are stored at several grid points at equal intervals and cubic spline interpolations using the not-a-knot ending conditions ([3]) are used to determine values between the grid points.

3. High performance vector libraries supplied by HP were used when possible. These libraries are optimized to perform well on the Series 715/75 workstation.

4. Reimann integration is used to integrate the cost function.

5. The general purpose ordinary differential equation solver DDEABM ([16]) which uses an Adams-Bashforth-Moulton method, is used to integrate all differential equations.

6. An inexact line search is used for each method. For each method, a step size of 1.0 is tried; if certain conditions are not satisfied (i.e., the Wolf conditions for NLP), the step size is reduced until the conditions are satisfied.

7. The general purpose program IDESIGN ([2]) was used to solve the problem for the NLP method. This program uses the sequential quadratic programming technique.

4.1. Constant flow system

An HVAC problem described by House et al. ([8]) is considered where a single zone system is controlled to minimize the energy consumption and a comfort measure; thus the cost functional can be written as

$$J = \int_0^T \Big(\alpha_1(u_1 + u_3\rho c_p(T_O - x_2))^2 + \alpha_2(x_2 - x_{\text{ref}})^2 + \alpha_3 u_1^2 + \alpha_4(u_2 - u_{\text{ref}})^2 + \alpha_5 u_2^3\Big)\, dt \tag{22}$$

where α_i are the weighting constants, and other variables are explained in the sequel. The state equations that model the temperature response of the heat exchanger and the zone and are given as

$$\dot{\mathbf{x}}_1(t) = \frac{u_2(T_1 - x_1)}{V_{\text{he}}} + \frac{u_1}{\rho c_p V_{\text{he}}} \tag{23}$$

$$\dot{\mathbf{x}}_2(t) = \frac{u_2(x_1 - x_2)}{V_{\text{ts}}} \tag{24}$$

where

$$T_1 = x_2 + (T_O - x_2)\frac{u_3}{u_2} \tag{25}$$

and the initial conditions are

$$x_1(0) = 20°\text{C}; \qquad x_2(0) = 20°\text{C}. \tag{26}$$

The state variable $x_1(t)$ is the temperature of air exiting the heat exchanger and $x_2(t)$ is the temperature of the thermal zone being controlled. The control variable $u_1(t)$ is the heat input, $u_2(t)$ is the airflow rate through the thermal zone being controlled and $u_3(t)$ is the inlet airflow rate from outside. The description and values of the parameters used in the above equations are summarized in Table 1. For a more detailed description of the system, see House et al. ([8]).

Description	Parameter	Value and units
Control Reference	u_{ref}	0.142 m^3/s
State Reference	x_{ref}	22.05° C
Air Density	ρ	1.195 kg/m^3
Specific Heat	c_p	1.005 $\times 10^3$ J/kg-°C
Volume of Heat Exchanger	V_{he}	25.5 m^3
Volume of Thermal Zone	V_{ts}	255.0 m^3
Input Temperature	T_0	20.0 ° C
Weighting Constant	α_1	8.983 $\times 10^{-12}$ \$/s-W^2
Weighting Constant	α_2	0.810$\times 10^{-4}$ \$/s-W^2
Weighting Constant	α_3	8.983 $\times 10^{-12}$ \$/s-°C
Weighting Constant	α_4	3.120 $\times 10^{-3}$ \$s/m^6
Weighting Constant	α_5	4.392 $\times 10^{-3}$ \$/s^2/m^9

Table 1. System parameters.

As a first example, the airflow rate is considered constant so that $u_3 = u_2 = u_{ref}$ leaving only u_1 as the control variable. The initial guess for u_1 is taken as 0.0 W and the terminal time is set to 5400 seconds. The simulation and optimization are executed for 41, 61 and 101 grid points and the results summarized in Table 2. The results show that DDP solves the problem in approximately half the CPU time compared to NLP. DDP also requires less than half the number of function evaluations.

Method	No. of Grid Points	Optimal Cost	No. of Iters.	Function Evals.	CPU Time, sec.
DDP	41	0.1843	3	10	0.24
	61	0.1852	3	10	0.34
	101	0.1856	3	10	0.57
NLP	41	0.1888	16	26	0.45
	61	0.1872	19	23	0.57
	101	0.1863	24	25	1.15

Table 2. Comparison of numerical results for constant flow system.

4.2. Variable flow system

As second example, same system is considered again but the airflow rate u_2 through the zone is also considered to be a control variable and u_3 is set to u_2 at all times. Initial values of u_1 and u_2 are set to 375.0 W and 0.284 m^3/s. The results are summarized in Table 3.

Method	No. of Grid Points	Optimal Cost	No. of Iters.	Function Evals.	CPU Time, sec.
DDP	41	0.1677	3	10	0.22
	61	0.1649	3	10	0.33
	101	0.1637	3	6	0.46
NLP	41	0.1668	22	95	1.01
	61	0.1651	20	22	0.76
	101	0.1639	24	24	1.58

Table 3. Comparison of numerical results for variable flow system.

For this case, DDP was up to 4 times faster than NLP. With 41 grid points, the NLP method had problems near the solution, and required 95 function evaluations. DDP yielded the optimal solution in 3 iterations against about 20 iterations using NLP. Also a lower optimal cost compared to Section 4.1 is obtained because of the additional flexibility of the variable airflow rate.

4.3. Variable flow and recirculation system

In addition to the variable airflow rate, air exiting the zone is allowed to recirculate so that the inlet airflow and the airflow through the zone need not be equal. However, to prevent air from flowing in the wrong direction in the recirculation loop, a constraint requiring $u_2 \geq u_3$ is included in the problem. Since this constraint does not include state variables, the DSA techniques described in Section 2 are not required. Initial values of u_1, u_2 and u_3 are set to 500.0 W, 1.0 m^3/s and 0.8 m^3/s respectively. The numerical results are summarized in Table 4.

Method	No. of Grid Points	Optimal Cost	No. of Iters.	Function Evals.	CPU Time, sec.
DDP	41	0.1607	7	14	0.67
	61	0.1602	6	13	0.80
	101	0.1585	6	10	1.18
NLP	41	0.1622	49	115	6.21
	61	0.1600	64	65	20.15
	101	0.1585	61	70	107.69

Table 4. Comparison of numerical results for variable flow and recirculation system

Numerical difficulties are encountered in solving this problem with the DDP technique because $H_{\mathbf{uu}}$ becomes singular when $x_2 = x_0$ which violates the sufficiency condition. In order to overcome this difficulty, an additional cost term $\alpha_6 u_3^3$ is added to the integrand of the cost function of (22) with $\alpha_6 = \alpha_5$. This added cost is negligible compared to the other cost terms.

DDP showed dramatically better performance in solving the modified problem. However, DDP is implemented without considering the constraint as it was known to be inactive at the solution. NLP, with its superior robustness, took the extra effort to make sure the constraints were not violated, substantially increasing the computational effort. Also for NLP the computational effort increases dramatically with the number of grid points. With more design variables and more active constraints, the computational effort to solve the quadratic subproblem within IDESIGN increases rapidly.

5. Mechanical and structural applications

In this section, two mechanical and structural optimal control problems are solved using the DDP and NLP algorithms of Sections 2 and 3. The numerical procedures described in Section 4 are used with some modifications:

1. Simpson's method is used to integrate the cost function with the integrand evaluated only at grid points where $\mathbf{x}$ and $\mathbf{u}$ are stored.
2. The general purpose ordinary differential equation solver LSODE ([6]) is used to integrate all differential equations. The option which uses an Adams-Bashforth-Moulton method is used for all examples in this section.
3. For all examples, $\mathbf{u}(t) = \mathbf{0}$ is used as the initial value.
4. The program IDESIGN was used only to solve the constrained example. A limited memory BFGS method described by Nocedal ([15]) is used to solve the unconstrained problems.
5. In order to approximate functions (such as $\mathbf{x}$, $\mathbf{u}$, α, etc.), values are stored at several grid points at equal intervals and cubic spline interpolations with the not-a-knot ending conditions ([3]) are used to determine values between the grid points. For functions with first derivatives available (such as x) the Hermite interpolation is used.

5.1. Nonlinear impact absorber

An airplane landing gear with active damping is modeled as a mass-spring-damper system with nonlinear elements. This unconstrained example was solved by Lin and Arora ([12]) using the DDP method. By letting the displacement be x_1 and the velocity be x_2, the state equations can be written in the first order form as

$$\dot{x}_1 = x_2 \tag{27}$$

$$\dot{x}_2 = -0.597x_1^2\text{sgn}(x_1) - 0.597x_2^2\,\text{sgn}(x_2) + u \tag{28}$$

where u is the control force and $\text{sgn}(x) = 1$ if $x \geq 0$ and $\text{sgn}(x) = -1$ if $x < 0$. The impact of the landing gear touching down can be modeled by setting the initial conditions to $x_1(0) = 0.0$ and $x_2(0) = 1.0$. The desired purpose is to minimize the acceleration while not allowing the position or control force from becoming too large; therefore the cost function is written as

$$J = \int_0^{12} \left(100\dot{x}_2^2 + x_1^2 + 0.005u^2\right) dt\,. \tag{29}$$

The DDP algorithm was implemented as described in Section 3 and the NLP method used a limited memory BFGS algorithm by Nocedal ([15]) with 10 previous correction vectors used to approximate the Hessian matrix. Two methods are used for NLP; in the first method the identity matrix is used as the initial guess for the Hessian matrix, and in the second method the identity matrix is scaled by the factor

$$\frac{(\mathbf{y}, \mathbf{s})}{(\mathbf{y}, \mathbf{y})} \tag{30}$$

where $(\mathbf{y}, \mathbf{y})$ denotes scalar product, $\mathbf{s}$ is the change in design variables from consecutive iterations and $\mathbf{y}$ is the change in the gradients from consecutive iterations. Numerical results with 11, 21, 41 and 101 grid points are summarized in Table 5.

Method	No. of Grid Pts.	Optimal Cost	No. of Iters.	Function Evals.	CPU Time, sec.
DDP	11	45.990	8	15	0.32
	21	44.083	8	15	0.34
	41	44.006	8	15	0.38
	101	44.005	7	14	0.40
NLP	11	45.864	23	38	0.49
(Scaling)	21	44.080	22	29	0.50
	41	44.005	20	28	0.50
	101	44.002	22	30	0.77
NLP	11	47.459	10	64	0.91
(No	21	45.354	31	142	2.17
Scaling)	41	44.005	62	451	7.14
	101	44.009	39	265	5.18

Table 5. Comparison of numerical results for nonlinear impact absorber.

DDP outperformed the scaled version of NLP by 53% to 92% in terms of CPU time and required fewer iterations and function evaluations. NLP without scaling performed very poorly, stopping at higher optimal solutions and requiring much more computational effort.

The optimal cost function also varied with the number of grid points. This difference is attributed mostly to the numerical error from the Simpson's rule. A more robust procedure may be required for the numerical integration of the cost function.

5.2. Constrained nonlinear impact absorber

This constrained example was solved by Lin and Arora ([12]) using the same methods described here; however, a better numerical implementation for NLP gives better results. The same cost function and state equations from Section 5.1 are used but, for this example, the position is constrained to move by no more than 1.0 unit, i.e.,

$$|x_1| - 1.0 \leq 0 \quad \text{for } 0 \leq t \leq 12\,. \tag{31}$$

To handle this constraint, two very different approaches were used. DDP uses an augmented Lagrangian approach to reduce the problem to a series of unconstrained problems described in detail by Lin and Arora ([11]). NLP enforces the constraint at several grid points; the same points that are used to store the functions ([17]). Since these constraints contain state variables, the AVM is used to determine total derivatives of all active constraints. The program IDESIGN, using a limited memory BFGS method with scaling, as described in Section 5.1, is used to solve the constrained NLP problem.

The numerical results are summarized in Table 6. The column for gradient evaluations in Table 6 denotes the total number of gradients computed for active constraints. For the problems with 21 and 41 grid points, DDP and NLP give similar results and computational effort. However, with 101 grid points NLP required 5.6 times more CPU time than DDP. The computational effort required for gradient evaluations and the solution of the quadratic subproblem in IDESIGN increases rapidly with the number of grid points for NLP. Also note that these results are quite different from those given by Lin and Arora ([11]). In that paper DDP was shown to be dramatically superior to NLP in terms of CPU time and converged to a better solution. The improved performance for NLP shown here is a result of better scaling (which reduces the number of iterations and function evaluations), and more efficient computation of the total derivatives of the constraints.

Method	No. of Pts.	Optimal Cost	Iters.	Fn. Evals.	Grad. Evals.	CPU Time.
DDP	21	50.410	8*	41	NA	1.05
(Multiplier	41	50.303	9*	36	NA	1.32
Method)	101	50.302	10*	37	NA	1.57
NLP	21	50.435	21	31	68	0.84
(IDESIGN)	41	50.301	21	29	140	1.57
	101	50.425	28	34	462	8.82

* This denotes the number of unconstrained optimizations.

Table 6: Comparison of numerical results for constrained nonlinear impact absorber.

6. Conclusions

Two numerical methods, DDP and NLP, are described and implemented for solving nonlinear optimal control problems. Several example problems are solved using the methods and their performance is compared. The major conclusions from this study are as follows:

NLP

1. NLP is more robust and general purpose compared to DDP.

2. Care must be taken to ensure an efficient implementation of the NLP algorithm. General purpose packages that provide good implementations are widely available. Scaling and sensitivity analysis were shown to be very important implementation issues.

3. Number of grid points, as implemented for this study, greatly affected the performance of the method. More robust methods are needed to calculate and store functions.

4. For constrained problems, gradient evaluation can dominate the computational effort. Use of multiplier methods or more efficient integration of sensitivity equations should be studied. Note that all sensitivity equations are linear time varying with the same Jacobian matrix. This could be exploited to improve performance.

5. Only first order partial derivatives are required.

DDP

1. When applicable, this method is quite efficient. For some examples, it was 5 times faster than NLP.

2. DDP is less robust than NLP. General purpose computer programs based on DDP are not available.

3. For good performance second order partial derivatives are required. If they are not available, approximations can be used which may compromise efficiency and accuracy.

4. The number of grid points had little effect on the performance of the method.

Choice of the above methods will depend on the application. If robustness and generality are required, NLP is very attractive. If efficiency is required and second order information is available, DDP is an attractive method.

References

[1] J. Arora; *Introduction to Optimum Design* , McGraw-Hill, New York, 1989.

[2] J. Arora and C. Tseng; *IDESIGN User's Manual: Version 3.5*, Optimal Design Laboratory, College of Engineering, The University of Iowa, Iowa City, 1987.

[3] K. Atkinson; *An Introduction to Numerical Analysis*, John Wiley & Sons, New York, 1989.

[4] E. Edge and W. Powers; Function-space quasi-Newton algorithms for optimal control problems, *J. Optim. Theo. Appl.* **20**, 1976, 455–479.

[5] P. Enright and B. Conway; Discrete approximations to optimal trajectories using direct transcription and nonlinear programming, *J. Guid. Cont. Dynam.* **15**, 1992, 994–1002.

[6] A. Hindmarsh; LSODE and LSODI, two new initial value ordinary differential equation solvers, *ACM-SIGNUM Newsletter* **15**, 1980, 10–11.

[7] J. House, T. Smith, and J. Arora; Optimal control of a thermal system, *ASHRAE Trans.* **97**, 1991, 991–1001.

[8] J. House, J. Arora, and T. Smith; Comparison of methods for design sensitivity analysis for optimal control of thermal systems, *Opt. Cont. Appl. Meth.* **14**, 1993, 17–37.

[9] D. Jacobson and D. Mayne; *Differential Dynamic Programming*, American Elsevier, N. Y., 1970.

[10] C. Kelley, E. Sachs, and B. Watson; Pointwise quasi-Newton method for unconstrained optimal control problems, II, *J. Optim. Theo. Appl.* **71**, 1991, 535–547.

[11] T. Lin and J. Arora; Differential dynamic programming technique for constrained optimal control, *Comp. Mech.* **9**, 1991, 27–53.

[12] T. Lin and J. Arora; Differential dynamic programming technique for optimal control, *Optimal Control: Applications and Methods*, to appear.

[13] D. Mayne; A second-order gradient method for determining optimal trajectories of nonlinear discrete-time systems, *Intern. J. Cont.* **3**, 1966, 85–95.

[14] S. Mitter; Successive approximation methods for the solution of optimal control problems, *Automatica* **3**, 1966, 135–149.

[15] J. Nocedal; Updating quasi-Newton matrices with limited memory, *Math. Comp.* **35**, 1980, 773–782.

[16] L. Shampine and H. Watts; *DEPAC - Design of a User Oriented Package of ODE Solvers* , SAND 79-2347, Sandia Laboratory, Albuquerque, 1979.

[17] C. Tseng and J. Arora; Optimum design of systems for dynamics and controls using sequential quadratic programming, *AIAA J.* **27**, 1989, 1793-1800.

Optimal Control of Metal Forging

Jordan M. Berg[1], Richard J. Adams[2],
James C. Malas III[3], and Siva S. Banda[4]
Wright Laboratory
Wright-Patterson Air Force Base OH 45433-7531, USA

The development of good material models, accurate nonlinear finite element codes, and computer-controlled presses makes practical the application of control techniques to metal forging. This paper considers the problem of selecting a ram velocity profile to produce a desired microstructure, given a specified die and preform geometry and forging temperature. Two approaches for doing so are successfully applied to a simple but representative problem. The first is based on classical numerical optimization techniques. The second is based on inverse neural networks, and offers potential savings in critical computations. A method for choosing the forging temperature is also presented.

1. Introduction

Forging has been used to shape and harden metal since ancient times. Traditionally forging has been an imprecise process, relying on experience and experiment rather than analysis. Despite the high performance of modern equipment and much greater understanding of fundamental material behavior this is still largely true. Typical industrial forge presses cannot provide accurate velocity or force profiles. Although precise computer-controlled presses are available, they are not in widespread use. This is due in large part to the lack of systematic design procedures for ram velocity profiles. This paper presents two such procedures. It begins with an overview of the forging process.

1.1. Metal forming

Metal forming is the process of mechanically deforming a solid workpiece into a desired shape. Forming is distinguished from other fabrication techniques such as casting, machining, or welding, by the production of large strains distributed throughout the workpiece. These strains cause microstructural changes that give the finished part desirable mechanical properties. In forging-the particular metal forming operation considered in this paper-the workpiece, or *billet*, is deformed by squeezing between two or more shaped, hardened, *dies*. Most commonly there are two dies, one stationary and the other moving. Figure 1 is a schematic.

Many of the mechanisms driving microstructural changes are activated, or at least influenced, by thermal energy. This leads to the classification of forging processes based on the billet temperature. In hot forging, the only type discussed in

[1]NRC Resident Research Associate, Flight Dynamics Directorate
[2]Captain, USAF, Flight Dynamics Directorate
[3]Materials Engineer, Materials Directorate
[4]Aerospace Engineer, Flight Dynamics Directorate

this paper, the billet is heated to between sixty and seventy-five percent of its absolute melting temperature. When the billet is heated but the dies are cold, the billet will cool rapidly during the forging. If the forging times are long this *die chilling* leads to an inhomogeneous final microstructure, which is almost certainly undesirable. Die chilling can be reduced by pre-heating the dies (*hot-die forging*), or eliminated entirely by maintaining the die/billet system at a uniform desired temperature in a furnace (*isothermal forging*) during the forging. The combination of high billet temperature and low strain-rates enables more difficult shapes to be forged [11], and in some cases, allows the forging to take place in the superplastic state [5]. Because the need for precise deformation rate control is often combined with the need for prolonged high temperature processing this paper focuses on isothermal forging. Isothermal forging is used for production of critical components in aerospace, automotive, and construction applications.

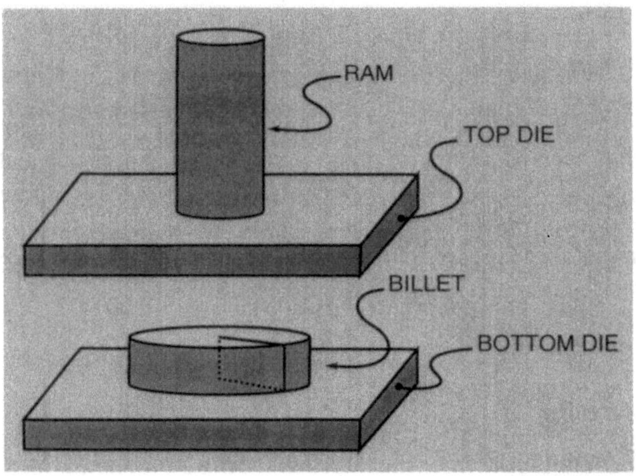

Figure 1. Forging schematic and nomenclature.

In addition to providing energy to drive desirable microstructural transitions, the high strains produced by forging can cause voids and fractures. Avoiding these problems may require a multi-step forging, with intervening heat treatment and machining. Each forging stage requires dedicated tooling, increasing production costs. Extensive machining is time-consuming and wasteful of material, also driving up costs. Precise control of material deformation rates minimizes unnecessarily high strains and so improves the economy of the process.

1.2. Microstructure

The mechanical properties of a metal alloy are strongly influenced by metallurgical factors at a microscopic level, such as grain size and dislocation density. Collectively these metallurgical properties are referred to as *microstructure*. When the characteristic distances, for example the average grain size, are small, the material is said to have a *fine* microstructure. When these distances are large the material is said to have a *coarse* microstructure. Finer microstructure generally corresponds to greater fracture toughness. For many forgings the final microstructure is as important as the final shape.

Three parameters dominate microstructural behavior. These are the scalar rate of deformation of an infinitesimal element, or *effective strain-rate*, $\dot{\varepsilon}$, the integrated effective strain-rate, or *effective strain*, ε, and the temperature, T. Together these quantities form the *thermomechanical state* at a point in the material. It is often possible, at least over some range, to characterize microstructural behavior as depending only on these variables.

1.3. Workability and processing maps

A number of factors constitute the workability of a material. Macroscopic cracking, of course, implies that the material has been worked beyond its limits. Microscopic voids are also unacceptable. Flow localization, in which deformation occurs only in narrow shear bands, results in undesirable inhomogeneous material properties and must be avoided. These are all examples of *material instabilities*. Necessary conditions for material stability are summarized in a *processing map* for the material [14], [21], [25]. Processing maps indicate regions in strain/strain-rate/temperature space where the material will be unstable. It is extremely desirable, if not imperative, that all points of a workpiece remain outside the unstable regions during forming. The processing map depends very strongly on the material under consideration. Producing a processing map for a new material is a difficult task, requiring analytical modeling and experimental measurement.

Because the processing map gives necessary but not sufficient conditions for material stability, remaining outside the unstable region does not guarantee a successful forging. Experimental verification is still necessary to validate the forging parameters.

1.4. Nonlinear FEM

Some analytical approaches to plasticity exist [7], but solutions to realistic problems, with complex die geometries and strain, strain-rate, and temperature dependence of the flow stress, require powerful numerical techniques. The most widely accepted means of predicting metal flow during forging is through nonlinear finite element methods (FEM). The theory underlying these methods is described by several authors [3], [13], [19], [29]. Studies comparing FEM output to experiment show excellent agreement [3], [10], [19], [22]. Grandhi, et al. have used the FEM program ALPID [18] to design ram velocity profiles based on linear quadratic optimal control [6]. This paper reports on the use of the FEM package *Antares* [2], a commercial descendant of *ALPID*, to apply nonlinear optimization techniques to the same problem.

The FEM code represents the billet as many small elements. The vertices of the elements are nodes, with associated positions and velocities. When the nodal positions are given the FEM uses energy considerations and a nonlinear root-finder to determine the nodal velocities. Thus, with the initial billet shape known, and all other inputs specified, the FEM algorithm generates the initial velocity field. This velocity field is applied over a time-step, Δt, to give the updated nodal geometry. The inputs are updated by Δt and the process is repeated. In this way the entire forging is simulated. Refinements to the procedure are possible, but this rough description suffices here.

1.5. Scope and objectives

For isothermal forging, the design parameters are the forging temperature, the ram velocity profile, the die shape, and the initial billet geometry. In this paper the die shape and undeformed billet geometry are considered to be fixed and specified. The temperature and strain-rate are chosen to give the best possible compatability between the final shape requirement and the desired microstructure. This is possible because the final strain is a function mainly of the geometries and not of the ram velocity or forging temperature. Once the strain-rate is given, the the ram velocity profile is selected to trade-off competing requirements in different parts of the deforming billet. Some typical design objectives might be preventing fracture in any part of the billet, achieving an "average" desired microstructure, or producing a desirable microstructural in a particular portion of the product. These goals can be mixed in a single cost function and weighted to reflect their importance. In this paper the representative goal is achieving a uniform, specified, microstructure.

The specific examples presented here are all axisymmetric, and the material is austenitic carbon steel. The dominant mechanism for microstructural transformation during forging is dynamic recrystallization [23]. The microstructure is specified by the recrystallized grain size, d, and the fraction of material recrystallized, χ. The static mechanisms that operate after the forging have not been included in this analysis.

2. Controls issues

One of the main problems making it difficult to address the forging problem from a control system viewpoint is the lack of formal actuators and sensors. This section discusses the current standard and some future possibilities.

2.1. Actuators

In the most general sense the metal forging process has many control inputs. These include the ram velocity, the die shapes, the starting billet geometry, the initial billet and die temperatures, heat inputs, initial billet microstructure, the mechanical and thermal properties of any die lubricants, etc. Of these, however, only the ram velocity and external heat inputs are available for adjustment during forging. The resolution available now for thermal control is coarse. The dies can be internally heated, and a furnace can be used to heat the surface of the billet. Although active temperature control during forging holds promise, in this paper it is assumed that such control is limited to regulating the average temperature about a desired constant value.

The steady-state ram velocity of a modern hydraulic press can be controlled quite precisely. The resolution claimed for some equipment is on the order of 10^{-3} in/sec, with a sampling rate of 200 Hz. The control bandwidth is about 10 Hz, and actual performance depends on the desired velocity profile. Press dynamics have been neglected in this paper.

2.2. Sensors

A hot, plastically deforming, billet is a hostile environment for sensor hardware. Some instrumentation is available, however. Thermocouples mounted in the dies can give an estimate of billet temperature, as can infrared or optical sensors in some cases. The ram force is also available in many hydraulic presses. Visual feedback has been used successfully to measure billet shape, but is limited to open-die forgings [12], [27]. Advanced sensors for industrial metal forming are an active area of study. Sensors have been proposed to measure microstructural properties directly, using ultrasonics [24] and x-rays [28]. A design for hot isostatic pressing by Meyer and Wadley uses direct measurement of microstructure [16]. Ultrasonic sensors were also used to determine temperature distributions in hot steel billets [1]. A host of other techniques based on electrical properties, optical properties, and magnetic properties are also under consideration for future development [17]. It is expected that feedback will be necessary for robust control of difficult forgings, but this paper concentrates on selecting open-loop ram velocity profiles.

3. Nonlinear numerical optimization

The ram velocity is a piecewise-continuous function, and so a search for the optimal solution takes place in an infinite-dimenional space. The dimension of the solution space is made finite by solving for the ram velocity only at discrete intervals of time or stroke. In this study the problem is further simplified by assuming it to be *memoryless*. That is, not only is the ram velocity/stroke curve discretized, but the optimal value is chosen at each point without allowing previous values to vary, and without considering the effects on subsequent points. Thus the problem is solved as a series of one-dimensional optimizations. One-dimensional optimization has the simplifying property that a minimum can be unambiguously bounded. The memoryless assumption is physically valid when inertial forces are small, as they typically are.

The representative objective selected for this study, based on previous work [6], [15], is achieving a specified strain-rate uniformly throughout the billet.

3.1. Brent's Method-Classical Numerical Optimization

The problem of choosing the best ram velocity at each step, given the forging temperature, is one-dimensional optimization problem. The cost function can be calculated at any point simply by running the FEM code. As a first cut it is natural to turn to a general-purpose one-dimensional optimization algorithm such as Brent's method [20]. Brent's method switches between parabolic interpolation and a golden search, depending on whether the parabolic fit is good. Each step reduces the size of an interval known to contain a function minimum. When the interval is within an acceptable tolerance the routine terminates. Figure 2 shows how Brent's method is integrated with *Antares*.

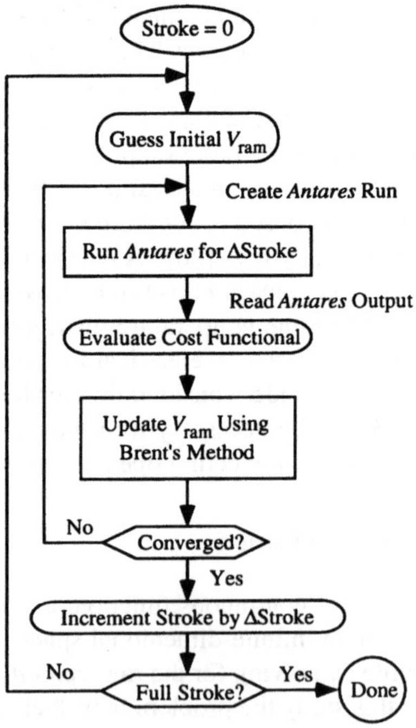

Figure 2. General scheme for optimizing ram velocity via Brent's method.

Brent's method requires that a minimum be bracketed initially between bounds. This is accomplished as follows. The initial guess supplied by the user, and a value 10% greater, are input to an algorithm from Press, et al. This algorithm simply moves "downhill" until the cost function starts to increase. In some circumstances this process could be costly, but the ram velocity is not expected to change drastically from one step to the next, so a reasonable initial guess should always be available. Brent's method is very safe, and performs reasonably well, but the metal forging problem is dominated by the high cost of function evaluations. It is important to choose convergence criteria carefully, and come up with a tight initial bracket, in order to reduce unnecessary computation. Even greater savings are possible by using derivative information. The derivatives are available, in principle, from the FEM code. The required modifications to *Antares*, however, are formidable.

3.2. Inverse neural networks

The deforming billet can be viewed as a mapping that relates two variables, stroke and ram velocity, to the average strain-rate of the material. This mapping is unknown, and in fact can only be studied through its input/output behavior (at least at current levels of modeling). Neural networks have proven to be an effective tool for approximating such mappings. Roughly speaking, a network is trained to approximate the finite-element model for a narrow operating condition. Then the numerical search for the optimal ram velocity is conducted on the network rather than the FEM. If a reasonably sized network can be trained to sufficient accuracy on a small set of training data the net result is a big savings in computation.

There are two primary considerations in designing a neural network, the network architecture and the approach to training. Both of these considerations are driven by the specific application. In general, neural networks are useful for a variety of operations [9]. This work uses networks as function approximators. A two layer neural network with a sigmoidal hidden layer and a linear output layer has been shown to be capable of approximating any static input-output mapping with a finite number of discontinuities to an arbitrary accuracy [8].

This study used backpropagation as the training method. A backpropagation learning rule adjusts the weights and biases to minimize the sum squared network error. The network error is the difference between the actual output of the network for a given set of inputs and the desired outputs corresponding to those inputs. The most commonly used approach to adjusting the weights and biases is a standard gradient descent algorithm. Other techniques that have been used to provide faster convergence results are conjugate gradient and quasi-Newton methods [26]. As in any numerical optimization technique, backpropagation is subject to the problem of local minima.

Neural networks have a number of properties which makes them particularly well suited for the forging problem. First of all, because they require only input/output data for training, the process can be treated as a "black box". This is a useful property due to the difficulty in coming up with a analytic model of the forging process. Neural networks have an information storage property which makes them reusable for similar problems. Once a network is found for one problem, it can be quickly retrained for a similar problem. Thus, a network that is trained to solve a forging problem provides a baseline for all similar forging problems.

In this work, a neural network training problem is formulated to calculate the ram velocity profile which minimizes the sum squared error between the average strain rate profile of the billet, as calculated by the FEM, and a desired average strain rate profile. The solution to the problem is found through a two step process.

Step 1. First, a training problem is formulated to model the mapping between the ram velocities at each stroke increment and the average strain rate within the billet. The network architecture for the model network is shown in Figure 3. It contains one hidden layer of sigmoidal functions and a linear output layer. V is the ram velocity and X is the value of stroke. A standard gradient descent backpropagation algorithm is used to minimize the sum squared error between the response of the FEM and the response of the model network to a representative set of velocities and values of stroke. The training model is illustrated in Figure 4.

Next, with the model network fixed, error backpropagation is used to find the ram velocity, at each value of stroke, which minimizes the sum squared error between the average strain rate, as calculated by the model network, and some desired average strain rate. Figure 5 illustrates the calculation of the ram velocity profile.

Step 2. The calculated values of ram velocity for each value of stroke from step 1 are used to drive the FEM to create a new set of training data. This data is then used to retrain the model network, see Figure 3. This retraining or fine tuning of the model network within a more limited operating window allows it to more accurately reflect the material behavior within the region of interest. The retrained model is then used to calculate a new ram velocity profile as shown in Figure 5. Step 2 is repeated until the average strain rate profile of the billet, as calculated by the FEM, accurately tracks the desired strain rate profile, or until the sum squared error between the two profiles ceases to decline with additional iterations.

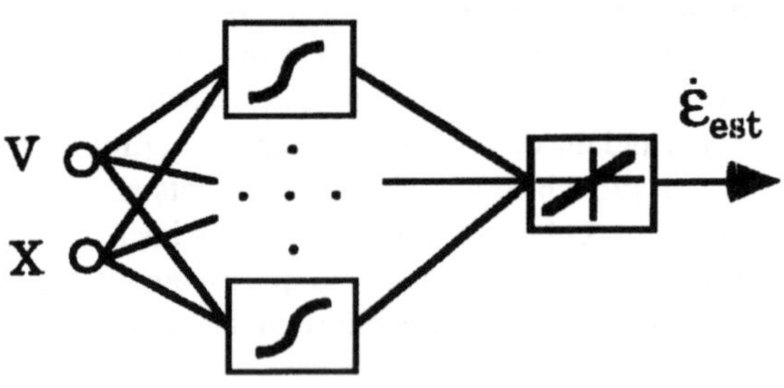

Figure 3. Model network architecture.

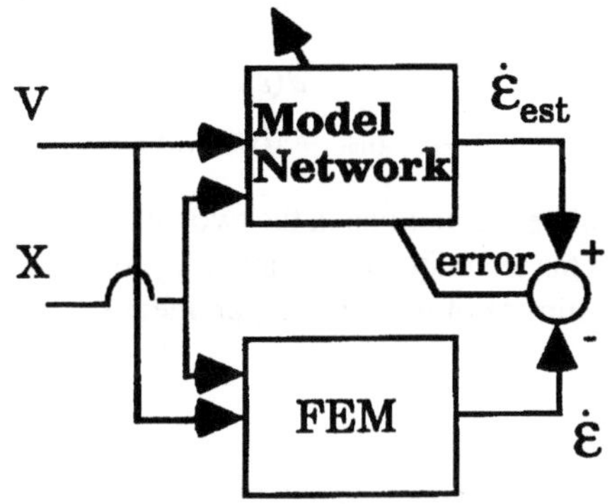

Figure 4. Neural network training model.

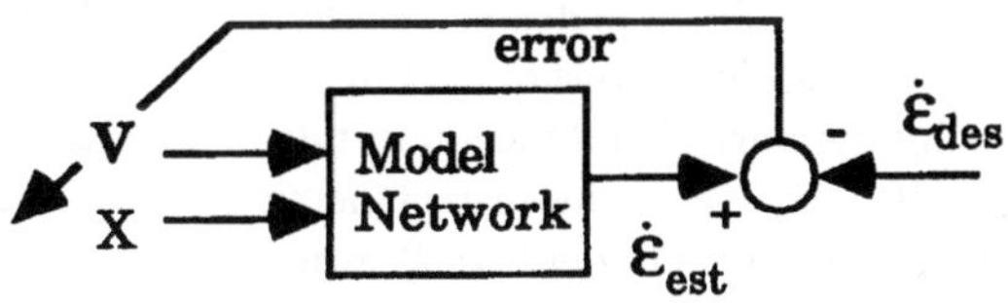

Figure 5. Neural network training model.

4. Selecting the forging temperature and strain-rate

Models can be formulated to relate microstructural parameters, which for the purposes of this paper will be grain size d and recrystallization fraction χ, to the thermomechanical state. These models are typically based on metallurgical principles, but depend largely on fitting experimental data. Writing the models in general,

$$\begin{aligned} d &= d(\varepsilon, \dot{\varepsilon}, T) \\ \chi &= \chi(\varepsilon, \dot{\varepsilon}, T) \end{aligned}$$

The final desired grain size can be specified as d_f. Generally full recrystallization is desired in the finished part, but this may not be possible or required after a single forging. It is necessary then to specify a minimum acceptable $\chi_{\min}$.

Two critical assumptions are made. Previous work [15] suggests the optimal strain-rate trajectory is close to constant. Based on this result the strain-rate is assumed to take a constant, though as yet unknown, value $\dot{\varepsilon}_s$ The second assumption is that the strain distribution in the finished forging is approximately independent of the forging temperaure and strain-rate profile. In this case the minimum strain in the finished forging, $\varepsilon_{\min}$, is an invariant of the process. Write,

$$\begin{aligned} d &= d_f = d(\varepsilon_{\min}, \dot{\varepsilon}_s, T) \\ \chi &= \chi_{\min} = d(\varepsilon_{\min}, \dot{\varepsilon}_s, T) \end{aligned}$$

These are two nonlinear equations in the two unknowns $\dot{\varepsilon}_s$ and T. They can be solved numerically. This technique is suitable for preliminary studies such as this one, and as a rough estimate of the forging parameters.

5. Results

The techniques described above were applied to a simple, but representative, sample problem. In this problem a steel cylinder one inch high and two inches in diameter is compressed between two flat plates until its height is reduced by sixty per cent. This type of forging, called *upsetting*, may be a first step in a more complex forging. Upsetting is often used to achieve a desired microstructure. Upsetting is also used experimentally to obtain material response data. When the die/billet interface is frictionless the problem is one of the few with an analytic solution. In that case the billet remains cylindrical and the strain-rate is homogeneous throughout, and equal to V_{ram}/H, where H is the instantaneous height of the cylinder. The final strain of each point in the cylinder is ln R, where R is the reduction in height of the cylinder.

When the die/billet friction is nonzero the billet geometry changes as the forging progresses, the strain-rates are non-uniform, and no analytical description of the problem exists. This paper reports two cases. One with no friction, for comparison with the analytical solution, and one with a realistic shear friction coefficient of 0.3.

5.1. Dynamic recrystallization model for austenitic steel

The microstructure is characterized by grain size, d, and recrystallization fraction, χ. The relation to the thermomechanical state, using equations due to Yada [4], is as follows:

$$\begin{aligned} d &= 22,600 \;\; Z^{-0.27} \\ \chi &= \begin{cases} 0 & \text{if } \varepsilon < \varepsilon_c \\ 1 - \exp\left(-0.693\left(\dfrac{\varepsilon - \varepsilon_c}{\varepsilon_{0.5}}\right)^2\right) & \text{if } \varepsilon \geq \varepsilon_c \end{cases} \end{aligned}$$

where d has units of microns, and Z is the Zener-Holloman parameter,

$$Z = \dot{\varepsilon} \exp\left(\frac{Q}{RT}\right)$$

Q is an activation energy with value,

$$Q = 267.1 \quad \text{kJ/mole}$$

and R is the gas constant,

$$R = 8.31 \times 10^{-3} \quad \text{kJ/mole.}$$

The temperature T must always be given in degrees Rankine. ε_c is the critical strain,

$$\varepsilon_c = 4.76 \times 10^{-4} \exp\left(\frac{8000}{T}\right)$$

and $\varepsilon_{0.5}$ is the strain required for 50% recrystallization,

$$\varepsilon_{0.5} = 1.144 \times 10^{-5} d_0^{0.28} \varepsilon \delta_1 / \delta_3^{\dot{}\,0.05} \exp\left(\frac{6420}{T}\right)$$

where d_0 is the initial grain size in microns.

FEM simulation shows that the minimum strain in the billet after upsetting is about 0.66. The recrystallization fraction corresponding to the minimum strain is set to 0.99. Realistic values for initial and final grain size [4] are 180 μm and 20 μm, respectively. Then the steps described in Section 4 give a forging temperature of 1547 degrees Fahrenheit and a desired strain-rate of 0.06 sec^{-1}.

5.2. Results using Brent's method

The full 0.6 inch stroke was discretized into 0.025 inch intervals. Brent's method was used to minimize a quadratic cost function,

$$J = \sum_{\text{all elements}} (\dot{\varepsilon}_i - 0.06)^2$$

at each stroke interval. Because of details of the implementation, the ram velocity associated with a stroke value is that of the previous interval. Initially the tolerance used to determine convergence was set to correspond to machine precision. At this level the algorithm required about 20 iterations per step to converge. The tolerance was later increased to reflect a more realistic level of precision-a ram velocity resolution of 10^{-6} in/sec. At this level, which is still atleast three orders of magnitude more precise than can actually be implemented, the algorithm required about 10 iterations per step for convergence. The initial guess was based on the analytic solution, and so was quite accurate. In every case the minimum was bracketed on the first try.

Accurate simulation of the upsetting requires the billet to be represented by over a thousand elements. This made the optimization run extremely slowly. To speed up the process a coarser mesh was used, corresponding to about 100 elements. The resulting ram velocity was then evaluated on a more accurate "truth model" to try to determine the effect of the coarser grid.

Case 1: Frictionless upsetting

In the case of frictionless upsetting the analytical solution is known. The optimal ram velocity is a linear function of stroke with an initial value of 0.06 in/sec, and a slope of -0.06 sec^{-1}. Figure 6 shows the ram velocity calculated by the optimization scheme versus the analytic solution. A constant ram velocity is also shown for comparison.

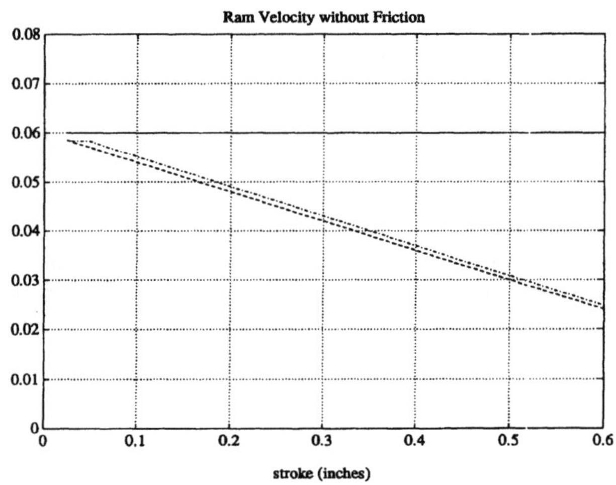

Figure 6. Ram velocities corresponding to three strategies: constant (solid), linear (dashed), and optimal (dash-dot) for frictionless upsetting.

The calculated velocity profile is very close to the analytic solution. The difference is due to inaccuracies in the FEM code. This is seen more clearly in Figure 7, showing the strain-rates calculated by *Antares*.

The ram velocity profiles must be evaluated using a more detailed simulation. This is done by using a finer grid and shorter stroke increment. Figure 8 shows the optimal trajectory applied to such a "truth model." Overall the performance of the numerical optimization indicates that it is behaving properly.

Case 2: Upsetting with realistic friction

To give a more realistic case, for which no analytic solution is known, a realistic die/billet interface shear friction coefficient of 0.3 was used. The calculated ram velocity is compared to the constant and linear velocities in Figure 9.

Figure 10 compares the resulting average strain-rates.

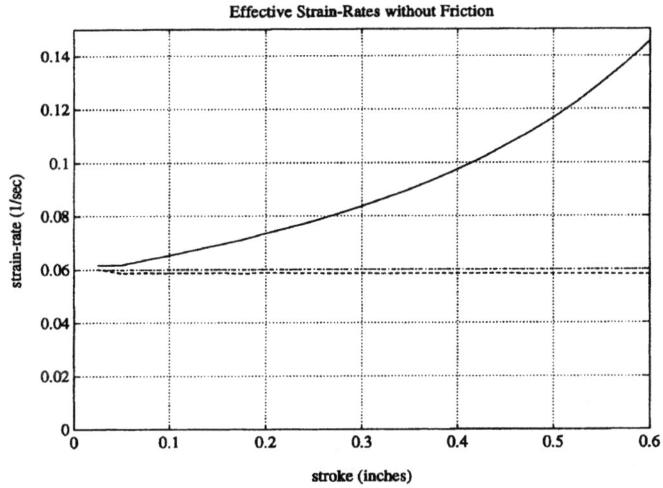

Figure 7. Strain-rates resulting from three strategies: constant (solid), linear (dashed), and optimal (dash-dot) for frictionless upsetting.

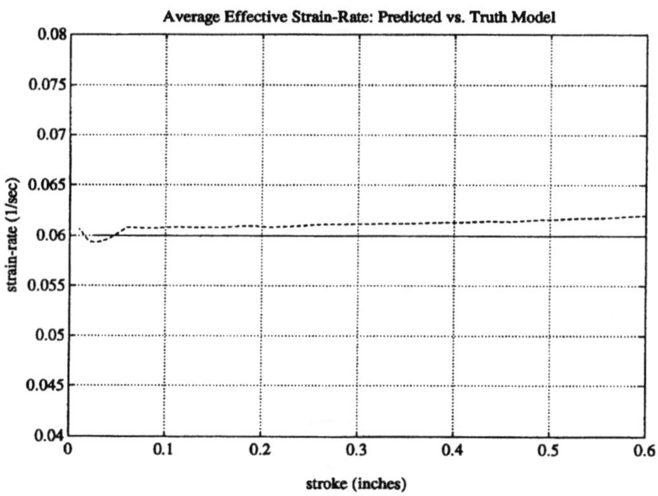

Figure 8. Effective strain-rate predicted using design model (solid) and effective strain-rate resulting from calculated ram velocity when applied to truth model (dashed).

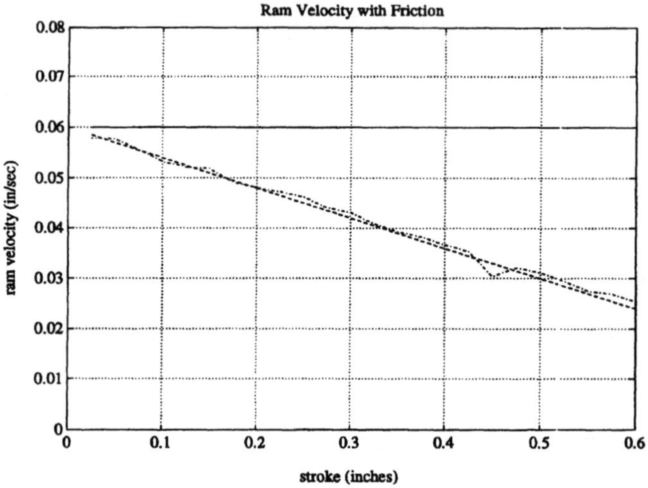

Figure 9. Ram velocities corresponding to three strategies: constant (solid), linear (dashed), and optimal (dash-dot) with realistic friction.

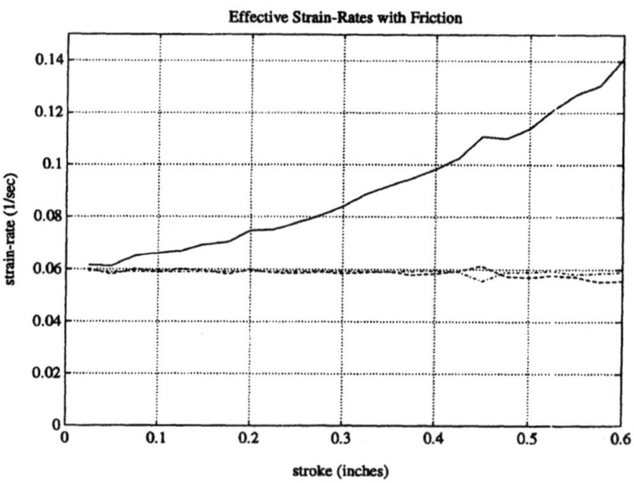

Figure 10. Average strain-rates corresponding to three strategies: constant (solid) linear (dashed), and optimal (dash-dot) with realistic friction.

The analytic solution without friction appears to be quite close to optimum for this case as well. Strain-rate distributions through the billet (for either the linear velocity profile, or the optimal-they are indistinguishable for practial purposes) are shown in Figure 11. Clearly either is much better than the constant velocity case, which follows a "rule-of-thumb" standard in the industry. This is apparent from Figure 12 which shows strain-rate distributions corresponding to constant ram velocity.

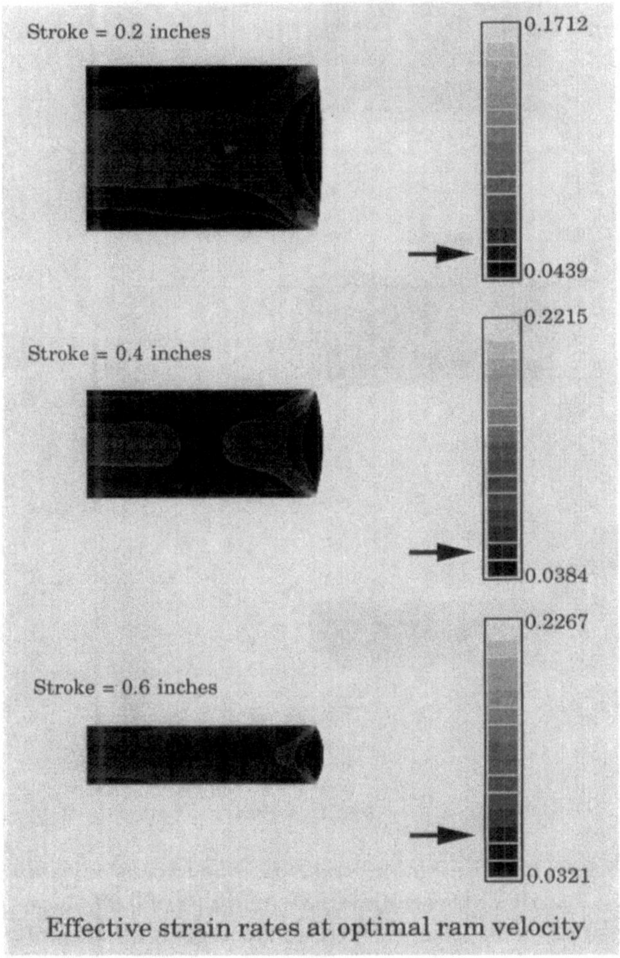

Figure 11. Strain-rate distributions corresponding to optimal ram velocity. Arrows indicate the target strain-rate of 0.06.

The strain-rate distributions are quite inhomogeneous when realistic friction is considered. Note, however, that despite the differences in average value between the optimal and the constant case, the distribution geometries are very similar. It is natural to conjecture that control ram velocity influences only the mean value of the strain-rate, but does not strongly affect the strain-rate contours.

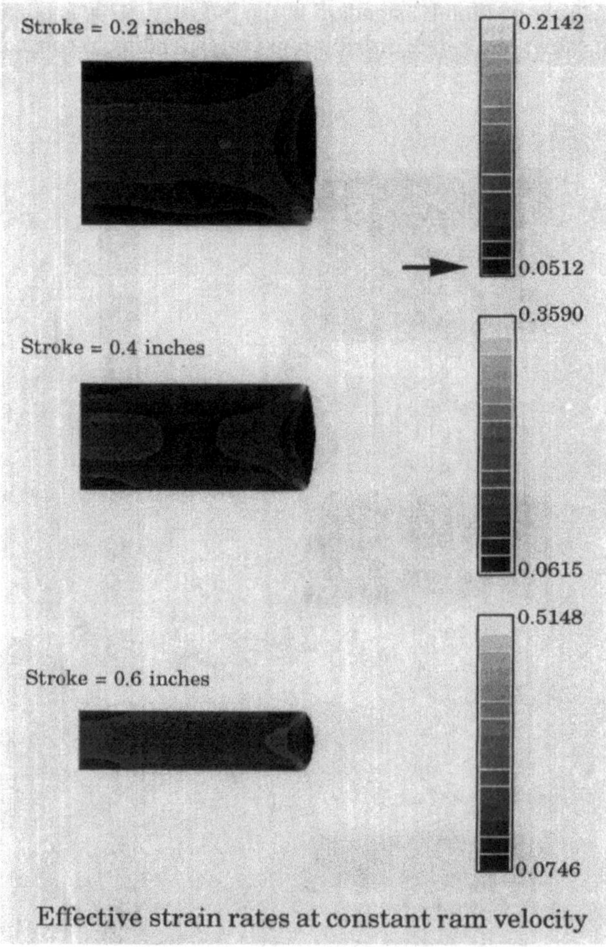

Figure 12. Strain-rate distributions corresponding to constant ram velocity. Arrows indicate the target strain-rate of 0.06. Note that on the lower two figures the target is off the scale.

Figure 13 shows the result of applying the calculated trajectory to a truth model. As in the frictionless case the performance degrades noticably as the forging progresses. Whether this is acceptable depends on the demands of the particular user. If not, a finer mesh and shorter stroke interval will be required.

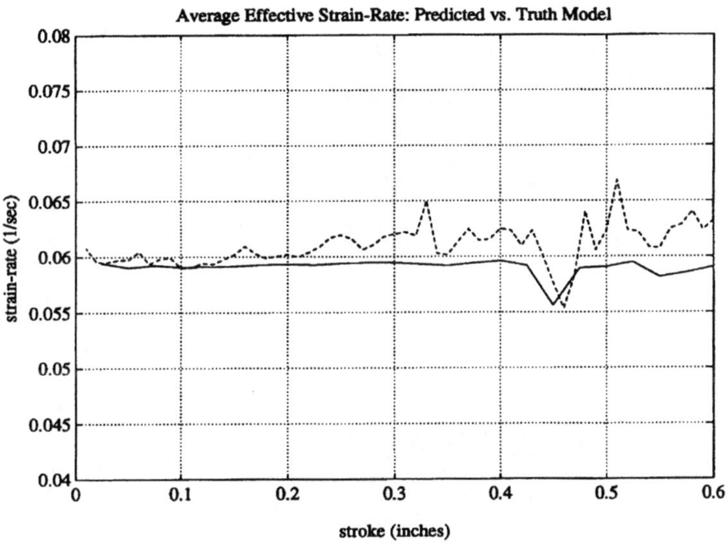

Figure 13. Effective strain-rate predicted using design model (solid) and resulting from calculated ram velocity when applied to truth model (dashed).

Recall that the effective strain distribution at each value of stroke was assumed to be independent of the stroke history. The results of this example provide a good test of that assumption. Figures 14 and 15 show the effective strain distributions corresponding to the constant ram velocity and the optimal ram velocity at stroke values of 0.4 inches and 0.6 inches, respectively. The distributions are very similar, although the two ram velocities are very different, suggesting that the assumption of path independence is valid.

5.3. Results using inverse neural network

The first step of the neural network approach is to generate the data that is required to train the model network. Ram velocity profiles are chosen which span the anticipated solution space. For the simple upsetting problem, it is known that in order to maintain a constant average strain rate the ram velocity must decrease as the stroke increases. The maximum value of ram velocity can be approximated as the initial height times the desired average strain rate. The ram velocity must always be positive, so its minimum value approaches zero in the limit. Three constant ram velocity profiles are used to generate the training data for step 1, 0.0 in/sec, 0.03 in/sec, and 0.06 in/sec. These are shown in Figure 16. The plots are labeled as simulation run 0, 1, and 2. The zero ram velocity profile is referred to as run 0 because it is a trivial case which does not require a FEM run to generate data.

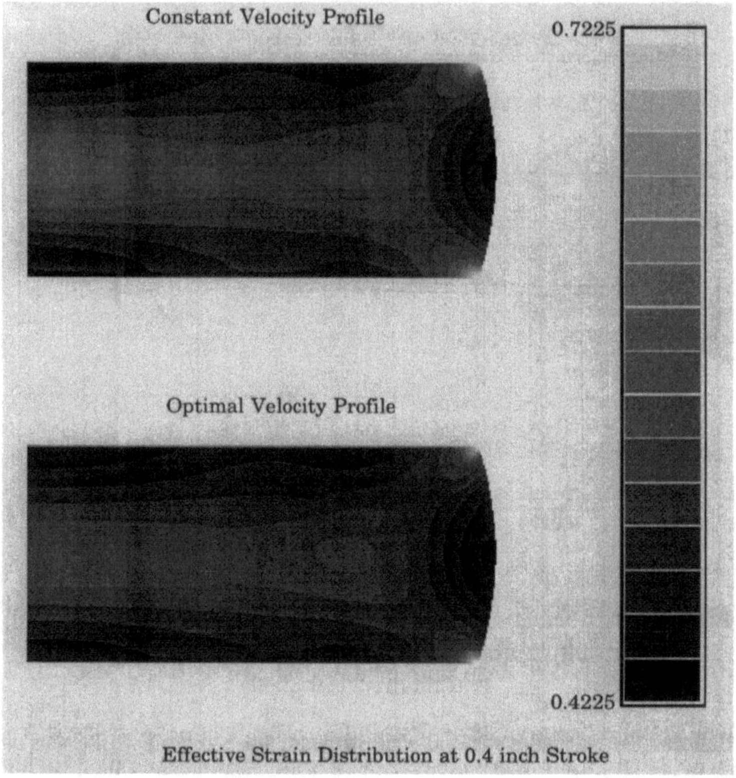

Figure 14. Comparison of effective strain distribution due to constant ram velocity to effective strain due to optimal ram velocity at stroke of 0.4 inches (67%).

Case 1: Frictionless upsetting

The ram velocity profiles for runs 1 and 2 are used to drive the FEM simulation and generate average strain rate output profiles for the frictionless upsetting problem. These are shown in Figure 17. These three sets of stroke, ram velocity, and average strain rate data are used to train a network with four sigmoidal nodes in the hidden layer. The model network is trained for 2000 iterations of backpropagation. This model is then fixed and error backpropagation is used to find the ram velocity profile which achieves the desired average strain rate output for the network. The result is shown in Figure 18 as the dotted line. This completes step 1 of the procedure.

The ram velocity profile resulting from step 1 is used to drive the FEM and generate a new set of training data. The resulting average strain rate profile is shown in Figure 19 as the dotted line, simulation run 3. The sum squared error between this profile and the desired average strain rate profile is $8.1210e^{-06}$. The data from run 3 is used to retrain the model network and recalculate a new ram velocity profile, the dashed line in Figure 18. The step is repeated. The new velocities are used to drive the FEM and generate new training data. The sum squared error resulting from the second training iteration is 1.6373e^{-06}. One more training iteration is performed to reduce the sum squared error to 5.9226e^{-07}. The final iteration is shown as the solid line in Figures 18 and 19. Figure 19 shows the average strain rate profiles converging on the desired solution of a 0.06. Figure 18 shows the ram velocities converging on the analytic solution for the frictionless upsetting problem, a straight line.

Case 2: Upsetting with realistic friction

The ram velocity profiles for runs 1 and 2 in Figure 16 are used to drive the FEM simulation and generate training data for the problem of upsetting with a realistic level of friction. These are shown in Figure 20.

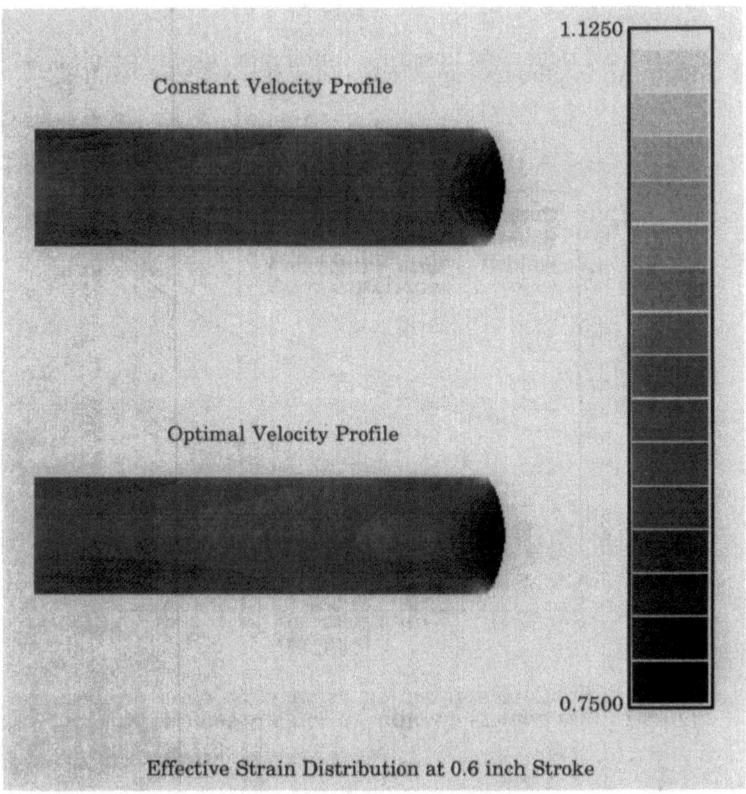

Figure 15. Comparison of effective strain distribution due to constant ram velocity to effective strain due to optimal ram velocity at stroke of 0.6 inches (100%).

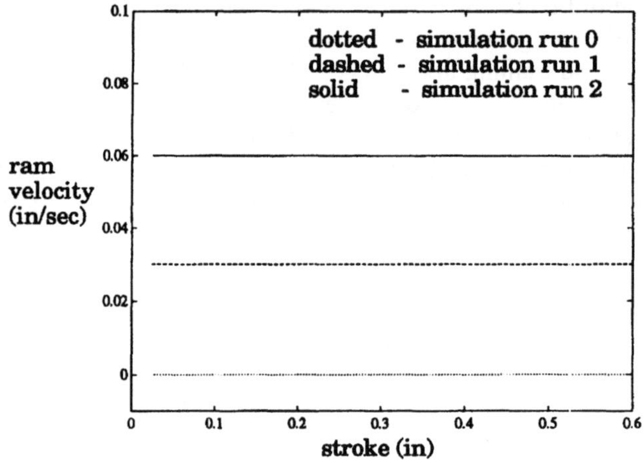

Figure 16. Frictionless upsetting, initial ram velocity profiles.

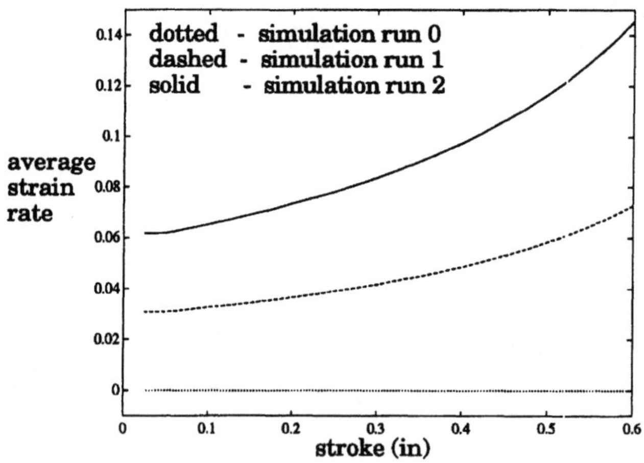

Figure 17. Frictionless upsetting, average strain rate profiles.

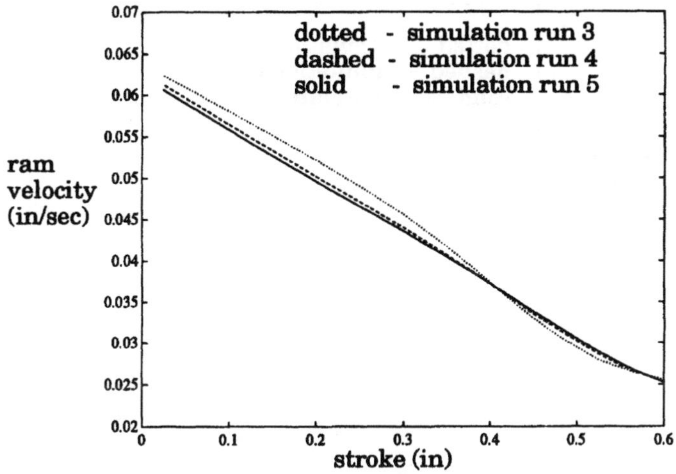

Figure 18. Frictionless upsetting, calculated ram velocity profiles.

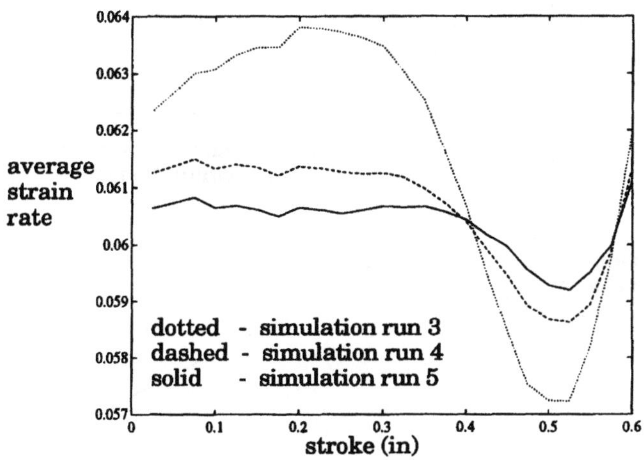

Figure 19. Frictionless upsetting, converging average strain rate profiles.

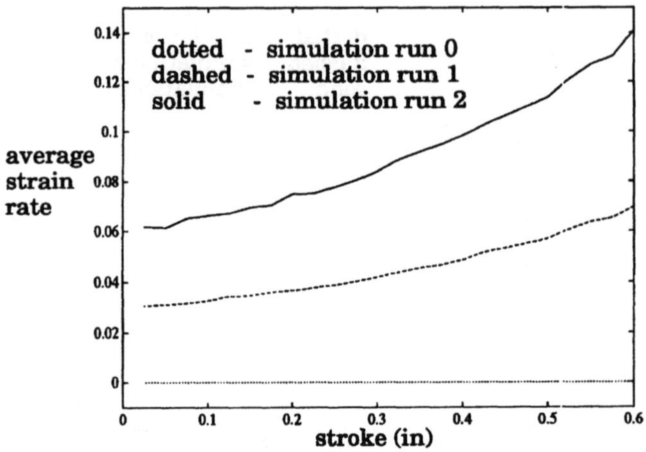

Figure 20. Upsetting with friction, average strain rate profile.

The average strain rate profiles are similar in scale to the frictionless case, except that there are ripples in the data caused by the effects of friction on the FEM simulation. The same procedure as in the frictionless case is used to train the model network and generate a ram velocity profile which minimizes the sum squared error between the model output and the desired value of average stain rate. The result is shown in Figure 21 as the dotted line.

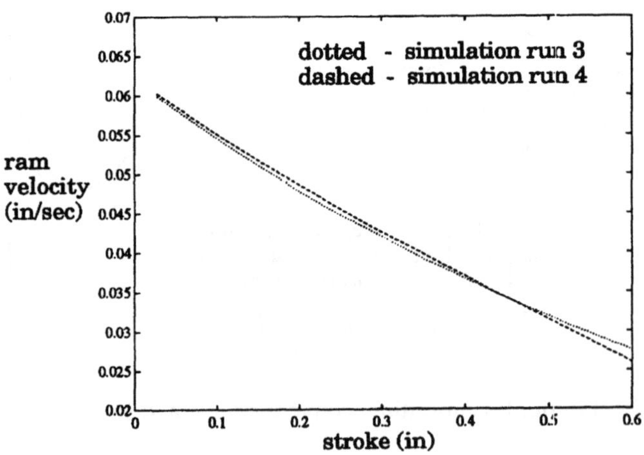

Figure 21. Upsetting with friction, calculated ram velocity profiles.

Again, the ram velocity profile resulting from step 1 is used to drive the FEM and generate a new set of training data. The resulting average strain rate profile is shown in Figure 22 as the dotted line, simulation run 3. The sum squared error between this profile and the desired average strain rate profile is 2.4542e^{-06}. The data from run 3 is used to retrain the model network and recalculate a new ram velocity profile, the dashed line in Figure 22. The sum squared error resulting from the second training iteration is 8.2598e^{-07}.

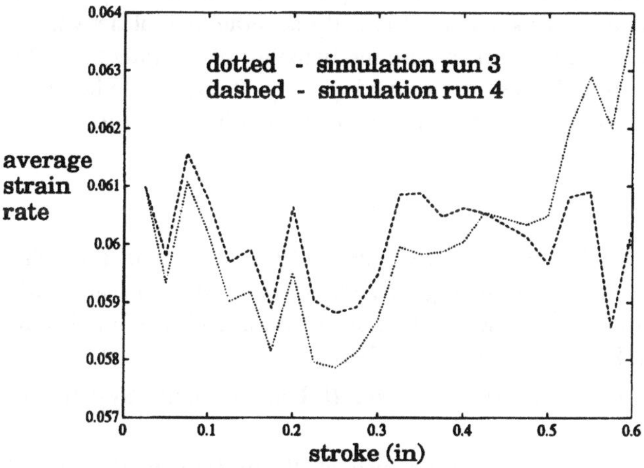

Figure 22. Upsetting with friction, converging average strain rate profiles.

Subsequent iterations show no further reduction in the sum squared error. This is due to the slightly erratic behavior of the data which results from the addition of friction to the FEM. For the case of upsetting with friction, it can be said that the best solution has been reached to the accuracy of our material model.

The ram velocity profiles generated by the inverse neural network technique are essentially the same as those generated by Brent's method. Therefore all the strain and strain-rate contours shown in the previous section apply here as well.

6. Conclusions

This paper outlined the problem of controlling the metal forging process. Off-the-shelf industrial equipment-computer-controlled presses and nonlinear finite element codes-can be combined with material models and processing maps in offline generation of ram velocity profiles. The resulting profiles are in some sense "optimal" and balance the competing demands of various parts of the billet.

A standard one-dimensional nonlinear optimization algorithm, Brent's method, was effective in generating optimum ram velocity profiles for two simple upsetting problems, with and without friction. In the frictionless case the algorithm recovered the known analytic solution. Perhaps suprisingly the addition of a realistic level of shear friction did not significantly change the result.

An approach using neural networks to calculate ram velocities in hot metal forging has also been examined. The difference between the classical method and this one is that the numerical optimization is performed on the neural network approximation to the FEM, rather than on the FEM directly. Since the neural network is far simpler than the FEM the optimization is much faster. The cost is the training required to produce a sufficiently accurate neural network. In this case a small network sufficed to capture the behavior of the rather simple forging geometry. Therefore the network trained quickly. It remains to be seen whether the trade-off will remain advantageous in more complicated forgings.

References

[1] G. Alers and H. Wadley; Noncontact ultrasonic sensors for high-temperature process control, in *Intelligent Processing of Materials and Advanced Sensors*, Ed. by H. Wadley, P. Parrish, B. Rath, and S. Wolf, The Metallurgical Society, Pennsylvania, 1987, 85–101.

[2] *Antares User's Manual Version 3.0*, United Energy Systems, Inc., Dayton, 1993.

[3] C. Chen; *Finite Element Analysis of Plastic Deformation in Metal Forming Processes*, Ph.D. dissertation, University of California, Berkeley, 1978.

[4] C. Devadas, I.. Samarasekeara, and E. Hawbolt; The thermal and metallurgical state of steel strip during hot rolling: part III. Microstructural evolution, *Metallurgical Transactions A* **22A**, 1991, 335–348.

[5] G. Gessinger; Isothermal forming: a low-cost method of precision forging, *Engineering*, 1979, 926–929.

[6] R. Grandhi, A. Kumar, A. Chaudhary, and J. Malas; State-space representation and optimal control of non-linear material deformation using the finite element method, *Int. J. Num. Meth. Engr.* **36**, 1993, 1967–1986.

[7] R. Hill; *The Mathematical Theory of Plasticity*, Oxford, London, 1950.

[8] K. Hornik; Approximation capabilities of multilayer feedforward networks, *Neural Networks* **1**, 1991, 251–257.

[9] D. Kemsley, T Martinez, and D. Campbell; A survey of neural network research and fielded applications, *Inter. J. Neural Networks* **2**, 1991, 123–132.

[10] S. Kobayashi, S. Oh, and T. Altan; *Metal Forming and the Finite-Element Method*, Oxford, New York, 1989.

[11] K. Kulkarni; Isothermal forging – from research to a promising new manufacturing technology, *Sixth North American Metalworking Research Conference Proc.*, 1978, 24–32.

[12] G. Larabee and K. Stelson; Digital image shape measurement for the monitoring and control of non-homogeneous, axisymmetric plastic deformation, *Proc. American Controls Conference*, Minnesota, **1** 1987, 729–734.

[13] C. Lee and S. Kobayashi; New solutions to rigid-plastic deformation problems using a matrix method, *Trans. ASME, J. Engr. Ind.* **95**, 1973, 865–873.

[14] J. Malas and V. Seetharaman; Using material behavior models to develop process control strategies, *JOM* **44**, 1992, 8–13.

[15] J. Malas, R. Irwin, and R. Grandhi; An innovative strategy for open loop control of hot deformation process, *J. Mater. Engrg. Perf.* **2**, 1993, 703–714.

[16] D. Meyer and H. Wadley; Model-based feedback control of ceformation processing with microstructure goals, *Metal. Trans. B* **24B**, 1993, 289–300.

[17] National Materials Advisory Board; *On-line Control of Metal Processing*, Report of the Committee on On-line Control of Metal Processing, NMAB-444, National Academy Press, 1989.

[18] S. Oh, G. Lahoti, and T. Altan; ALPID – a general purpose FEM program for metal forming, in *Proceedings NAMRC-IX*, State College, 1981, 83–92.

[19] S. Oh; Finite element analysis of metal forming processes with arbitrarily shaped dies, *Int. J. Mech. Sci.* **24**, 1982, 479–493.

[20] W. Press, B Flannery, S. Teukolsky, and W. Vetterling; *Numerical Recipes: The Art of Scientific Computing*, Cambridge, Cambridge, 1986.

[21] Y. Prasad, H. Gegel, S. Doraivelu, J. Malas, J. Morgan, K. Lark, and D. Barker; Modelling of dynamic material behaviour in hot deformation forging of Ti-6242, *Metal. Trans. A* **15A**, 1984, 1883–1892.

[22] J. Price and J. Alexander; Specimen geometries predicted by computer model of high deformation forming, *Inter. J. Mech. Sci.* **21**, 1979, 417–430.

[23] W. Roberts; Dynamic changes that occur during hot working and their significance regarding microstructural development and hot workability, in *Deformation, Processing, and Structure*, Ed. by G. Krauss, American Society for Metals, OH, 1984, 109–184.

[24] M. Rosen; Ultrasonic sensors for characterization of phase transformations, in *Intelligent Processing of Materials and Advanced Sensors*, Ed. by H. Wadley, P. Parrish, B. Rath, and S. Wolf, The Metallurgical Society, Pennsylvania, 1987, 29–48.

[25] S. Semiatin and G. Lahoti; The occurrence of shear bands in isothermal, hot forging, *Metal. Trans. A* **13A**, 1982, 275–288.

[26] E. Sontag; Some topics in neural networks and control, Report LS93-02, Siemens Corporate Research, Inc., 1993.

[27] R. Srinivasan, C. Hartley, B. Raju, and J. Clave; Measurement of neck development in tensile testing using projection moire, *Opt. Engrg.*, 1982, 655–662.

[28] J. Winter and R. Green; Real-time monitoring of microstructural transformations using synchotron and flash X-ray diffraction, in *Intelligent Processing of Materials and Advanced Sensors*, Ed. by H. Wadley, P. Parrish, B. Rath, and S. Wolf, The Metallurgical Society, Pennsylvania, 1987, 61–75.

[29] O. Zienkiewicz and R. Taylor; *The Finite Element Method, Fourth Edition, Volume 1, Basic Formulation and Linear Problems*, and *Volume 2, Solid and Fluid Mechanics, Dynamics, and Non-linearity*, McGraw-Hill, London, 1991.

The Application of Sparse Least Squares in Aerospace Design Problems

John T. Betts
Mathematics and Engineering Analysis
Research and Technology Division
Boeing Computer Services
P.O. Box 24346, MS 7L-21
Seattle WA 98124-0346, USA

This paper describes the extension of a general purpose method for solving large sparse nonlinear programming problems to problems characterized by nonlinear least squares objective functions. The method incorporates a sequential quadratic programming algorithm which is solved using a multifrontal method for sparse symmetric indefinite linear equations in conjunction with a Schur-complement technique for solving the quadratic programming subproblem. To demonstrate the utility of the tool the algorithm is used to solve both linear and nonlinear least squares problems which arise in trajectory design. An approach for approximating tabular data which minimizes the curvature subject to interpolation and monotonicity constraints requires the solution of a large sparse linear least squares problem. Finally, the utility of the tool for solving nonlinear least squares problems which arise from prescribed path trajectory design is demonstrated.

1. A sparse nonlinear least squares algorithm

1.1. The NLP problem

Let us begin the discussion with an overview of the general nonlinear programming problem (NLP). It is desired to find an n-vector $\mathbf{x}$ to minimize

$$f(\mathbf{x}) \tag{1}$$

subject to

$$\overline{\mathbf{b}}_\ell \leq \begin{bmatrix} \mathbf{c}(\mathbf{x}) \\ \mathbf{x} \end{bmatrix} \leq \overline{\mathbf{b}}_u \tag{2}$$

where $\mathbf{c}(\mathbf{x})$ is an m-vector, with $f(\mathbf{x})$ and $\mathbf{c}(\mathbf{x})$ having continuous second derivatives.

An NLP is considered *sparse* when most of the elements in the $m \times n$ Jacobian matrix

$$\mathbf{G} = \nabla \mathbf{c}(\mathbf{x}) \tag{3}$$

and the Hessian of the Lagrangian

$$\mathbf{H}_L(\mathbf{x}, \boldsymbol{\lambda}) = \nabla^2 f(\mathbf{x}) - \sum_{i=1}^{m} \lambda_i \nabla^2 c_i \tag{4}$$

are zero. Typically less than 1% of the elements are nonzero. In these expressions $\boldsymbol{\lambda}$ is an m-vector of Lagrange multipliers.

1.2. Optimization strategy

A general purpose algorithm for solving large sparse nonlinear programming problems is described in [4], [5], and [6]. A fundamental feature of the approach is to compute a sequence of estimates for the variables according to the formula

$$\overline{\mathbf{x}} = \mathbf{x} + \alpha \mathbf{p} \tag{5}$$

where $\mathbf{p}$ is referred to as the search direction, and the scalar step length α is adjusted at each iteration to produce a sufficient reduction in an augmented Lagrangian merit function suggested by [8].

Each search direction solves a quadratic programming (QP) subproblem: i.e. minimize

$$\mathbf{g}^{\top}\mathbf{p} + \frac{1}{2}\mathbf{p}^{\top}\mathbf{H}\mathbf{p} \tag{6}$$

subject to

$$\mathbf{b}_{\ell} \leq \begin{bmatrix} \mathbf{G}\mathbf{p} \\ \mathbf{p} \end{bmatrix} \leq \mathbf{b}_{u} \tag{7}$$

where g is the gradient of the objective function. The QP subproblem is well posed provided the *projected Hessian* is positive definite for all active set changes. Unfortunately the projected Hessian $\mathbf{Z}^{\top}\mathbf{H}\mathbf{Z}$ is not readily available for large sparse problems and even if it could be formed in general it would be dense. Furthermore for most iterations the correct active set is unknown. Consequently we correct the defective QP subproblem by modifying the *full Hessian*

$$\mathbf{H} = \mathbf{H}_{L} + \tau(|\sigma| + 1)\mathbf{I} \tag{8}$$

where σ is the Gerschgorin bound for the most negative eigenvalue of $\mathbf{H}_L$. The Levenberg parameter, $0 \leq \tau \leq 1$ is chosen such that the projected Hessian is positive definite which can be inferred from calculations performed during the QP solution process. In addition at each iteration the Levenberg parameter is modified using a trust region strategy such that $\tau \to 0$.

1.3. Solving the sparse QP subproblem

The efficient solution of the sparse quadratic programming subproblem (6)-(7) is achieved using a technique based on the Schur-complement method of Gill, Murray, Saunders, and Wright [10]. The fundamental idea of this method is to solve the large sparse symmetric indefinite Kuhn-Tucker (KT) system

$$\begin{bmatrix} \mathbf{H} & \mathbf{G}^{\top} \\ \mathbf{G} & \mathbf{0} \end{bmatrix} \begin{bmatrix} -\mathbf{p} \\ \boldsymbol{\pi} \end{bmatrix} \equiv \mathbf{K}_0 \begin{bmatrix} -\mathbf{p} \\ \boldsymbol{\pi} \end{bmatrix} = \begin{bmatrix} \mathbf{g} \\ \mathbf{0} \end{bmatrix}. \tag{9}$$

This linear system is factored only *once* using the *multifrontal* algorithm developed by Duff, Ashcraft, Grimes, Lewis, Peyton and Pierce [1], [2]. Subsequently changes to the QP active set can be computed using a *solve* with the previously factored KT matrix $\mathbf{K}_0$ and a solve with a small dense "Schur-complement" matrix. As a part of the solution process it is also possible to compute the inertia of the matrix $\mathbf{K}_0$ and update the inertia as the active set changes. It is then possible to infer that the projected Hessian is positive definite if the inertia of $\mathbf{K}$ is $[n, m, 0]$. If at some point during the QP solution process the inertia of $\mathbf{K}$ is incorrect it is necessary to terminate the QP solution process and modify the Hessian using the Levenberg parameter as in (8).

1.4. The nonlinear least squares problem

In contrast to the general nonlinear programming problem as defined in (1)-(2), the *Nonlinear Least Squares Problem* (NLS) is characterized by an objective function

$$f(\mathbf{x}) = \frac{1}{2}\mathbf{r}^\top(\mathbf{x})\mathbf{r}(\mathbf{x}) = \frac{1}{2}\sum_{i=1}^{\ell}\mathbf{r}_i^2 \tag{10}$$

where $\mathbf{r}(\mathbf{x})$ is an ℓ-vector of *residuals*. As for the general NLP it is necessary to find an n-vector $\mathbf{x}$ to minimize $f(\mathbf{x})$ while satisfying the constraints (2). The $\ell \times n$ *Residual Jacobian* matrix $\mathbf{R}$ is defined by

$$\mathbf{R}^\top = [\nabla\mathbf{r}_1, \ldots, \nabla\mathbf{r}_\ell] \tag{11}$$

and the gradient vector is

$$\mathbf{g} = \mathbf{R}^\top\mathbf{r} = \sum_{i=1}^{\ell} r_i \nabla r_i\,. \tag{12}$$

And finally the Hessian of the Lagrangian is given by

$$\mathbf{H}_L(\mathbf{x}, \boldsymbol{\lambda}) = \sum_{i=1}^{\ell} r_i \nabla^2 r_i - \sum_{i=1}^{m} \lambda_i \nabla^2 c_i + \mathbf{R}^\top\mathbf{R} \tag{13}$$

$$\doteq \mathbf{V} + \mathbf{R}^\top\mathbf{R}\,. \tag{14}$$

The matrix $\mathbf{R}^\top\mathbf{R}$ is referred to as the *normal matrix* and we shall refer to the matrix $\mathbf{V}$ as the *Residual Hessian.*

1.5. Sparse least squares

Having derived the appropriate expressions for the gradient and Hessian of the least squares objective function one obvious approach is to simply use these quantities as required within the framework of a general NLP. Unfortunately there are a number of well known difficulties which are related to the normal matrix $\mathbf{R}^\top\mathbf{R}$.

First, it is quite possible that the Hessian $\mathbf{H}_L$ may be dense even when the residual Jacobian $\mathbf{R}$ is sparse which can occur for example when there is one dense row in $\mathbf{R}$. Even if the Hessian is not completely dense, in general formation of the normal matrix introduces "fill" into an otherwise sparse problem. Secondly it is well known that the normal matrix approach is subject to ill-conditioning even for small dense problems. In particular, formation of $\mathbf{R}^\top\mathbf{R}$ is prone to cancellation and the condition number of $\mathbf{R}^\top\mathbf{R}$ is the square of the condition number of $\mathbf{R}$. It is for this reason that utilization of the normal matrix is contraindicated. For small dense least squares problems the preferred solution technique is to introduce an orthogonal decomposition for the residual Jacobian without forming the normal matrix. Unfortunately these techniques do not readily generalize to large sparse systems when it is necessary to repeatedly modify the active set (and therefore the factorization). A survey of the various alternatives is found in [11]. In order to ameliorate the difficulties associated with the normal matrix, while maintaining the benefits of the general sparse NLP Schur-complement method a "sparse tableau" approach has been adopted.

If we proceed formally to derive the quadratic objective function for the QP subproblem, combining (6) with (14) yields

$$\begin{aligned} \mathbf{g}^\top\mathbf{p} + \frac{1}{2}\mathbf{p}^\top\mathbf{H}\mathbf{p} &= \mathbf{g}^\top\mathbf{p} + \frac{1}{2}\mathbf{p}^\top\left[\mathbf{V} + \mathbf{R}^\top\mathbf{R}\right]\mathbf{p} \\ &= \mathbf{g}^\top\mathbf{p} + \frac{1}{2}\mathbf{p}^\top\mathbf{V}\mathbf{p} + \frac{1}{2}\mathbf{p}^\top\mathbf{R}^\top\mathbf{R}\mathbf{p}\,. \end{aligned} \tag{15}$$

This form suggests introducing new variables $\mathbf{w}$ and constraints $\mathbf{w} = \mathbf{R}\mathbf{p}$. With this identification made we then solve an augmented QP subproblem for the variables $\mathbf{q}^\top = (\mathbf{p}, \mathbf{w})^\top$, i.e., minimize

$$\frac{1}{2}(\mathbf{p}, \mathbf{w})^\top \begin{pmatrix} \overline{\mathbf{V}} & \mathbf{0} \\ \mathbf{0} & \mathbf{I} \end{pmatrix} \begin{pmatrix} \mathbf{p} \\ \mathbf{w} \end{pmatrix} + \mathbf{g}^\top\mathbf{p} \tag{16}$$

subject to

$$\begin{pmatrix} \mathbf{0} \\ \mathbf{b}_\ell \end{pmatrix} \le \begin{pmatrix} \mathbf{R} & -\mathbf{I} \\ \mathbf{G} & \mathbf{0} \\ \mathbf{I} & \mathbf{0} \end{pmatrix} \begin{pmatrix} \mathbf{p} \\ \mathbf{w} \end{pmatrix} \le \begin{pmatrix} \mathbf{0} \\ \mathbf{b}_u \end{pmatrix}. \tag{17}$$

As in the general NLP problem it is necessary that the quadratic programming subproblem be well posed. Consequently to correct a defective QP subproblem the *residual Hessian* is modified

$$\overline{\mathbf{V}} = \mathbf{V} + \tau(|\sigma| + 1)\mathbf{I} \tag{18}$$

where σ is the Gerschgorin bound for the most negative eigenvalue of $\mathbf{V}$. Notice that it is not necessary to modify the entire Hessian matrix for the augmented problem as given in (16) but simply the portion which can contribute to directions of negative curvature. As for the general NLP, we choose the Levenberg parameter, $0 \le \tau \le 1$ such that the projected Hessian of the augmented problem is positive definite, and modify the Levenberg parameter, at each iteration such that ultimately $\tau \to 0$. Observe that for linear residuals and constraints the residual Hessian $\mathbf{V} = 0$ and consequently $|\sigma| = 0$. Thus in the linear case no modification is necessary unless $\mathbf{R}$ is rank deficient. Furthermore for linearly constrained problems with small residuals at the solution as $\|\mathbf{r}\| \to 0$, $|\sigma| \to 0$ so the modification to the Hessian is "small" even when the Levenberg parameter $\tau \ne 0$. And finally for large residual problems, i.e. even when $\|\mathbf{r}^*\| \ne 0$ the accelerated trust region strategy used for the general NLP method adjusts the modification $\tau \to 0$ which ultimately leads to quadratic convergence.

2. Minimum time to climb trajectory example

In order to illustrate the behavior of the sparse least squares algorithm let us consider a series of problems derived from a well-known trajectory optimization problem. The original minimum time to climb problem was presented by Bryson, et.al. [7] and has been the subject of many analyses over the past 25 years. The basic problem is to choose the optimal control function $\alpha(t)$ (the angle of attack) such that an airplane flies from a point on a runway to a specified final altitude, as quickly as possible. In its simplest form, the planar motion of the aircraft is described by the following set of ordinary differential equations:

$$\dot{h} = v \sin\gamma \tag{19}$$

$$\dot{v} = \frac{1}{m}\left[T\cos(\alpha) - D\right] - \frac{\mu}{(R_e + h)^2}\sin\gamma \tag{20}$$

$$\dot{\gamma} = \frac{1}{mv}\left[T\sin(\alpha) + L\right] + \cos\gamma\left[\frac{v}{(R_e + h)} - \frac{\mu}{v(R_e + h)^2}\right] \tag{21}$$

$$\dot{w} = \frac{-T}{I_{sp}} \tag{22}$$

where h is the altitude, v the velocity, γ the flight path angle, w the weight, $m = w/g_0$ the mass, μ the gravitational constant, and R_e the radius of the earth.

The aerodynamic forces on the vehicle are defined by the expressions

$$D = \frac{1}{2}C_D S\rho v^2 \tag{23}$$

$$L = \frac{1}{2}C_L S\rho v^2 \tag{24}$$

$$C_L = c_{L\alpha}(M)\alpha \tag{25}$$

$$C_D = c_{D0}(M) + \eta(M)c_{L\alpha}(M)\alpha^2 \tag{26}$$

where D is the drag, L is the lift, C_L and C_D are the aerodynamic lift and drag coefficients respectively, with S the cross reference area of the vehicle, and ρ the atmospheric density. Although the results presented utilized a cubic spline approximation to the 1962 Standard Atmosphere, qualitatively similar results can be achieved with a simple exponential approximation to $\rho(h)$.

The following constants complete the definition of the problem:

$$\begin{array}{lcllcl}
h(0) & = & 0. & h(t_f) & = & 65600.0 \\
v(0) & = & 424.260 & v(t_f) & = & 968.148 \\
\gamma(0) & = & 0. & \gamma(t_f) & = & 0. \\
w(0) & = & 42000.0 & S & = & 530. \\
I_{sp} & = & 1600.0 & \mu & = & .14076539 \times 10^{17} \\
g_0 & = & 32.174 & R_e & = & 20902900\,.
\end{array}$$

2.1. Tabular data

As with most real aircraft, the aerodynamic and propulsive forces are specified in tabular form. For the sake of completeness the data as it appeared in the original reference is given in Tables 1 and 2. There are a number of significant points which characterize the tabular data representation. First both the thrust and aerodynamic table values are given to limited numeric precision (i.e. approximately two significant figures). Perhaps the most obvious explanation for the limited precision is that the data probably was originally obtained from experimental tests and truncated at the precision of the test equipment. Unfortunately the statistical analysis (if any) of the original test data has long since been forgotten. Consequently, it is common to assume that the table values are "exact" and correct to full machine precision. A second difficulty which is evident in the bivariate thrust data is that apparently data is missing from the corners of the table. Of course the data is "missing" because a real aircraft can never fly in these regimes (e.g. at mach = 0, and h = 70000 ft). In fact for most experimentally obtained data it can be expected that data will be missing for unrealistic portions of the domain. In view of these realities it is common to linearly interpolate and never extrapolate the tablular data. Figure 1 illustrates a linear treatment of the thrust table.

2.2. Cubic spline interpolation

While a linear treatment of tabular function may be adequate for simply evaluating the functions it is totally inappropriate if the functions are to be used within a trajectory simulation and/or optimization. The principle difficulty stems from the fact that most numerical optimization and integration algorithms assume that the functions are continuously differentiable to at least second order. Thus, just to propagate the trajectory using an integration algorithm such as Runge-Kutta, or Adams-Moulton, it is necessary that the right hand side of the differential equations (19)-(22) have the necessary continuity. Similar requirements are imposed when a numerical optimization algorithm is used to shape the trajectory, since the optimization uses second derivative (Hessian) information to construct estimates of the solution.

Thrust T(M,h) (thousands of pounds)

M	Altitude h (thousands of ft)									
	0	5	10	15	20	25	30	40	50	70
0.0	24.2									
0.2	28.0	24.6	21.1	18.1	15.2	12.8	10.7			
0.4	28.3	25.2	21.9	18.7	15.9	13.4	11.2	7.3	4.4	
0.6	30.8	27.2	23.8	20.5	17.3	14.7	12.3	8.1	4.9	
0.8	34.5	30.3	26.6	23.2	19.8	16.8	14.1	9.4	5.6	1.1
1.0	37.9	34.3	30.4	26.8	23.3	19.8	16.8	11.2	6.8	1.4
1.2	36.1	38.0	34.9	31.3	27.3	23.6	20.1	13.4	8.3	1.7
1.4		36.6	38.5	36.1	31.6	28.1	24.2	16.2	10.0	2.2
1.6				38.7	35.7	32.0	28.1	19.3	11.9	2.9
1.8						34.6	31.1	21.7	13.3	3.1

Table 1. Propulsion data.

M	0	.4	.8	.9	1.0	1.2	1.4	1.6	1.8
$c_{L\alpha}$	3.44	3.44	3.44	3.58	4.44	3.44	3.01	2.86	2.44
c_{D0}	.013	.013	.013	.014	.031	.041	.039	.036	.035
η	.54	.54	.54	.75	.79	.78	.89	.93	.93

Table 2. Aerodynamic data.

The most direct way to achieve the required continuity is to approximate the data by a tensor product cubic B-spline of the form

$$T(M,h) = \sum_{i=1}^{n_1} \sum_{j=1}^{n_2} c_{i,j} B_i(M) B_j(h) \,. \tag{27}$$

In order to utilize this approximation it is necessary to compute the coefficients $c_{i,j}$. The simplest way to compute the spline coefficients is to force the approximating function to *interpolate* the data at all of the data points. However for a unique interpolant, data is required at all points on a rectangular grid. Since data is missing in the corners of the domain typically this difficulty is resolved by adding "fake" data in the corners using an "eyeball" approach. It is common to ignore the limited precision of the data and simply treat data as though it was of full precision. Finally by using divided difference estimates of the derivatives at the boundary of the region, the spline coefficients are uniquely determined by solving a sparse system of linear equations. Figures 2 and 3 illustrate the cubic interpolating spline obtained using this approach.

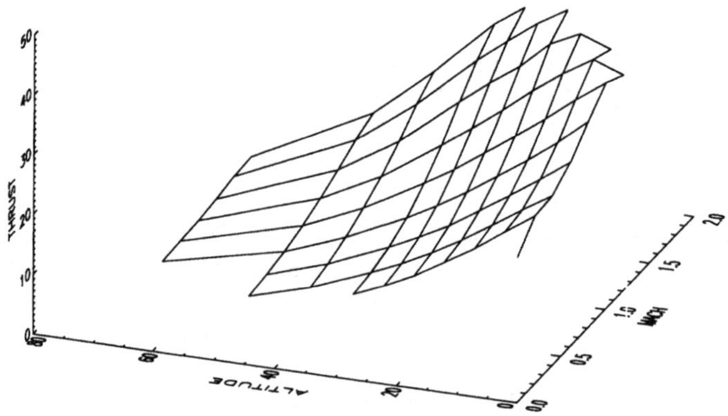

Figure 1. Original thrust data.

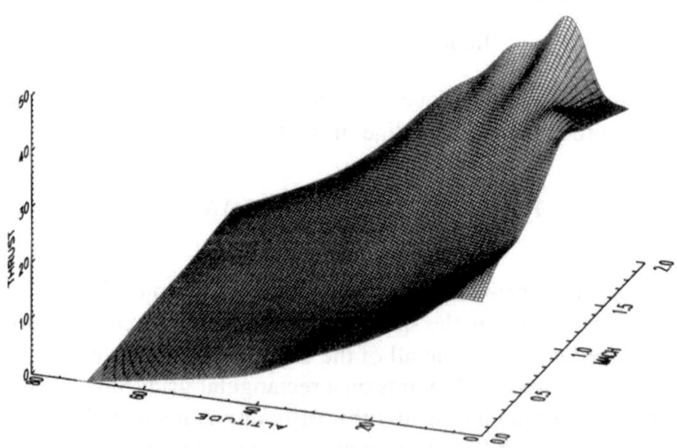

Figure 2. Cubic spline interpolant for thrust data.

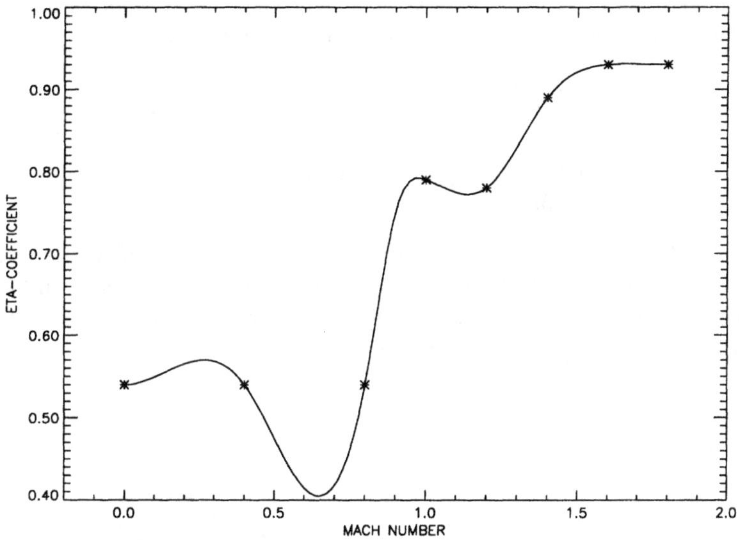

Figure 3. Cubic spline interpolant for aerodynamic data.

2.3. Constrained smoothing spline

The cubic spline interpolant does provide C^2 continuity as needed for proper behavior of integration and optimization algorithms. Unfortunately the approximation produced by simply interpolating the raw data does not necessarily reflect the qualitative aspects of the data. In particular it is common for the interpolant to introduce "wiggles" which are not actually present in the tabular data itself. This is clearly illustrated along the boundaries of the thrust surface in Figure 2 and especially in the aerodynamic data for $M \leq .8$ as in Figure 3. In the case of the latter it is obvious that the interpolant does not reflect the fact that η is constant for low Mach numbers. A second drawback of the cubic interpolant is the need for data at all points on a rectangular grid when constructing an approximation in more than one dimension.

A technique for eliminating the oscillations in the approximation is to carefully select the spline knot locations by inspection of the data, with fewer knots than data points. In this case it is no longer possible to interpolate the data because the number of coefficients is less than the number of interpolation conditions, i.e. the system is overdetermined. However it is reasonable to determine the spline coefficients $c_{i,j}$ such that

$$f(c_{i,j}) = \sum_{k=1}^{\ell} \left[T(M_k, h_k) - \hat{T}_k \right]^2 \tag{28}$$

is minimized. Furthermore it is possible to introduce constraints on the slope of the spline approximation to reflect monotonicity of the data; i.e.

$$\left[\frac{\partial T}{\partial M}\right] \geq 0 \qquad \text{if} \left[\hat{T}_{k+1} - \hat{T}_k\right] \geq 0 \tag{29}$$

$$\left[\frac{\partial T}{\partial M}\right] \leq 0 \qquad \text{if} \left[\hat{T}_{k+1} - \hat{T}_k\right] \leq 0 \tag{30}$$

for $M \in [M_k, M_{k+1}]$. Similar constraints can be imposed to reflect monotonicity in the h-direction. The coefficients which satisfy these conditions can be determined by solving a sparse constrained linear least squares problem. The resulting approximations are illustrated in Figure 4 and Figure 5.

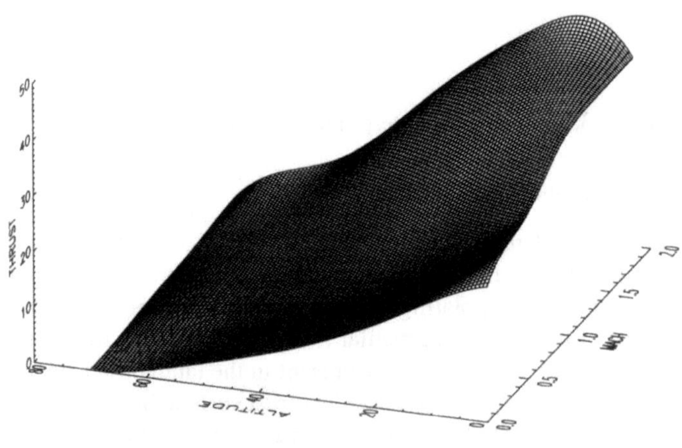

Figure 4. Constrained smoothing spline for thrust data.

2.4. Minimum curvature spline

By imposing monotonicity constraints and minimizing the error between the data and the approximating function it is possible to achieve one of the major goals, namely constructing a C^2 function which eliminates the "wiggles". Unfortunately the location of the knots in the spline approximation must be chosen such that the coefficients are determined by minimizing the least square error. In fact, special care must be taken not to locate knots in regions where data is missing, since this will result in a rank deficient least squares problem. In essence the knots must be located such that local constraint and data uniquely define the spline coefficients.

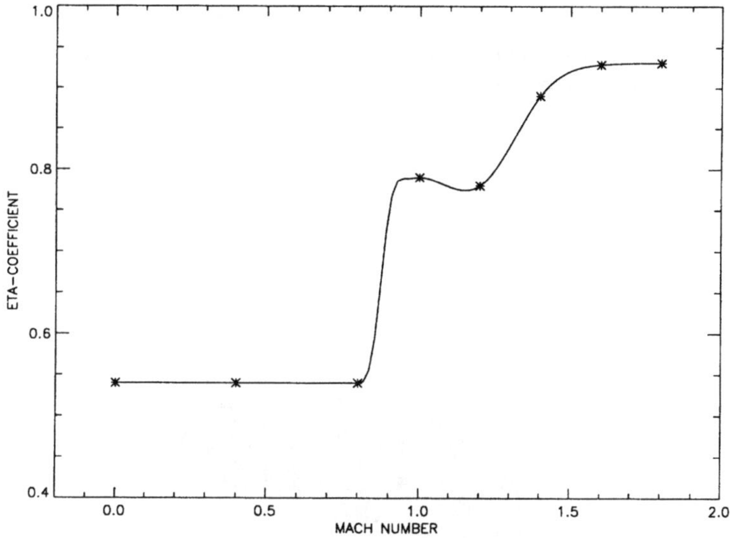

Figure 5. Constrained smoothing spline for aerodynamic data.

To resolve these deficiencies we introduce a rectangular grid with k_1 values in the first coordinate and k_2 values in the second coordinate. We require that all data points lie on the rectangular grid. However not all grid points need have data (i.e. data can be missing). Then let us consider introducing a quartic spline with; (a) double knots at the data points in order insure C^2 continuity and, (b) single knots at the midpoints of each interval. Then let us determine the spline coefficients $c_{i,j}$ which minimize the curvature

$$f(c_{i,j}) = \sum_{k=1}^{L} \left[\frac{\partial^2 T}{\partial M^2}(M_k, h_k) \right]^2 + \left[\frac{\partial^2 T}{\partial h^2}(M_k, h_k) \right]^2 \tag{31}$$

and satisfy the data approximation constraints

$$\hat{T}_k - \epsilon \leq T(M_k, h_k) \leq \hat{T}_k + \epsilon \tag{32}$$

for all data points $k = 1, \ldots, \ell$, where ϵ is the data precision. In order to insure full rank in the Hessian of the objective we evaluate the curvature at points determined by the knot interlacing conditions [9]. As with the smoothing spline we introduce the slope constraints (29) and (30) on the approximation in order to reflect the monotonicity of the data. As before, these coefficients can be determined by solving a sparse constrained linear least squares problem. The resulting approximations are illustrated in Figure 6 and Figure 7. For this particular fit the least squares problem had 900 variables, with 988 constraints of which 77 were equalities. In general the number of variables n for a bivariate minimum curvature fit is $n = 9k_1k_2$. For the general case the minimum curvature formulation requires imposition of m constraints where $10k_1k_2 \le m \le 14k_1k_2$ and the exact number depends on the number of algebraic sign changes in the slope of the data. In addition if interpolation is required the number of equality constraints m_e is $m_e \le k_1k_2$.

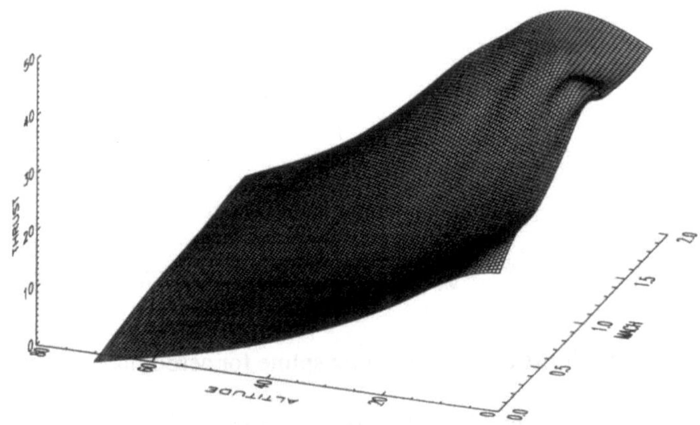

Figure 6. Minimum curvature spline for thrust data.

2.5. Nonlinear least squares example

The previous sections have described a number of alternatives for treating the tabular data needed to define the minimum time to climb trajectory problem (19)-(26). When this data is utilized the solution of the trajectory optimization problem can be computed in a number of ways. A direct transcription method as described in [3], [5], and [6], was used to solve the problem and the results are illustrated in Figure 8.

One obvious attribute of the solution which has been the subject of considerable interest is the fact that there is a dive during part of the climb. Essentially the aircraft dives to pick up energy thereby reaching the final altitude sooner. In order to illustrate a more benign climb trajectory let us suppose that the desired climb path is described by

$$\hat{h}(t) = c_0 + c_1 t + c_2 t^2 + c_3 t^3 \tag{33}$$

where the coefficients are chosen such that $\hat{h}(0) = \dot{\hat{h}}(0) = 0$ and $\hat{h}(t_f) = h_f$, $\dot{\hat{h}}(t_f) = 0$ for a specified final time $t_f = 350$ and free final velocity v_f. Let us now attempt to choose the control to minimize the deviation from the desired path, that is,

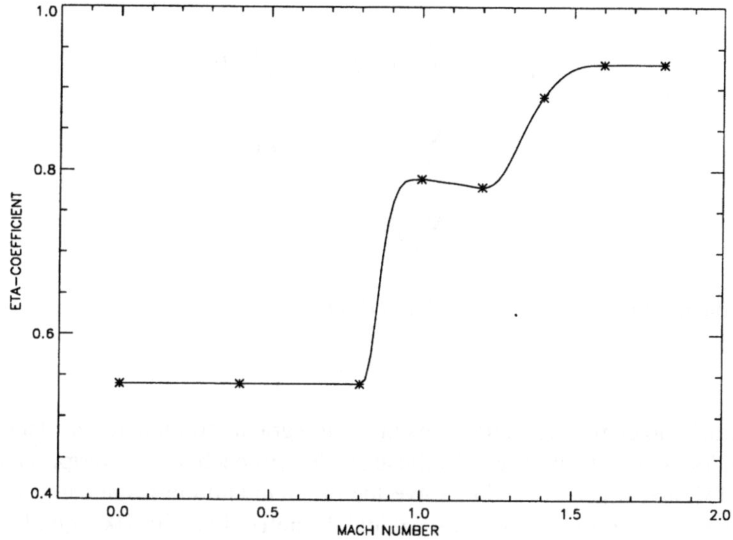

Figure 7. Minimum curvature spline for aerodynamic data.

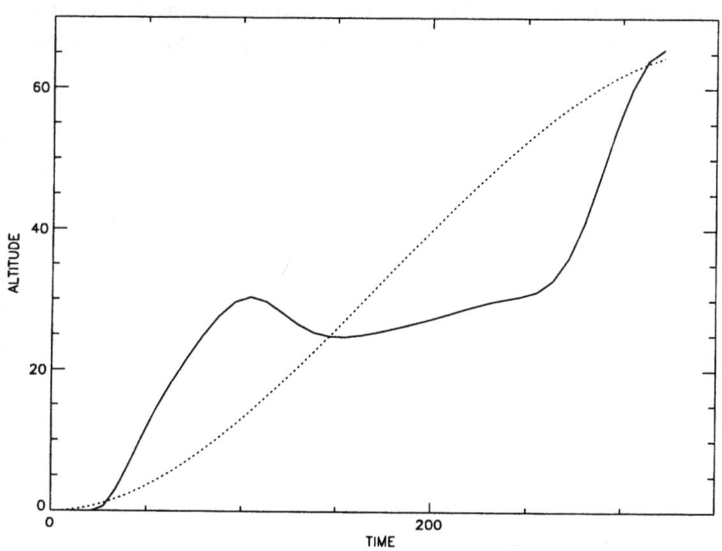

Figure 8. Minimum time to climb trajectory.

$$
\begin{aligned}
f &= \int_0^{t_f} \left[h(t) - \hat{h}(t)\right]^2 dt & (34)\\
&\approx \sum_{k=0}^{n_g} a_k \left[h(t_k) - \hat{h}(t_k)\right]^2 & (35)\\
&= \sum_{k=0}^{n_g} r_k^2 & (36)
\end{aligned}
$$

where the nonlinear *residuals* are defined by

$$r_k = \sqrt{a_k}\left[h(t_k) - \hat{h}(t_k)\right]. \tag{37}$$

The actual choice for the coefficients in the integral approximation a_k depends on the quadrature rule being used. To illustrate the approach the transcription method in [3], [5], and [6], was used to solve the problem and the results are illustrated in Figure 9. For this example a trapezoidal discretization with 100 equally spaced grid points was used, which results in a sparse nonlinear least squares problem with 500 variables, 396 nonlinear constraints, and 86 degrees of freedom at the solution. Using a linear initial guess for the state and control variables the solution was obtained in 110.5 seconds on a SUN IPX workstation. During the iteration the trajectory functions were evaluated 304 times, and the Hessian matrix was computed 11 times.

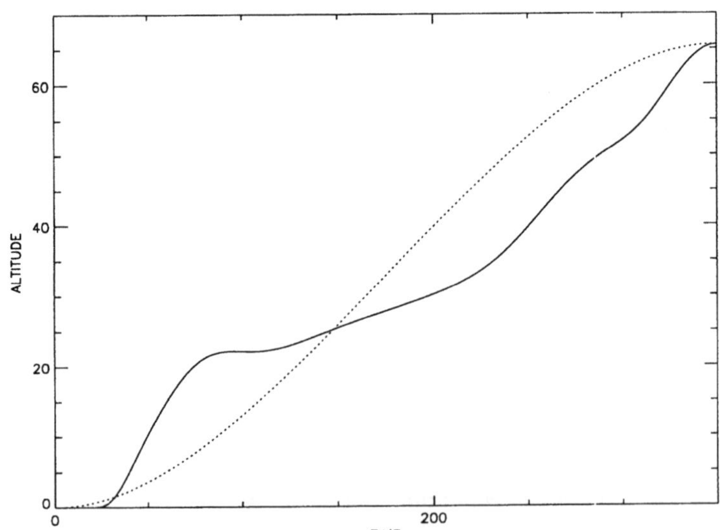

Figure 9. Minimum deviation climb.

3. Summary and conclusions

This paper describes an algorithm for solving general sparse nonlinear least squares problems. The method represents an extension of a previously described sparse nonlinear programming technique based on a Schur-complement implementation of a sequential quadratic programming approach. The detrimental affects of the normal matrix on matrix sparsity and conditioning are avoided by utilizing a sparse tableau formulation which results in an augmented quadratic programming subproblem. A fundamental feature of the method is to modify the residual Hessian matrix using a diagonal matrix whose size is determined by the Gershgorin bound for the most negative eigenvalue and a Levenberg parameter which is adjusted utilizing an accelerated trust region strategy. Since projected and/or reduced Hessian approximations are not explicitly constructed the algorithm is efficient even when the number of degrees of freedom is large.

The new algorithm is illustrated on both linear and nonlinear least squares examples which result from solving the minimum time to climb trajectory optimization problem. A technique for constructing a minimum curvature tensor product spline approximation to tabular data requires solving a large sparse linear least squares problem. The data approximation exhibits the desired continuity and differentiability, without introducing "wiggles" that are not actually in the original data. The technique is applicable even when the data is not available on a uniform rectangular grid. The monotonicity and precision of the data are treated directly by formulating linear inequality constraints. Finally the knot locations for constructing the approximation are defined without manual interaction. The nonlinear least squares capability is illustrated by solving a minimum deviation trajectory optimization problem.

Acknowledgement

The author wishes to acknowledge the valuable contributions of Mr. Richard Mastro during the development and implementation of the minimum curvature spline approximation procedure. The author also gratefully acknowledges the valuable interaction with Dr. Paul Frank concerning the implementation of the sparse least squares capability.

References

[1] C. Ashcraft; A vector implementation of the multifrontal method for large sparse, symmetric positive definite linear systems, Technical Report ETA-TR-51, Boeing Computer Services, 1987.

[2] C. Ashcraft and R. Grimes; The influence of relaxed supernode partitions on the multifrontal method, Technical Report ETA-TR-60, Boeing Computer

Services, 1988.

[3] J. Betts; Using sparse nonlinear programming to compute low thrust orbit transfers, *J. Astro. Sci.* **41**, 1993, 349–371.

[4] J. Betts and P. Frank; A sparse nonlinear optimization algorithm, *J. Optim. The. Appl.* **82**, 1994.

[5] J. Betts and W. Huffman; The application of sparse nonlinear programming to trajectory optimization, *J. Guid. Cont. Dynam.* **15**, 1992.

[6] J. Betts and W. Huffman; Path constrained trajectory optimization using sparse sequential quadratic programming, *J. Guid. Cont. Dynam.* **16**, 1993, 59–68.

[7] A. Bryson, M. Desai, and W. Hoffman; Energy-state approximation in performance optimization of supersonic aircraft, *J. Aircraft* **6** 1969.

[8] P. Gill, W. Murray, M. Saunders, and M. Wright; Some theoretical properties of an augmented Lagrangian merit function, Report SOL 86-6, Department of Operations Research, Stanford University, 1986.

[9] C. de Boor; *A Practical Guide to Splines*, Springer, New York, 1978.

[10] P. Gill, W. Murray, M. Saunders, and M. Wright; A Schur-complement method for sparse quadratic programming, Report SOL 87-12, Department of Operations Research, Stanford University, 1987.

[11] M. Heath; Numerical methods for large sparse linear least squares problems, *SIAM J. Scien. Stat. Comp.* **5** 1984, 497–513.

Algorithms for Flow Control and Optimization*

J. Borggaard, J. Burkardt, J. Burns, E. Cliff, M. Gunzburger,
H. Kim, H. Lee, J. Peterson, A. Shenoy, and X. Wu
Air Force Center for Optimal Design and Control
Virginia Tech
Blacksburg VA 24061-0531, USA

1. Introduction

The work on flow control and optimization at the Air Force Center for Optimal Design and Control (CODAC) has focused on *new tools for improved aerodynamic design* and the *optimal and feedback control of fluid flows.*

The aerodynamic performance of aerospace systems (or of their components) and of test facilities can often be enhanced by the judicious tailoring of their shape or of other parameters that determine the flows. Computational tools that combine modern computational fluid dynamics and optimization methods can aid in the design of these objects and result in enhanced performance and reduced development costs. Efficient optimization schemes require information about both the flow and its sensitivity to design changes. The use of standard finite-difference quotient approximations to the sensitivities requires the expenditure of far too large amounts of computational resources. As an effective alternative, we use the continuum model to derive sensitivity equations which are a system of linear partial differential equations that may be solved, in an approximate manner, for the flow sensitivities. We have successfully applied this method to several problems. This approach is now the basis for several joint ventures with Air Force Laboratories and industrial collaborators.

The development of convergent computational schemes for optimal control of fluid flows is an extremely complex undertaking. The problems are highly nonlinear and often the control is applied by using wall shapes, temperature, or mass fluxes as control mechanisms. A general framework for constructing finite dimensional approximating optimal control problems has been developed. This framework is applicable not only to flow control, but to nonlinear problems in many other settings as well. Our efforts in this direction have focused on building mathematical models of the physical problems, invoking a minimum of assumptions about the physical phenomena, and then rigorously analyzing these mathematical models. After this first step, we develop and analyze discretization methods for determining approximate solutions of these problems, develop computer codes implementing the algorithms, first for the purpose of showing the efficacy of these methods, and ultimately, to solve problems of practical interest. We have also considered various algorithmic issues in the feedback control of fluid flows.

*The work reported on here was supported by the Air Force Office of Scientific Research under AFOSR URI Grant Number F49620-93-1-0280.

In this paper, we first give a brief summary of the various projects that CODAC personnel have been involved in. The references cited may be consulted for details about these projects and the results that have been obtained. We then give a more detailed discussion of two of these projects.

2. Summary of CODAC projects

Optimal design of a two-dimensional forebody simulator [2], [3]

This problem was motivated by the design of the Free-Jet Test Facility at the Arnold Engineering Development Center (AEDC) that is used for testing aircraft engines under various flight conditions. The test cells in this facility can house a jet engine, but are too small to contain a full airplane forebody and engine. Thus the effect of the forward fuselage on the engine inlet flow conditions must be "simulated" in order to achieve an accurate test. One approach to solving this problem is to replace the actual forebody by a smaller object, called a "forebody simulator" (FBS) and determine the shape of the FBS that produces the best flow match at the engine inlet. In developing practical computational methods for such optimal design problems, one often relies on cascading simulation software into optimization algorithms. Most optimization algorithms require gradient (sensitivity) information which, in this case, describes how sensitive the cost function (flow variables) is (are) to changes in the shape of the FBS, boundary conditions, and wind tunnel operating conditions. In our study, we used the sensitivity equation method (SEM) to generate optimal FBS designs for a model two-dimensional problem. The combined flow/sensitivity code is a modification of the PARC analysis code used at AEDC. We used a quasi-Newton/trust region algorithm with gradient information obtained by the SEM method. Timing tests for twenty design parameters show that sensitivities may be determined by the SEM method in only 38% of the CPU time required by the finite difference quotient method.

One-dimensional Euler duct design problem [1]

A one-dimensional design problem has been used as a test vehicle for studying various issues that arise in the optimal design of compressible fluid flows that involve shock waves. The model problem consists of a steady-state Euler flow in a one-dimensional duct having variable cross-sectional area. The goal of the optimal design problem is to find the cross-sectional area that minimizes the discrepancy between the flow and a desired flow. With appropriate choices for the inlet and outlet conditions and constraints on the variation of the duct cross-sectional area, this flow contains a shock wave. Thus, this problem is complex enough to capture the difficulties presented by shock waves in a flow. On the other hand, the problem can be solved analytically so that, e.g., the cost functional can easily be determined; furthermore, one can use several numerical schemes to cheaply approximate solutions of this problem. This simple example shows that the numerical approximation of the cost functional can generate artificial local minima. The basic problem comes from trying to match discontinuous flow solutions. Thus, optimization strategies employed to solve flow design problems containing shock waves have to account for this phenomenom. One method for correcting this problem is to remove the artificial local minima by smearing out the shock wave through the addition of dissipation to the flow; this can be easily accomplished by using a highly dissipative flow solution algorithm. Once one has successfully avoided the artificial local minima and "reached a vicinity" of the desired global minimum, one can switch to a less dissipative flow solver in order to obtain a more accurate approximation to the optimal design. Another method for avoiding the artificial local minima problem is to use coarse grids during the initial stages of the optimization process to take advantage of the observation that the number of local minima increase with grid refinement; again, once "near" the global minimum one can switch to a finer grid in order to achieve greater accuracy. (A third method is discussed in Section 3 in connection with shock fitting schemes for the optimal design of a two-dimensional nozzle.)

Optimal control approach to the one-dimensional nozzle design problem [18]

The one-dimensional Euler nozzle design problem is also considered from an optimal control point of view. The fluid velocity and the cross-sectional area are the states, and the variation of the area, i.e., the slope of the area distribution function, is the control. The problem, then, is to determine how to control the area distribution in order to minimize the deviation of the obtained velocity from a desired profile. A priori bounds are placed on the allowable size of the control. The solution of this optimal control problem consists of singular arcs and arcs such that the control takes on one of its bounding values. Along the singular arcs, the area distribution assumes that of the desired flow. Approximate optimal solutions are found by the adjoint equations method and a subsequent discretization.

Analysis and computation of a shape control problem for the Navier-Stokes equations [4], [11]

Shape controls are of great interest in the control of fluid flows. For example, the problem of the optimal design of wings is a shape optimal control problem. There have been substantial analyses of linear shape control problems for partial differential equations. However, for the nonlinear Navier-Stokes equations of incompressible flow, no rigorous analyses have been previously carried out. We have given the first such analyses in the context of a two-dimensional flow in a channel-like domain. Our rigorous analyses include showing, in the context of a drag minimization problem, that optimal solutions exist, showing that the Lagrange multiplier rule may be used to enforce constraints, deriving an optimality system from which optimal states and co-states may be deduced, deriving the shape gradient, and finally, deriving error estimates for finite element approximations of optimal states. We have also carried out extensive numerical experiments with shape control problems for incompressible viscous flows. In this study, we examined various issues related to local minima, functional regularization, non-feasible solutions, sensitivities, and the efficiency of computations.

Optimal design of a two-dimensional nozzle [19]

We consider an optimal design problem for a two dimensional flow in a nozzle. The goal is to find an outflow boundary condition and shape of the nozzle such that a target flow is best matched. The design parameters are defined through a parameterization of the shape of the nozzle and the outflow boundary condition. Then, the inverse design problem is an optimization problem with respect to the design parameters which requires the minimization of a cost functional. We solve the optimization problem by combining a very accurate numerical shock-fitting scheme to solve the flow problem, a careful numerical integration of the cost functional, and a BFGS/trust region optimization algorithm. In this approach the artificial local minima problem discussed above are avoided by smoothing the cost functional; the addition of artificial dissipation in the flow solver is not needed. This project is described in greater detail in Section 3.

Finite dimensional approximation of a class of nonlinear optimal control problems [8], [10], [13]

The need to solve optimization and control problems arises in many settings. Although in some cases these problems may be easily solved, either analytically or computationally, in many other cases substantial difficulties are encountered. For example, candidate optimal states and controls may belong to infinite dimensional function spaces and one may have to minimize a nonlinear functional of the state and control variables subject to nonlinear constraints that take the form of a system of partial differential equations whose solutions are in general not unique. Our goal, which we have reached, was to construct, analyze, and apply a framework for the approximate solution of many such problems. The setting for our framework is a class of nonlinear control or optimization problems which is general enough to be of use in numerous applications. The major steps in the development and analysis of our framework are as follows: 1. define an abstract class of nonlinear control or optimization problems; 2. show that, under certain assumptions, optimal solutions exist; 3. show that, under certain additional assumptions, Lagrange multipliers exist that may be used to enforce the constraints; 4. use the Lagrange multiplier technique to derive an optimality system from which optimal states and controls may be deduced; 5. define algorithms for the finite dimensional approximation of optimal states and controls; and 6. derive estimates for the error in the approximations to the optimal states and controls. Two of the key ingredients used to carry out the above plan are the Tikhomorov version of the Lagrange multiplier theory and the Brezzi-Rappaz-Raviart theory for the approximation of a class of nonlinear problems. In both of these theories, certain properties of compact operators on Banach spaces play a central role. We point out that the nonuniqueness of solutions of the nonlinear constraint equations makes it appropriate to employ

Lagrange multiplier principles. After having developed and analyzed the abstract framework, we applied it to some specific, concrete problems. In each case, we used the abstract framework to analyze the concrete problems by merely showing that the latter fit into the former. The particular applications we considered are: 1. control problems in structural mechanics having geometric nonlinearities that are governed by the von Kármán equations; 2. control problems in superconductivity that are governed by the Ginzburg-Landau equations; 3. control problems for incompressible, viscous flows that are governed by the Navier-Stokes equations; 4. a coupled solid/fluid heat control problem; and 5. control problems for a Ladyzhenskaya model for viscous flows. In considering these applications, we purposely chose different types of controls in order to illustrate how these could be accounted for within the abstract framework. In all cases, approximations were effected through the use of finite element methods.

Analysis of optimal boundary control problems for the Navier-Stokes equations [9]
The control of fluid flows has emerged as an important area of technological and scientific research. From a rigorous mathematical point of view, much of that study has been directed at stationary problems, or at time dependent problems with specially structured controls, mostly of the distributed type. We have addressed the interesting problem of boundary control problems for the two and three dimensional Navier-Stokes equations of unsteady, viscous, incompressible flow. The allowable controls are quite general. So far, we have carried out the analyses of the existence of optimal solutions and of the existence of Lagrange multipliers and have derived the optimality system for a drag minimization problem.

Optimal control of an electrically conducting fluid [15], [16]
Another example of optimal control problems that we have considered deals with the study of electrically conducting fluids. Specifically, we want to match the electric current density by controlling the electric potentials on the boundary of the flow domain. Such a control may be easily effected by attaching to the flow boundary appropriate electric sources with required electric potentials or currents. Optimal control problems are precisely defined and the existence of an optimal solution is proved by standard techniques. Using the Lagrange multiplier technique, the constrained optimization problem is converted into an unconstrained one and an optimality system of equations is derived. Optimal solutions can be obtained as solutions of the optimality system. We have also examined a new solution algorithm for solving the discrete state equations which is a large system of nonlinear equations. The objective of the method is to require the solution of a much smaller system of nonlinear equations than the one obtained by standard methods while retaining optimal accuracy. The basic outline of the algorithm is to use, e.g., Newton's method, to first solve the nonlinear system of equations on a coarse mesh. One then solves an appropriately defined linear problem on the desired finer mesh. Under a uniqueness condition, this two-level procedure is shown to produce an op-

timally accurate solution at a substantially reduced cost.

Feedback control of Kármán vortex shedding [12], [14]

The control of the forces, e.g., lift and drag, exerted on a submerged obstacle by the fluid that surrounds it is important in many applications. Here we focus on the problem of controlling the lift in the plane, unsteady, incompressible, viscous flow around a circular cylinder. Flows about a cylinder at even moderate values of Reynolds numbers, exhibit an unsymmetric, periodic shedding of vortices. As a result, the lift force exerted by the fluid on the cylinder is periodic in time. Here, our goal is to show the efficacy of a simple feedback control law for the reduction of the magnitude of the lift. The control mechanism used to attempt to reduce the size of the lift oscillations is the injection and suction of fluid through orifices on the surface of the cylinder. The amount of fluid injected or sucked through the orifices is determined, using a simple feedback law, from the pressure sensed at various stations on the cylinder. Thus the sensor determines the pressure at various stations on the cylinder and the actuator injects or sucks fluid through orifices also on the cylinder. Other sensing and actuating mechanisms could also be used, e.g., the vorticity or stress components on the cylinder for sensing and rotation or shape modification of the cylinder for actuating. The type of feedback laws used in our study are based on the observation that differences in the pressure on the top and bottom halves of the cylinder should give an indication of the asymmetric behavior of the lift. Such pressure differences are then used to determine how much fluid to inject or suck through the orifices. No attempt is made to systematically design an optimal feedback law, i.e., one that in some sense does the best possible job in meeting the objective. However, the computational results we have obtained indicate that the simple feedback law we use here is quite effective in reducing the size of the oscillations in the lift. This project is described in greater detail in Section 4.

Optimal feedback control for Navier-Stokes equations [5], [6], [7], [17]

Active feedback control of fluid flows has many important applications in engineering sciences, both in military situations and in industrial processes. The intrinsic difficulties caused by, among other things, the complex nonlinearity of the governing equations, lead to difficult computational and theoretical problems. Our initial investigations are therefore limited, but our hope is that this investigation will shed some light toward the development of computational algorithms for the problems of flow control, and also serve as a guide to turbulence control. The control problems we considered are not restricted to bounded domains and include flows in exterior domains and flow in unphysical but important space-periodic domains. During the first phase of our work, we study flows in domains with rather simple geometrical configurations. For example, in the case of exterior domains, we have considered the flow past a rotating cylinder with the rotation rate as a control variable. We have also used the driven cavity as a test model, where the control is applied at the bottom boundary of the domain. For these problems, the control is applied at the boundary so that it appears as a boundary condition for the unsteady Navier-Stokes problem. One example for the space periodic domain is the well-known Kolmogorov flow. This particular flow not only provides a reasonable first test for linear feedback control laws, but it also simplifies the mathematical framework for theoretical analysis. For the periodic domain problems, a distributed control is applied. The control objectives we have considered thus far include controlling vortex shedding, enhancing lift-to-drag performance, and stabilizing unsteady flows. All these objectives are usually formulated as minimization problems for a suitable cost functional. The main goal of this phase of the project is the development of numerical methods for the aforementioned flow control problems, and to provide efficient algorithms to implement these methods. Although in our studies we have only considered laminar flows, our goal is to develop feedback controls for turbulent flows.

3. Optimal design of a two-dimensional nozzle

We consider an inviscid compressible flow with a shock wave in a two-dimensional nozzle. The flow is described by the Euler equations

$$P\frac{\partial \mathbf{U}}{\partial x} + Q\frac{\partial \mathbf{U}}{\partial y} = 0, \tag{1}$$

where

$$\mathbf{U} = \begin{pmatrix} p \\ u \\ v \\ \rho \end{pmatrix}, \quad P = \begin{pmatrix} u & \rho a^2 & 0 & 0 \\ 1/\rho & u & 0 & 0 \\ 0 & 0 & u & 0 \\ 0 & \rho & 0 & u \end{pmatrix}, \quad Q = \begin{pmatrix} v & 0 & \rho a^2 & 0 \\ 0 & v & 0 & 0 \\ 1/\rho & 0 & v & 0 \\ 0 & 0 & \rho & v \end{pmatrix},$$

ρ is density, p is pressure, a is the speed of sound, and u and v are velocity components in the x and y directions, respectively.

At the inflow boundary, a supersonic boundary condition is given:

$$p(0, y) = p_0(y), \quad u(0, y) = u_0(y), \quad v(0, y) = v_0(y), \quad \rho(0, y) = \rho_0(y), \tag{2}$$

where $p_0(y)$, $u_0(y)$, $v_0(y)$, and $\rho_0(y)$ are given functions satisfying $(u_0 \cos\alpha + v_0 \sin\alpha) < -\sqrt{\rho_0/\gamma p_0}$ and where $(\cos\alpha, \sin\alpha)$ is the unit outward normal vector of the inflow boundary.

At the outflow boundary, a subsonic boundary condition is given:

$$u(1, y) = u_1(y), \tag{3}$$

where $u_1(y)$ is a given function.

The shock wave is considered as an interior boundary separating the domains on each side of it. The boundary conditions at the shock boundary are the shock jump conditions. The nozzle is assumed symmetric about the centerline so that only the lower half of the nozzle is used in the computation. On the lower wall, a no penetration boundary condition is given:

$$u \cos\theta + v \sin\theta = 0\,, \tag{4}$$

where $(\cos\theta, \sin\theta)$ is the outward normal to the wall.

The flow problem is solved by time marching, i.e., we consider the system

$$\frac{\partial \mathbf{U}}{\partial t} + P\frac{\partial \mathbf{U}}{\partial x} + Q\frac{\partial \mathbf{U}}{\partial y} = 0,$$

with initial conditions satisfying the shock jump conditions and consistent with the boundary conditions. We then march this system to a steady state to obtain a solution of (1)–(4).

A highly accurate numerical scheme is used in the computation. The main idea of the scheme is to map the nozzle into two fixed domains such that the shock position in the mapped domain remains fixed; then, a second order finite difference scheme is applied to the transformed equations.

We now consider the inverse design problem. Given a target flow in a nozzle, external constraints on candidate optimal solutions may have to be imposed. For example, in a wind-tunnel setting, it may be necessary to use some existing hardware components. Thus, we shall consider a problem where the outflow area must be larger than that for the target flow so that the target flow cannot be matched exactly.

Let $\widehat{D}$ denote the target nozzle and let $\widehat{\mathbf{U}} = (\widehat{p}, \widehat{u}, \widehat{v}, \widehat{\rho})$ be the given target flow in $\widehat{D}$. Let D denote a nozzle having its two end dimensions specified and let $y = A(x)$ denote the lower wall boundary of D. The corresponding flow in D is denoted by $\mathbf{U} = (p, u, v, \rho)$. We consider the following optimization problem:

given $\widehat{\mathbf{U}}$ and $\widehat{D}$, find $A(x)$ and u_{out} such that

$$\begin{aligned} \mathcal{J}(A, \mathbf{u}_{out}) &= \omega_1 \int_{\widehat{D}} |p - \widehat{p}|^2 ds + \omega_2 \int_{\widehat{D}} |u - \widehat{u}|^2 ds \\ &\quad + \omega_3 \int_{\widehat{D}} |v - \widehat{v}|^2 ds + \omega_4 \int_{\widehat{D}} |\rho - \widehat{\rho}|^2 ds \end{aligned} \tag{5}$$

is minimized,

where u_{out} is the x-component of the outflow velocity, and $\omega_i, i = 1, \cdots, 4$, are weights that can be chosen, if desired, to better match some components of $\widehat{\mathbf{U}}$ than others. Note that the integration is taken over the target domain $\widehat{D}$; however, since the boundary of D may intersect the boundary of $\widehat{D}$, the flow field $\mathbf{U}$ may not be defined everywhere on the $\widehat{D}$. In this case, we assume that the flow field $\mathbf{U}$ can be extrapolated to the whole domain $\widehat{D}$. The intersection of the lower boundaries of D and $\widehat{D}$ can be avoided by a careful choice for the parameterized function $A(x)$.

We assume that the lower boundary $y = A(x)$ is described by the following function:

$$A(x) = A_l + (A_r - A_l)(1 + \tanh(\tan((x - 0.5)\pi) - q_1))/2\,, \tag{6}$$

where A_l and A_r are the y coordinates of $A(x)$ at the left and right ends and q_1 is a parameter. The outflow boundary condition for the x-component of the velocity is assumed to be of the form

$$u_{out}(y) = q_2 + 3(u_c - q_2)(\frac{y + A_r}{2A_r})^2 - 2(u_c - q_2)(\frac{y + A_r}{2A_r})^3\,, \tag{7}$$

where u_c is the centerline velocity and q_2 is a parameter. Thus the cost functional $\mathcal{J}$ in (5) may be viewed as a function of q_1 and q_2, i.e., $\mathcal{J} = \mathcal{J}(q_1, q_2)$.

The target domain $\widehat{D}$ is defined through the choices $A_l = 0.25$, $A_r = 0.375$, $q_1 = 0$ in (6); the target flow in $\widehat{D}$ is determined by the the outflow boundary condition (7) with $A_r = 0.375$, $u_c = 0.6$, and $q_2 = 0.5$. Throughout, the inflow x-velocity is a fixed cubic polynomial.

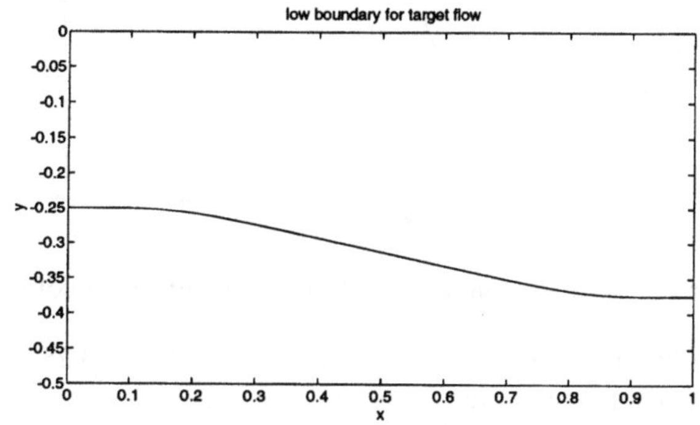

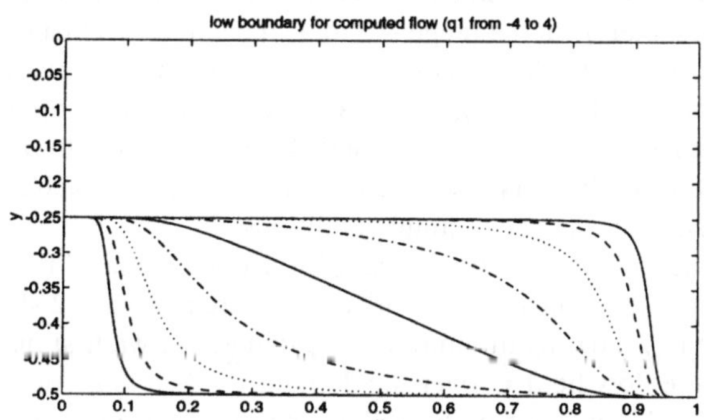

Figure 1. Shape of the nozzle vs. q_1.

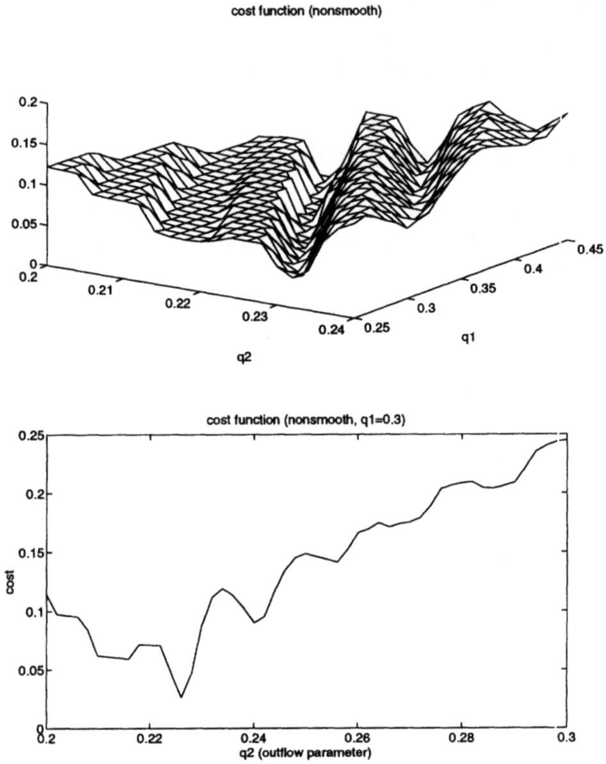

Figure 2. The approximate cost functional before smoothing.

Candidate optimal domains are given by (6) with $A_l = 0.25$ and $A_r = 0.5$ with q_1 as a free design parameter. Note that since the values of A_r are different for the target and candidate optimal flows, one cannot expect that exact matching is possible. Figure 1 shows the shape of nozzle for the target flow and for various values of the parameter q_1 in the range -4 to 4. It can be seen that over that range a wide variation in nozzle shapes is possible. At the inflow boundary, the same boundary conditions as for the target flow are given. Candidate optimal flows are constrained to satisfy (7) with $u_c = 0.6$ and $A_r = 0.5$ and with q_2 a design parameter. The initial condition is a function that is linear along each grid line in the x-direction and satisfies the shock jump conditions.

In the computation, the cost functional $\mathcal{J}$ must be evaluated by numerical integration, and this aspect must be carefully considered. Figure 2 shows a plot of the cost functional $\mathcal{J}$ near its minimum. To evaluate $\mathcal{J}$, we first made a bilinear interpolation of the flow field $\mathbf{U}$ over the domain $\widehat{D}$, then numerically integrated over $\widehat{D}$. Although this integration procedure seems reasonable, one sees that the resultant approximate cost functional is badly behaved in the sense that it has many local minima. An optimization procedure will have great difficulty for such a cost functional. The cause of this problem is the movement of the shock wave across mesh lines of the grid. Recall that we are evaluating the cost functional by integrating over the domain $\widehat{D}$ of the target flow. The shock position for the flow computed at other values of the parameters q_1 and q_2 is not well-adapted to the $\widehat{D}$ grid and, as these parameters change, the computed shock wave will cross cell boundaries in the $\widehat{D}$-grid.

To solve this problem we smooth the flow fields $\mathbf{U}$ and $\widehat{\mathbf{U}}$ by smearing the shock over three points. This can be done by averaging the dependent variables. Once this smoothing is done, the same integration procedure is performed for the smoothed fields. As a result, we now have a well-behaved cost functional as is shown in Figure 3; the circles denote the parameters at each iteration of the optimization procedure described below. Note that we compute the flow using a second order accurate shock-fitting scheme and the smoothing is introduced only in the evaluation of the cost functional.

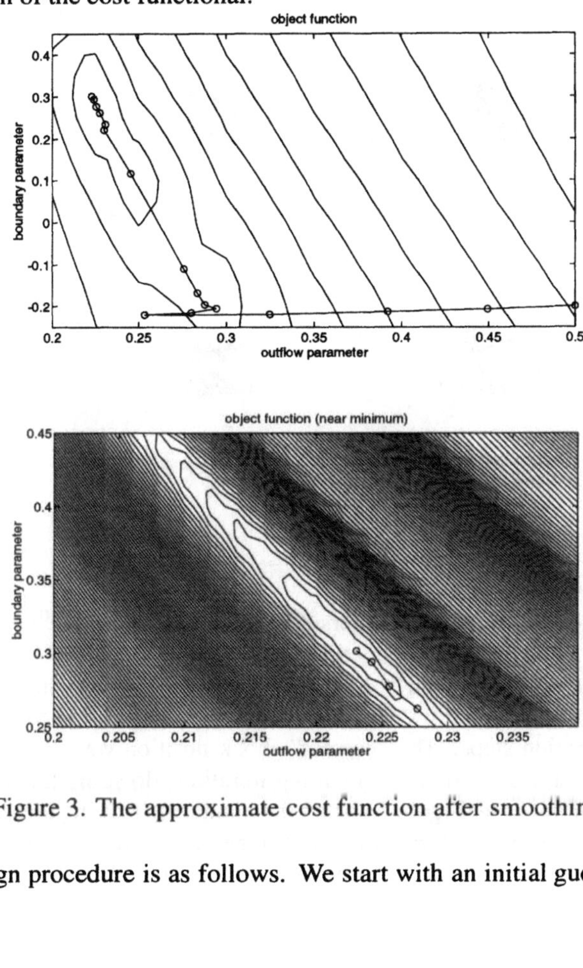

Figure 3. The approximate cost function after smoothing.

The design procedure is as follows. We start with an initial guess for the design parameters q_1 and q_2. The flow problem is solved for these parameters, then the objective functional is computed. The output is sent to an optimizer. The optimizer then produces new parameters which are sent back to the flow solver. This procedure is repeated until an optimal solution is obtained. The optimization algorithm is a BFGS/trust region scheme. This scheme requires gradient information for the functional which was obtained by a forward finite difference quotient approximation.

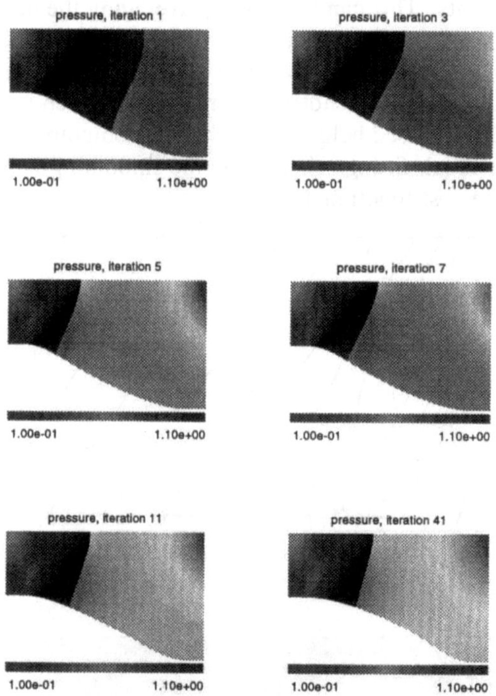

Figure 4. Convergence history for the pressure.

The initial guess for the parameters employed is $\mathbf{q}^0 = (q_1^0, q_2^0) = (-0.2, 0.5)$ and the initial value of the cost functional is $\mathcal{J}(\mathbf{q}^0) = 0.179969$. The converged optimal parameters are $\mathbf{q}^* = \mathbf{q}^{41} = (0.301607, 0.222969)$ and the corresponding value of the cost functional is $\mathcal{J}(\mathbf{q}^*) = 0.0107119$. Figure 4 shows the pressure at different iteration steps. The "correct" shock position was picked up quickly in about 11 iterations, while the remaining iterations do some fine tuning of the smooth parts of the flow. Figure 5 shows the comparison of the pressure between the converged flow and the target flow. From the graph, we can see that the match is very good. Figure 6 shows a close matching between the converged flow and the target flow along the centerline.

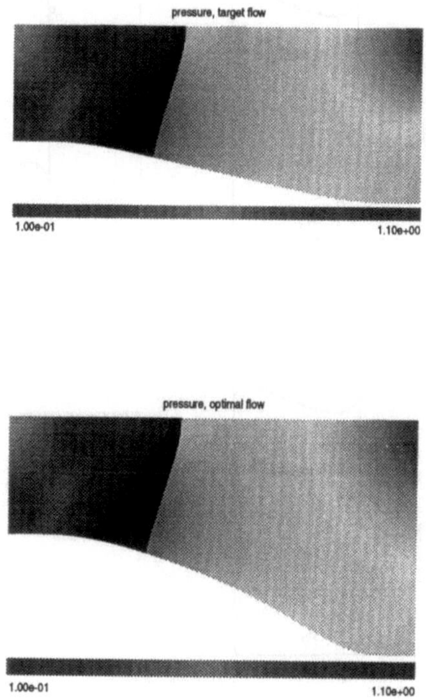

Figure 5. Optimal and target pressures.

If the cost functional $\mathcal{J}$ is redefined so that the integration takes place along the centerline, we eliminate the problem resulting from having the target and candidate optimal flows being defined on different domains. On the other hand, since we are only trying to match along the centerline, one cannot expect a good match at other points of the flow. The initial guesses for the parameters are chosen to be $\mathbf{q}^0 = (q_1^0, q_2^0) = (0.0, 0.4)$ so that $\mathcal{J}(\mathbf{q}^0) = 0.245003$. The converged optimal parameters are $\mathbf{q}^* = \mathbf{q}^{24} = (0.183405, 0.235287)$ with $\mathcal{J}(\mathbf{q}^*) = 0.005869$.

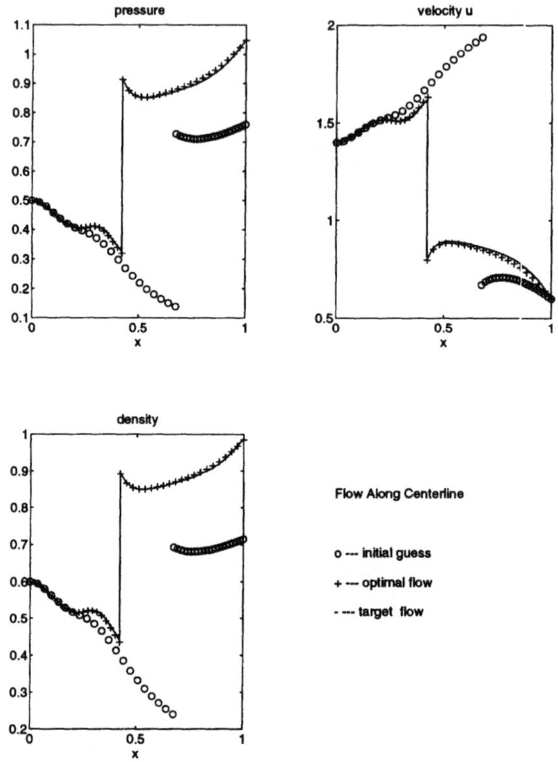

Figure 6. Comparison between the target and optimal flows.

4. Feedback Control of Kármán Vortex Shedding

The control of flows has a long and rich history. Some of these past efforts have been devoted to conducting experiments and simulations with different controlling mechanisms, without any attempt to optimally design or to actively change these mechanisms. For example, one may consult textbooks for detailed expositions of such efforts in the area of boundary layer control. Of course, there has been great success in the design and application of active controls, especially in the area of aerodynamic controls. These efforts, for the most part, use simple models to account for the effect of the flow on the submerged object.

Flows about a cylinder at even moderate values of Reynolds numbers, e.g., $Re > 47$, exhibit an unsymmetric, periodic shedding of vortices. As a result, the lift force exerted by the fluid on the cylinder is periodic in time. Here, our goal is to show the efficacy of a simple feedback control law for the reduction of the magnitude of the lift. The control mechanism used to attempt to reduce the size of the lift oscillations is the injection and suction of fluid through orifices on the surface of the cylinder. The amount of fluid injected or sucked through the orifices is determined, using a simple feedback law, from the pressure sensed at various stations on the cylinder. Thus, in the language of feedback control, the *sensor* determines the pressure at various stations on the cylinder and the *actuator* injects or sucks fluid through orifices also on the cylinder. Other sensing and actuating mechanisms could also be used, e.g., the vorticity or stress components on the cylinder for sensing and rotation or shape modification of the cylinder for actuating.

The type of feedback laws used in our study are based on the observation that differences in the pressure on the top and bottom halves of the cylinder should give an indication of the asymmetric behavior of the lift. Such pressure differences are then used to determine how much fluid to inject or suck through the orifices. No attempt is made to systematically design an "optimal" feedback law, i.e., one that in some sense does the best possible job in meeting the objective. However, the computational results we have obtained indicate that the simple feedback law we use here is quite effective in reducing the size of the oscillations in the lift.

We have investigated the control problem by numerically solving the two dimensional Navier-Stokes equations

$$\frac{\partial \mathbf{u}}{\partial t} + \mathbf{u} \cdot \nabla \mathbf{u} + \nabla p = \frac{1}{Re} \nabla^2 \mathbf{u} \qquad \text{and} \qquad \nabla \cdot \mathbf{u} = 0 \,,$$

where $\mathbf{u}$ and p are the velocity and kinematic pressure (pressure divided by density), respectively. The above equations are nondimensionalized using the cylinder diameter d, the free stream velocity U, and the kinematic viscosity of the fluid ν. The Reynolds number Re is defined by $Re = Ud/\nu$. Initial and boundary conditions on the cylinder surface are imposed on the velocity:

$$\mathbf{u}(\mathbf{x}, 0) = \mathbf{0} \quad \text{in the computational domain } \Omega$$

and

$$\mathbf{u}(\mathbf{x}, t) = \mathbf{g} \quad \text{on the cylinder surface } \Gamma \text{ and for } \; t > 0 \,.$$

For the uncontrolled flow about the cylinder, the cylinder surface is a solid wall so that in this case $\mathbf{g} = \mathbf{0}$. When controls are applied, this boundary condition becomes inhomogeneous on the portions of Γ covered by the injection and suction orifices, i.e., $\mathbf{g} \neq \mathbf{0}$ at these orifices.

In order to avoid complications with boundary conditions at infinity, a specific finite domain was defined. The origin of the (x, y) coordinate system is located at the center of the cylinder and the cylinder has a unit diameter. The computational domain we use is the rectangle $-5 \leq x \leq 15$ and $-5 \leq y \leq 5$ with the cylinder excluded. The geometry of the domain is sketched in Figure 7. The exterior boundary of the computational domain consists of the four sides of the rectangle; again, see Figure 7. At the inflow boundary Γ_i the velocity is set to the uniform value at infinity, i.e., $u = 1$ and $v = 0$, where u and v denote the x and y components of the velocity vector, respectively. At the outflow Γ_o a vanishing "stress" boundary condition is imposed. Specifically, if $\mathbf{t} = -p\mathbf{n} + (1/Re)\partial\mathbf{u}/\partial n$, then on Γ_o we set $t_1 = t_2 = 0$, where t_1 and t_2 respectively denote the x and y components of $\mathbf{t}$. The vector $\mathbf{t}$ is not actually the stress vector; however, it has been found that imposing such an outflow condition is often more effective than requiring the true stress to vanish. Moreover, it is a more convenient boundary condition to use with the particular form of the viscous term appearing in the Navier-Stokes equations given above. On the top Γ_t and bottom Γ_b of the rectangle we impose the mixed conditions $v = 0$ and $t_1 = 0$. The boundary conditions are also indicated on the sketch of the computational domain given in Figure 7.

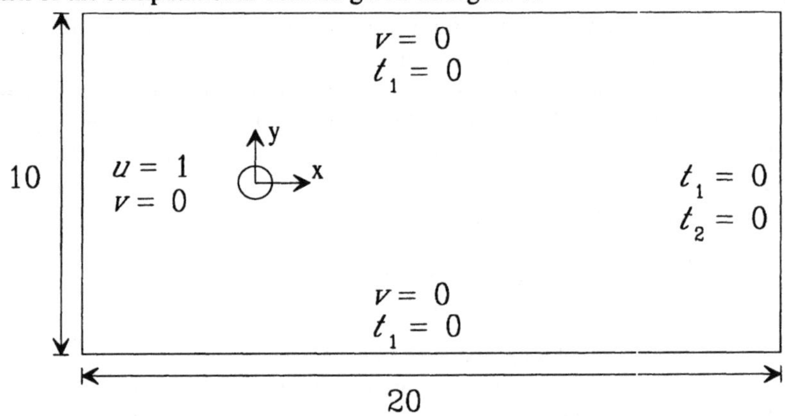

Figure 7. Computational domain for flow around a cylinder.

For the uncontrolled case, the problem we wish to discretize is then defined by the above equations with boundary conditions on Γ_i, Γ_o, Γ_b, and Γ_t.

The spatial discretization is effected using the Taylor-Hood finite element pair based on a triangulation of the computational flow domain. A typical triangulation is depicted in Figure 8; it consists of 1735 triangles and 3735 velocity nodes. The velocity (pressure) field is approximated using continuous piecewise quadratic (linear) polynomials; the same grid is used for both fields.

The semi-implicit single-step Euler method is employed for the time discretization. This time-stepping algorithm is defined as follows: given $\mathbf{u}^{m-1}$, the pair ($\mathbf{u}^m$, p^m) is obtained by solving the linear system

$$\frac{\mathbf{u}^m - \mathbf{u}^{m-1}}{\delta} + (\mathbf{u}^{m-1} \cdot \nabla)\mathbf{u}^m + \nabla p^m - \frac{1}{Re}\nabla^2\mathbf{u}^m = 0 \qquad \text{and} \qquad \nabla \cdot \mathbf{u}^m = 0\,,$$

where δ denotes the time step. Of course, these are supplemented by the appropriate boundary and initial conditions. This scheme has been previously analyzed and is found to be unconditionally stable.

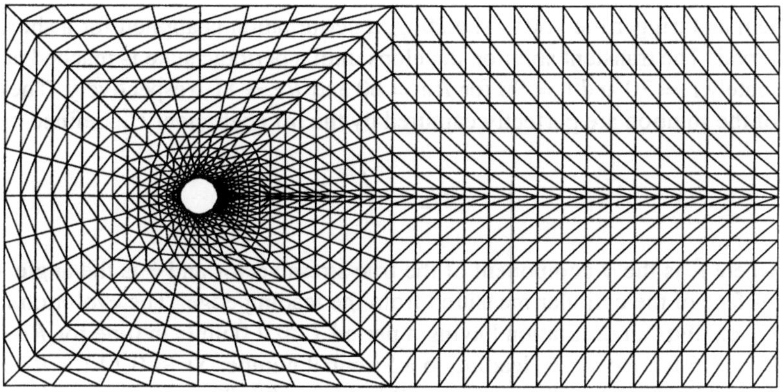

Figure 8. Mesh for flow around a cylinder.

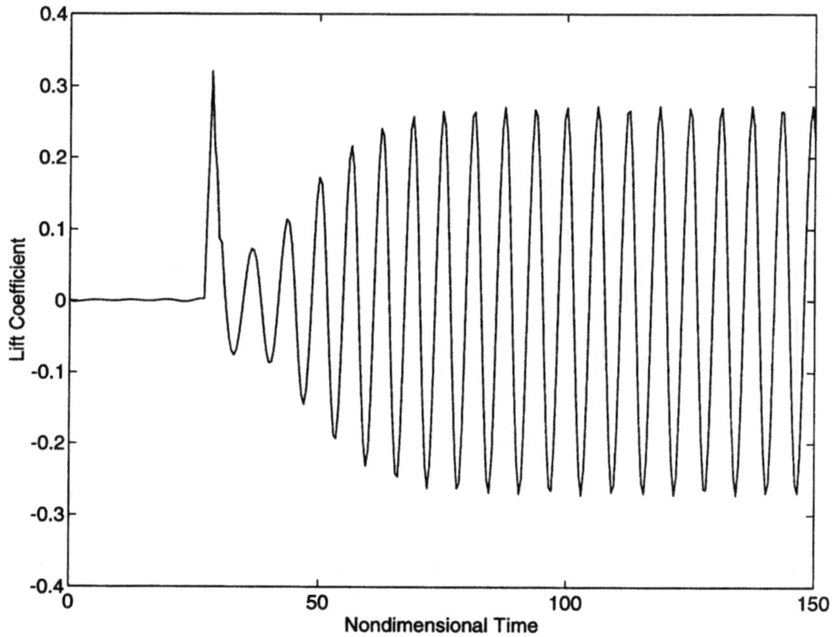

Figure 9. Uncontrolled lift coefficient for $Re = 80$.

Using the computational scheme described above, we have simulated the uncontrolled problem at $Re = 60$ and $Re = 80$. To break the symmetry in the problem and trigger the vortex shedding, a perturbation is applied for a short time interval. The transient solution develops until a periodic state is reached at approximately $t = 70$ and $t = 90$ for $Re = 60$ and $Re = 80$, respectively. These phases of the solution are perfectly captured by the lift coefficient C_L defined by

$$C_L = \int_0^{2\pi} \tilde{t}_2(\theta) d\theta,$$

where $\tilde{t}_2$ is the y component of the true stress vector and θ is the angle along the surface of the cylinder measured from the leading edge. The evolution in time of C_L exhibits an oscillation having a period $T \approx 7.0$ units in time for $Re = 60$ and $T \approx 6.5$ for $Re = 80$, leading to a Strouhal number $St \approx 0.14$ for $Re = 60$ and $St \approx .15$ for $Re = 80$; see Figure 9 for $Re = 80$. These values are in agreement with experimental results and other numerical simulations.

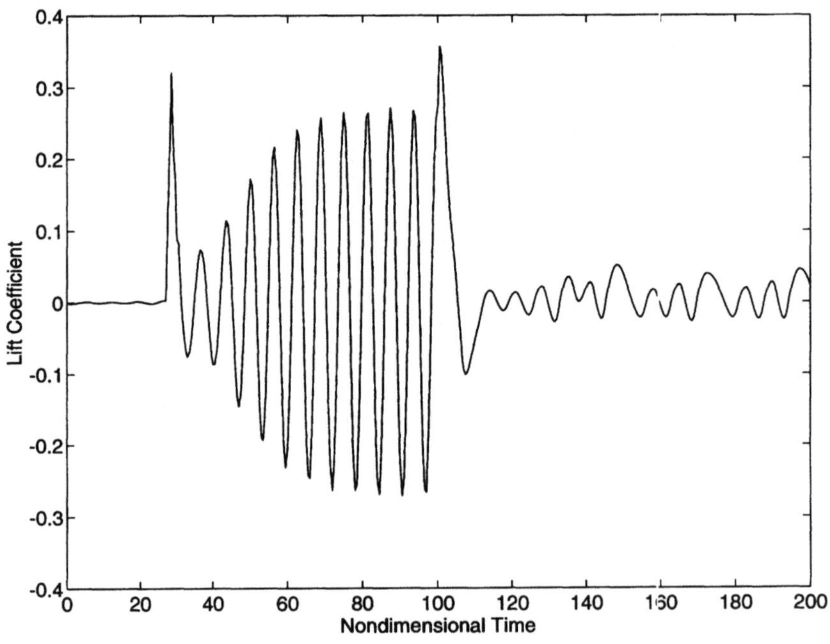

Figure 10. Controlled lift coefficient for $Re = 80$.

We have used a variety of number and location of blowing and suction orifices along the back side of the cylinder. One effective arrangement is to have two suction slots centered at $\pm 23\pi/32$ on the back-side of the cylinder and a single blowing slot centered at π. We place the sensors at symmetric locations on the front-side of the cylinder and we sense the pressure at these locations. The feedback law is chosen as follows:

$$\mathbf{u} = \min\{\alpha|p(\theta_1) - p(\theta_2)|, \beta\}\, \mathbf{g}(\theta).$$

Here, α and β are constants, with β being used to limit the size of allowable controls; $\mathbf{g}$ is a given velocity profile at the orifices. We have performed computational simulations for numerous cases but report only the case of $\alpha = 80$ and $\beta = 2$. The control is turned on at $t = 100$. The results are shown in Figure 10 from which one can conclude that the simple feedback law can be quite effective in reducing the size of the oscillations in the lift.

References

[1] J. Borggaard; On the presence of shocks in domain optimization of Euler flows, in *Flow Control* (Ed. by M. Gunzburger) Springer, 1995, to appear.

[2] J. Borggaard and J. Burns; A sensitivity equation approach to shape optimization in fluid, in *Flow Control* (Ed. by M. Gunzburger) Springer, 1995, to appear.

[3] J. Borggaard, J. Burns, E. Cliff, and M. Gunzburger; Sensitivity calculations for a 2D, inviscid, supersonic forebody problem, in *Identification and Control of Distributed Parameter Systems*, (Ed. by H. Banks, R. Fabiano, and K. Ito), SIAM, Philadelphia, 1993, 14–24.

[4] J. Burkardt, M. Gunzburger, and J. Peterson; Computational experiences in shape control for incompressible viscous flows, to appear.

[5] J. Burns and Y. Ou; Effect of rotation rate on the forces of a rotating cylinder: simulation and control, ICASE Report No. 93-11, ICASE, Hampton, 1993.

[6] J. Burns and Y. Ou; Active control of vortex shedding, AIAA Report No. 94-0182, AIAA, Washington, 1994.

[7] J. Burns and Y. Ou; Feedback control of driven cavity problem using LQG designs, in *Proc. 33rd IEEE Conference on Decision and Control*, IEEE, 1994, to appear.

[8] Q. Du, M. Gunzburger, and L. Hou; Analysis and finite element approximation of optimal control problems for a Ladyzhenskaya model for stationary, incompressible, viscous flows, *Comp. Appl. Math.*, to appear.

[9] A. Fursikov, M. Gunzburger, and L. Hou; Boundary control of the unsteady Navier-Stokes equations: the two-dimensional case, to appear.

[10] M. Gunzburger and L. Hou; Finite dimensional approximation of a class of constrained nonlinear optimal control problems, to appear.

[11] M. Gunzburger and H. Kim; Analysis of an optimal shape control problem for the Navier-Stokes equations; to appear.

[12] M. Gunzburger and H. Lee; Feedback control of fluid flows, *Proc. 14th IMACS World Congress*, Atlanta, 1994.

[13] M. Gunzburger and H. Lee; Analysis, approximation, and computation of a coupled solid/fluid temperature control problem, *Comput. Methods Appl. Mech. Engrg.* **118**, 1994, 133–152.

[14] M. Gunzburger and H. Lee; Feedback control of Karman vortex shedding, to appear.

[15] L. Hou and J. Peterson; Matching electric current in electrically conducting fluid flows by using an optimal control formulation, to appear.

[16] W. Layton, K. Lenferink, and J. Peterson; A two-level Newton finite element method for approximating electrically conducting incompressible fluid flows, to appear.

[17] Y. Ou and S. Sritharan; On the robustness of the Navier-Stokes global attractor, *Proc. 14th IMACS World Congress*, 1994.

[18] A. Shenoy and E. Cliff; An optimal control formulation for a flow matching problem, *Proc. 5th AIAA/USAF/NASA/ISSMO Symposium on Multidisciplinary Analysis and Optimization*, Panama City, 1994, 520–528.

[19] X. Wu, X., E. Cliff, and M. Gunzburger; An optimal design problem for a two dimensional flow in a duct, to appear.

A Generalized Reduced Gradient Algorithm for Large-scale Trajectory Optimization Problems*

Kathryn E. Brenan and Wayne P. Hallman
The Aerospace Corporation

This paper describes the design of a "sparse" nonlinear programming algorithm for a "large scale" flight optimization and sizing code. This program is designed to solve trajectory optimal control problems with path constraints. The direct transcription technique is applied to discretize an optimal control problem into a sparse, large scale parameter optimization problem for subsequent numerical solution by a nonlinear programming algorithm. This paper focuses on the design of a generalized reduced gradient algorithm to exploit the sparsity structure of the collocation equations.

1. Introduction

In a trajectory optimal control problem, a trajectory is determined which simultaneously minimizes a performance index and satisfies the equations of motion, additional path constraints and boundary conditions for the vehicle. The control profile along the trajectory, as well as the state profile, must be determined. In general, these constraints form a nonlinear system of differential-algebraic equations [8] [9].

Over the past 35 years, many methods have been developed to solve these trajectory problems. One such technique, known as the *direct transcription method*, has been implemented in the trajectory optimization and sizing code FONSIZE [22]. In the direct transcription method, a discretization based on a collocation formula is applied to the differential equations and mission constraints to obtain a parameter optimization problem.

In theory, any nonlinear programming (NLP) algorithm can be used to solve the parameter optimization problem. However, it is critical to the efficiency and ultimate success of this approach to employ an NLP algorithm designed for *sparse, large-scale* parameter optimization problems. The sparsity has two origins: the collocation method and the inherent sparse character of the trajectory problems (i.e., each variable is involved in relatively few constraints). It is important to exploit the sparsity properties in order to increase efficiency of the solution of linear systems required by the NLP algorithm and to reduce storage requirements.

One commercial code capable of efficiently handling sparse, large-scale optimization problems is MINOS [21]. There also is a proprietary NLP code [5] based on a multifrontal algorithm developed at Boeing Computer Services. This multifrontal NLP code has been employed by Betts and Huffman [7] in their trajectory optimization program. It is believed that neither of these NLP codes is designed specifically to exploit the almost block diagonal (ABD) structure of the collocation

*This work was supported in part by an Aerospace Sponsored Research Grant.

equations. McDonnell Douglas also is solving trajectory optimization problems via the Optimal Trajectories by Implicit Simulation program [17] [18] and an NLP code NZSPARSE [23]. Both the multifrontal NLP code and the NZSPARSE code employ a sequential quadratic programming technique, while MINOS employs a projected augmented Lagrangian algorithm. We take a different strategy by basing our sparse NLP code on a generalized reduced gradient (GRG) algorithm. In addition, our sparse NLP code is specifically designed to exploit the structure of the collocation equations, while other NLP codes are for more general sparse, large-scale problems.

In Section 2, we describe a class of trajectory optimal control problems, the direct transcription method, and the resulting parameter optimization problem. Our sparse GRG algorithm is described in Section 3. Next a typical trajectory optimization and sizing problem is studied in Section 4. Finally, in Section 5, we discuss future enhancements for the sparse GRG algorithm.

The following notation will be used throughout this paper. All vectors and matrices are assumed to be real-valued, with $\mathcal{R}^{n\times m}$ denoting the set of all real $n \times m$ matrices, and $\mathcal{R}^n$ the set of all real column vectors of dimension n. The superscript T will be used to denote the transpose of a matrix or a vector. The notation

$$a \left\{ \begin{array}{c} \geq \\ = \end{array} \right\} b$$

is used to denote the fact that the relation may be an equality ($a = b$) or an inequality ($a \geq b$). Finally, to denote the complement of a set ρ of indices with respect to another set β, where $\rho \subseteq \beta$, we use the notation $\beta \setminus \rho$.

2. Direct transcription method

The FONSIZE program has been developed to perform trajectory optimization and sizing of vehicles. It has been used successfully to solve a wide range of flight mechanics problems for single-stage-to-orbit, two-stage-to-orbit, as well as conventional launch vehicles.

Trajectory optimal control problem

Determine the control $u(t)$, the corresponding trajectory $x(t)$, the final time t_f, and a set of design parameters p to minimize the performance index

$$\varphi(t_f, x(t_f), u(t_f), p) + \int_{t_o}^{t_f} \chi(t, x(t), u(t), p)dt$$

such that

$$\begin{array}{rcll} x'(t) & = & f(t, x, u, p), & t \in [t_o, t_f] \\ \phi(x(t_o), p) & \left\{ \begin{array}{c} \geq \\ = \end{array} \right\} & 0, & \end{array}$$

$$\psi(x(t_f), p) \quad \begin{Bmatrix} \geq \\ = \end{Bmatrix} \quad 0,$$

$$c(t, x(t), u(t), p) \quad \begin{Bmatrix} \geq \\ = \end{Bmatrix} \quad 0, \qquad t \in [t_o, t_f]$$

where $x \in \mathcal{R}^n$, $u \in \mathcal{R}^m$, $p \in \mathcal{R}^\ell$, and $\varphi \in \mathcal{R}^1$, $\chi \in \mathcal{R}^1$, $\phi \in \mathcal{R}^s$, $\psi \in \mathcal{R}^q$, and $c \in \mathcal{R}^r$.

In the direct transcription method, the optimal control problem is discretized using only the differential equations for the state, the state boundary conditions, and the path constraints. The differential equations and associated boundary conditions for the adjoint variables, the control optimality conditions, and the transversality conditions are not explicitly included. However, these conditions are satisfied indirectly via the NLP algorithm which bases its convergence criteria on conditions representing discretized approximations to these first order necessary conditions. In the FONSIZE code, the Hermite-Simpson (HS) collocation scheme is used to enforce the differential equations at the nodes and centers of each subinterval (i.e., at the collocation points) and path constraints at the nodes.

The HS method is based on representing the state and control variables by piece wise polynomials in which coefficients are determined by function and derivative values at a sequence of nodes. Hermite cubic polynomials are employed to represent the state variables, while the controls may be approximated by piecewise constant, linear, quadratic, or cubic polynomials. For simplicity, assume that cubic polynomials are utilized for both the state and control variables. Suppose the interval $[t_0, t_f]$ is partitioned into N subintervals such that $t_0 < t_1 < \ldots < t_N = t_f$. Define $h_i = t_i - t_{i-1}$ for $i = 1, \ldots, N$. Let x_i and u_i represent the approximate state and control values, respectively, at the nodes t_i. Introduce variables w_i to represent weighted derivatives of the controls at the nodes. The parameter optimization problem can now be stated.

Parameter optimization problem

Determine the control vector $[u_0, w_0, u_1, \ldots, u_N, w_N]$, the corresponding state vector $[x_0, x_1, \ldots, x_N]$, the final time t_f, and the design parameters p to minimize the cost function

$$\varphi(t_f, x_N, u_N, p) + \sum_{i=0}^{N-1} \int_{t_i}^{t_{i+1}} \chi(t, x(t), u(t), p)dt \tag{1}$$

such that the HS defects

$$\Delta_i^{HS} = 0, \qquad \text{for } i = 1, 2, \ldots, N \tag{2}$$

the initial and terminal conditions, and the path constraints

$$c(t_i, x_i, u_i, p) \begin{Bmatrix} \geq \\ = \end{Bmatrix} 0, \qquad \text{for } i = 0, 1, \ldots, N \tag{3}$$

are satisfied where

$$
\begin{aligned}
\Delta_i^{HS} &= x_i - x_{i-1} - \frac{h_i}{6}\left(f_p(t_{i-1}, x_{i-1}, u_{i-1}) + 4 f_p(\hat{t}_i, y_i, v_i) + f_p(t_i, x_i, u_i)\right) \\
y_i &= \frac{1}{2}(x_{i-1} + x_i) + \frac{h_i}{8}\left(f_p(t_{i-1}, x_{i-1}, u_{i-1}) - f_p(t_i, x_i, u_i)\right), \\
v_i &= \frac{1}{2}(u_{i-1} + u_i) + \frac{1}{8}(w_{i-1} - w_i),
\end{aligned}
$$

$f_p(t, x, u) = f(t, x, u, p)$, and $\hat{t}_i = (t_{i-1} + t_i)/2$. The integrals in (1) may be discretized by a quadrature rule. Equations (2) and (3) form a nonlinear system of equations for the unknown variables x_i, u_i, and w_i at the nodes as well as for the final time t_f and the design parameters p.

3. A sparse GRG algorithm

In designing an NLP algorithm appropriate for the sparse, large-scale parameter optimization problems arising from the collocation of optimal control problems, we have selected as our starting point The Aerospace Corporation code NLP3 [15]. The NLP3 code, based on the GRG method [1], has evolved over the last twenty years into a very reliable and robust optimization algorithm. It has been used successfully to solve hundreds of trajectory optimization problems arising from shooting strategies. In the shooting approach, the parameter optimization problems tend to be relatively small and dense. In the direct transcription approach, the parameter optimization problems are very large, but the Jacobian of the constraints has a sparse block structure. We have redesigned the NLP3 algorithm to exploit the sparsity structure of the Jacobian.

The NLP3 algorithm determines the vector z of dimension n_z which solves the NLP problem:

$$
\begin{aligned}
&\min \Phi(z) \\
\text{subject to: } c_i(z) &= 0, \ i = 1, 2, \ldots, m_e \\
c_i(z) &\geq 0, \ i = m_e + 1, m_e + 2, \ldots, m.
\end{aligned}
$$

It is assumed that there are more variables than active constraints, and that all functions are twice continuously differentiable. For the trajectory optimization problems considered here, the vector z includes the discretized states x, the controls u, and the weighted derivatives w of the controls at all the nodes, as well as the final time t_f and the design parameters p.

The NLP3 algorithm utilizes an *active set strategy*. Thus, on each cycle of the algorithm, a basis of active constraints is identified. This basis β includes all equality constraints (such as the HS defects) and a subset of the inequality constraints. The reduced gradient method then is applied to solve the equality constrained optimization problem, treating all the basis constraints as equality constraints. The (inequality) constraints not in the basis are monitored during this constrained optimization. If one of the constraints not in the basis becomes violated in minimizing Φ, then it is added to the basis.

The NLP3 code is a *feasible point* method; i.e, the algorithm follows the constraint surface to the solution. After an initial feasible point is determined, the variables are factored into two sets, z_γ and z_ζ. The *constraint-solving variables*, z_γ, are used to satisfy the basis constraints. By construction, the number of constraint-solving variables equals the number of basis constraints. The remaining *optimization variables*, z_ζ, are used for minimizing the objective function Φ. After the equality constrained optimization problem is solved with the current basis, tests are performed to determine if any inequality constraint can be released from the basis in order to further reduce the objective function Φ. If such a constraint is identified, then a new basis is formed and another cycle of the algorithm is performed. If no such constraint is found, then the final point satisfies the necessary conditions for a local optimum.

Our focus in the redesign of NLP3 has been on the efficient solution of large, sparse linear systems. We have identified four locations in the NLP3 code where a dense linear solver could be replaced by a sparse linear solver. In each of these locations, a system of linear equations involving the Jacobian of the active constraints with respect to the constraint-solving variables $G_{\gamma\beta}$, or its transpose $G_{\gamma\beta}^T$ must be solved. This matrix is not only sparse, but it has a block-type structure which can be exploited by a specialized linear solver routine. Since there are four linear systems involving this matrix, we also reuse the "LU" factorization of the matrix whenever possible.

The four places in the NLP3 code which have been modified to use the sparse linear solver are:

1. **Return to the constraint surface**
   The subroutine SRCHDR computes the search direction s_γ for the constraint solving process using a linear model of the basis constraints with respect to the constraint-solving variables. Specifically, the following Gauss-Newton linear system must be solved for the search direction:

   $$G_{\gamma\beta}^T s_\gamma + c_\beta(\bar{z_\gamma}, \bar{z_\zeta}) = 0. \tag{4}$$

   The constraint-solving variables then can be updated as $z_\gamma = \bar{z_\gamma} + \rho s_\gamma$ where the nominal value of ρ is one.

2. **Calculation of Lagrange multipliers and reduced gradient**
   In the subroutine REDUCG, the Lagrange multipliers λ are determined as the solution to the linear system

   $$G_{\gamma\beta}\lambda + \frac{\partial\Phi}{\partial z_\gamma} = 0. \tag{5}$$

The Lagrange multipliers then are used in the computation of the reduced gradient of the objective function

$$\nabla_r \Phi(z_\zeta) = \frac{\partial \Phi}{\partial z_\zeta} + G_{\zeta\beta}\lambda \tag{6}$$

where $G_{\zeta\beta}$ is the Jacobian of the basis constraints with respect to the optimization variables.

3. **Calculation of the step length bounds**
The subroutine SOLVER solves the equality constrained NLP problem with the current basis β:

$$\begin{aligned} &\min \Phi(z) \\ &c_i(z) \;=\; 0, \quad i = 1, 2, \ldots, m_z \end{aligned}$$

where m_z is the number of constraints in the basis ($m_z \geq m_e$). In addition, a set of (nonactive) inequality constraints

$$c_i(z) \geq 0, \quad i = m_z + 1, \ldots, m$$

is monitored. Using a linear model of each nonactive inequality constraint, one can compute the distance d to the root along the search direction s_ζ with respect to the optimization variables along which the reduced objective function (i.e., restricted to the constraint surface) decreases:

$$G_{\gamma\beta}^T\, d = G_{\zeta\beta}^T s_\zeta .$$

The remaining details of the computations of the step lengths are given in [15].

4. **Factorization of variables**
The new factorization of variables algorithm to be described in Section 3.2 also requires solving a linear system involving $G_{\gamma\beta}^T$.

Note that the coefficient matrix in each linear system described above is either $G_{\gamma\beta}$ or its transpose. Therefore, if the "LU" matrix factorization for $G_{\gamma\beta}$ has been saved, then only the forward and backward substitution processes are performed for the solution of subsequent linear systems.

3.1. Initial feasible point strategy

The first step in the GRG algorithm is to find a feasible point where all of the equality and inequality constraints are satisfied. The technique used in NLP3 determines a minimum one-norm search direction based on a linear model of the constraints. The linear constraint model about a base point $\bar{z}$ is given by

$$\tilde{c}_i(\bar{z}+s) = c_i(\bar{z}) + \left(\frac{\partial c_i}{\partial z}\right)^{\mathsf{T}} s.$$

The minimum one-norm search direction which satisfies the linear constraint model is determined by solving the optimization problem

$$\begin{aligned} &\min \|s\|_1 \\ \text{subject to: } \quad &\tilde{c}_i(\bar{z}+s) = 0, \quad i = 1, \ldots, m_e \\ &\tilde{c}_i(\bar{z}+s) \geq 0, \quad i = m_e+1, \ldots, m. \end{aligned}$$

This problem can be formulated as a linear programming problem with $2n_z$ variables and m constraints. The LINPRG code [24] [25] is used to solve this linear programming problem. This algorithm does not exploit the Jacobian matrix structure for the collocation problem, but does use sparse storage techniques. The nonzero elements of the basis inverse are stored using the product form of the basis.

Each time the LINPRG code is called to determine the search direction, the basic variable indices at the solution are saved and used as a starting basis for the next search direction computation. This warm start procedure has been found to make a significant reduction in the number of linear programming iterations. One other modification to this warm start procedure has been implemented inside the LINPRG code. After the basis is inverted from a warm start, the signs of the basic variables are checked. If some basic variable is negative, then this variable is replaced by the corresponding nonbasic variable which is its negative. Due to the special form of the linear programming problem, such a variable always exists. By adjusting the initial basis in this fashion, a basic feasible solution is generated where all basic variables are non-negative.

This strategy has worked well on small dense NLP problems but we have found that it frequently fails when we are searching for an initial feasible point for a large-scale NLP problem. In the future, we plan to investigate these difficulties and either modify the existing algorithm to be more efficient or replace it by a new strategy. Currently, when this strategy fails, the NLP code reverts to a backup strategy which computes a search direction as the minimum two-norm solution to the Gauss-Newton equations [15]. In this case, the least squares solution to a typically nonsquare ($m \times n_z$) linear system is determined by a solver based on the (dense) QR factorization of the coefficient matrix. Since this matrix is in modified ABD form, in the future we may design a QR algorithm to exploit it [20].

3.2. Factorization algorithm

The factorization of the variables into constraint-solving z_y and optimization variables z_ζ is a trademark of a GRG algorithm. The original NLP3 algorithm determines this factorization from the minimum one-norm estimate of the Lagrange multipliers by solving a linear programming (LP) problem. For small dimension problems, this technique is not a computational burden. In adapting NLP3 for larger sparse optimization problems, an alternate, more efficient factorization technique is needed which avoids solving a linear programming problem.

Given an active constraint basis β with corresponding Jacobian $G_{*\beta}$, a factorization of the variables simply is a choice of an index set γ such that $G_{\gamma\beta}$ is nonsingular. For collocation problems, natural candidates for constraint-solving variables are the discrete state variables x, while the control variables u can be used for optimization variables when no path constraints are active. The new LU factorization technique for collocation problems is built around the sparse linear solver. This method is initialized by specifying the index set δ as the indices of the discrete state variables x, and the index set ρ as the indices of the corresponding defect constraints for the differential equations. Let $\rho \subseteq \beta$, $\delta \subseteq \{1, \ldots, n_z\}$, and $G_{\delta\rho}$ be nonsingular. The factorization method is outlined below:

1. Factorize $G_{\delta\rho}$ using the sparse linear solver.
2. Choosing a constraint $j \,\varepsilon\, \beta \setminus \rho$, compute
$$d_i = |g_{ij} - G_{i\rho}(G_{\delta\rho})^{-1}G_{\delta j}| \text{ for all } i \,\varepsilon\, \{1, \ldots, n_z\} \setminus \delta.$$
3. Choose $d_{\hat{i}} = \max |d_i|$; if $d_{\hat{i}} = 0$ stop.
4. Update $\rho = \rho \cup j$ and $\delta = \delta \cup \hat{i}$.
5. Continue until $\rho = \beta$.

When the factorization process is complete, the constraint-solving variables are identified as $z_\gamma = z_\delta$ for the final δ index set. Note that in step 3, if $d_{\hat{i}} = 0$, then the matrix $G_{*\beta}$ does not have full rank. In this case, the method is terminated since there are no valid factorizations. In practice, we assume that this matrix has full rank, so this test should not be satisfied.

When the constraint basis changes during the optimization iteration procedure, the factorization of variables must be changed. An efficient way to modify the factorization simply is to update the previous factorization. There are three possible ways the basis could have changed:

1. **Add constraint**: Perform one loop of the factorization method.
2. **Delete constraint**: Delete the variable associated with constraint j from γ.
3. **Add and delete constraint**: Combine above steps.

Note that when a constraint is deleted, the inequality constraint leaving the basis must have been added during a previous cycle, and thus there is a constraint-solving variable associated with this constraint. In the final case, the constraint first is deleted from the basis, and then the new constraint is added.

In practice, a factorization restart mechanism is needed if the matrix $G_{\gamma\beta}$ becomes ill-conditioned during the iteration process. This restart mechanism can be based on the condition number estimate of the matrix $G_{\gamma\beta}$ which can be obtained from the sparse LU factorization. Currently, we do not have this computation implemented but anticipate doing so in the future [19].

Note that for the trajectory optimization problems of interest here, we have fixed the state variables x as constraint-solving variables. This decision does in fact affect the rate of convergence of a GRG algorithm, as it determines the subspace in which the reduced gradient is computed [3]. For large parameter optimization problems it is difficult, if not impossible, to evaluate the effect of this decision on the rate of convergence. However, the advantages to fixing some of the constraint-solving variables are clear. First, assuming the initial state vector is known, there is a natural one-to-one correspondence between the HS defects and the discrete state variables. Therefore, the work required in factorizing the variables is reduced significantly. Second, the identification of the block structure of the basis Jacobian $G_{\gamma\beta}$ is simplified. Still, in spite of these advantages, we may wish to revisit this strategy in the future simply because we do not know if this choice is appropriate for all trajectory optimization problems, in particular path-constrained problems.

3.3. Jacobian matrix

The *external* ordering of variables and constraints by FONSIZE leads to a Jacobian matrix whose sparsity structure is difficult to exploit during the solution of linear systems by NLP3. The sparsity in the Jacobian occurs because the state and control variables at a node affect only the equations in the adjacent intervals. Similarly, the evaluation of a path constraint at a node involve only variables "local" to that segment. Hence, by reordering the variables and constraints to reflect the natural ordering of the nodes, the sparse block structure of the Jacobian can be "improved." NLP3 works with this new *internal* ordering. In Figure 1, a portion of the Jacobian matrix of a typical trajectory test problem is shown in internal order.

The large-scale FONSIZE program automatically reorders the variables and constraints so the Jacobian matrix will have a modified almost block diagonal form [13]. Specifically, the variables are reordered as

$$[x_0, u_0, w_0, x_1, u_1, w_1, \ldots, x_N, u_N, w_N, p].$$

The path constraints are interspersed with the collocation equations in a natural order according to the ordering of nodes. A dense column block on the right will occur only for each parameter p_i ($i = 1, \ell$) designated to be a constraint-solving variable. The sparse NLP3 algorithm works with this internal ordering of variables and constraints.

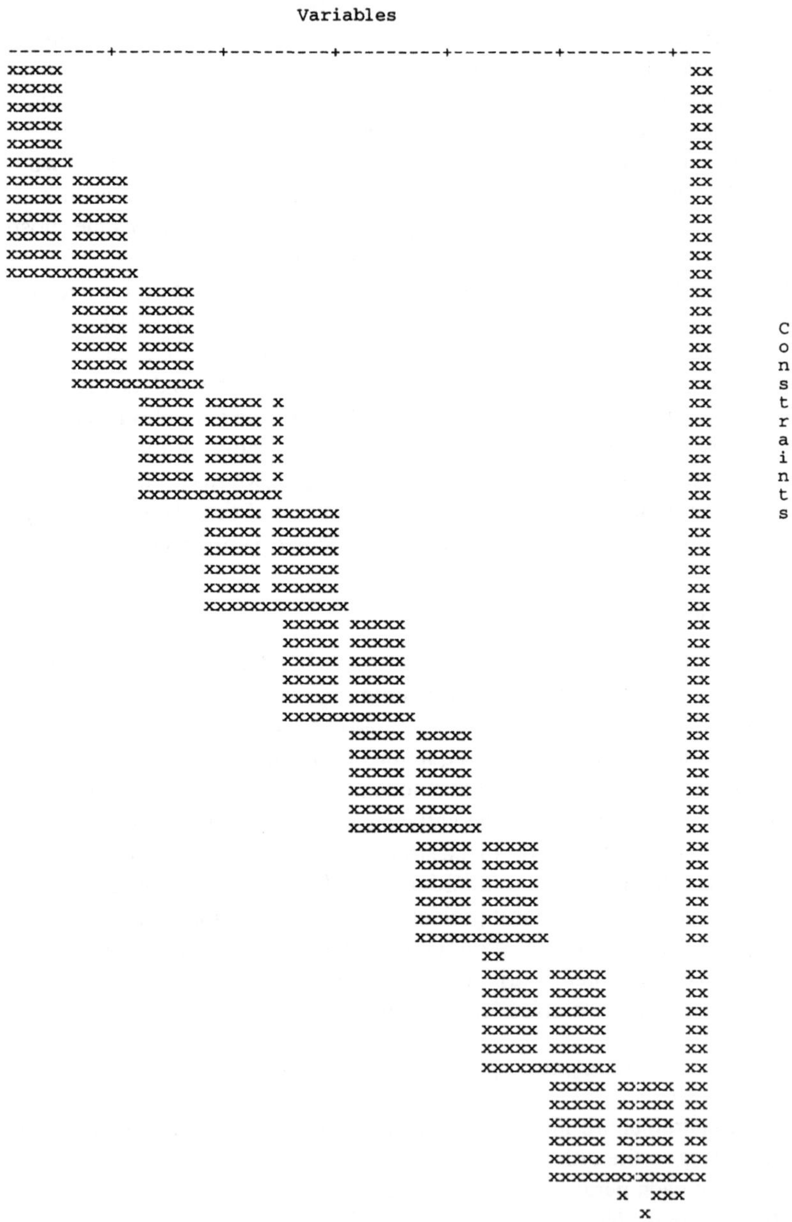

Figure 1. Sparsity structure of the Jacobian.

The identification of the block structure of the Jacobian matrix $G_{\gamma\beta}$ is handled automatically by the new NLP3 algorithm. For example, suppose during the optimization process that a (path) constraint is deleted from the basis. Then a constraint solving variable must be switched to an optimization variable. The Jacobian of the new basis with respect to the new set of constraint-solving variables will contain one less row and column. A similar situation occurs when a constraint and variable are added to the basis and constraint-solving set.

Each nonzero element of the Jacobian matrix is approximated by a central finite difference formula. The user may request either dense or sparse finite differencing of the Jacobian matrix. Sparse finite differencing exploits the problem structure by perturbing groups of variables simultaneously, thereby reducing the number of function evaluations required to compute the gradients [12]. The dense option still is available but is useful primarily if a problem is suspected in the sparse finite difference computation. Note that the sparse differencing process might be improved further by tailoring it to our specific collocation algorithm (i.e., we know the nonzero structure of the Jacobian matrix a priori, but currently we do not use that information in the sparse differencing option).

In spite of the sparse differencing option, the evaluation of the Jacobian matrix is an expensive process. Thus the frequency of its evaluation will be an important factor in the overall execution cost of solving a problem. Therefore, during the constraint-solving phase (i.e., during the solution of the Gauss-Newton equations (4)), a new option for controlling the gradient evaluation process has been defined to avoid updating the Jacobian matrix and thus to eliminate its LU factorization and computation of the sparse block structure. This option may be viewed as a *modified* Newton method. A significant time savings can be realized by using the modified Newton option as the Jacobian matrix is very large in general, and reasonable convergence rates still can be achieved with an "old" Jacobian. The Jacobian is evaluated at the system level and refactorized at each new feasible point.

3.4. Sparse linear solver

Our sparse linear solver is a package for efficiently solving linear systems ($\mathbf{Ax} = \mathbf{b}$ and $\mathbf{A}^T\mathbf{x} = \mathbf{b}$), when the coefficient matrix $\mathbf{A}$ is given in *almost block diagonal form* (ABD) and the last column block is nonzero. In [13], a $\kappa \times \kappa$ matrix $\mathbf{A}$ is defined to be ABD if there exists a strictly increasing sequence

$$1 = e_1 < \ldots < e_{r+1} = \kappa + 1$$

so that, for $i = 1, \ldots, r$, rows e_i through $e_{i+1} - 1$ have nonzero entries only in column α_i through β_i for some strictly increasing sequence $\{\alpha_i\}$ and some nondecreasing sequence $\{\beta_i\}$. In our work, we have a *modified* ABD form in which there are nonzero entries in the last column block as well. Specifically, in the modified ABD form, there exists a positive integer k with $0 < k \leq \kappa$ such that other nonzero entries can take place only from column k to column κ. We based our sparse linear solver on the SOLVEBLOK software package developed by DeBoor and Weiss [13] for ABD linear systems.

First, the solver performs Gaussian elimination with row pivoting on the coefficient matrix **A** to generate the LU-factorization of **A**. Then, the solution is obtained by using forward and backward substitutions in both cases. This algorithm has been proven to be stable in the same sense that Gaussian elimination applied to a dense system is stable when applied with partial pivoting. The primary advantages of this approach, namely the reduction in storage requirements and computational effort, are derived by exploiting the zero structure of the coefficient matrix. A detailed description of our linear solver for modified ABD linear systems is given in [11].

Note that a matrix may be ABD in many ways, and that this concept becomes computationally important only if the property is present in sufficient strength (i.e., with many blocks). In general, linear systems arising from collocation techniques are very strongly ABD. Note also that we could consider an ABD matrix to be banded. However, applying a band linear solver would not take full advantage of the ABD form. Specifically, a typical band solver would require more storage than the ABD algorithm and would perform unnecessary computations with zero elements. A band algorithm applied to an ABD matrix also would generate more fill-in due to pivoting [2].

The sparse linear solver routine just described allows for a general nonzero column block corresponding to the parameters p. In the future, we may wish to solve optimal control problems which have integral path constraints or other coupling constraints that link variables across several segments. These constraints would introduce a row block of nonzero elements, destroying the modified ABD form. In this case, we shall need to be able to solve linear systems with a full border. We are currently investigating one such technique known as Mixed Block Elimination [14] but will not discuss this research here.

4. Numerical results

In this section, a sample trajectory optimal control problem is solved with the new large-scale FONSIZE program using the sparse NLP3 algorithm. For comparison, we also will give results when the problem is solved using the dense NLP3 algorithm as well as the original FONSIZE/NLP2 program. NLP2 [6], an Aerospace code designed for small dense parameter optimization problems, is based on a projected gradient algorithm. All test cases have been executed using double precision on a HP 9000/755 workstation. Execution timing results were obtained by calling the HP system routine ETIME with I/O turned off. Convergence tolerances were chosen such that NLP2 and NLP3 produced 11-12 digits of accuracy in the optimal objective function value. The gradients required for both NLP2 and NLP3 were computed using sparse central finite differences.

A two-stage expendable launch vehicle, launched vertically from the Kennedy Space Center into a 28.5 deg, 100 nmi circular orbit [22] [4], is studied. The coupled trajectory optimization and sizing problem is to determine the propellant tank section lengths which minimize the weight of the ascent propellants, necessary to deliver a payload weighing 81,947.0 lb. The trajectory is divided into seven phases, six of which have unknown time durations. The control variable, angle of attack, is approximated with piecewise linear polynomials across the collocation mesh. Inequality path constraints are imposed which place bounds on the control variable and the dynamic pressure. Three terminal constraints dictate the vehicle's final radius, velocity, and inertial flight path angle. Also there are two design constraints prescribing the engine mixture ratios in terms of the weights of the stage 1 and stage 2 LO_2 and LH_2 tanks. There are four design variables, namely the lengths of the propellant tank sections for the LO_2 and LH_2 tanks in both stages. For more details, see [10].

The parameter optimization problem contains 198 variables (165 state variables, 23 control variables, 6 additional time variables, and 4 design variables), and 191 constraints (165 collocation equations, 21 path constraints, 3 terminal conditions, and 2 sizing constraints). At the solution there are 19 degrees of freedom.

Using an initial feasible point to start, we solved this problem with the NLP3 code exploiting the sparsity, including the new factorization of variables algorithm and the modified Newton option. For comparison, we also solved it with the original NLP3 code (using dense linear algebra) and the FONSIZE/NLP2 program. Table 1 displays these results: (NFE) the total number of function evaluations, (FCALLS) the number of function evaluations excluding those due to the Jacobian matrix, (GCALLS) the number of Jacobian matrix evaluations, (CCALLS) the number of calls to the sizing algorithm, and execution time in seconds. The objective function values agree to 11-12 digits for all cases. More extensive testing, including results for infeasible starting points, has been documented in [4] [10].

Comparison of NLP Algorithms

	Dense NLP3	**Sparse NLP3**	**NLP2**
NFE	2691	1332	527
FCALLS	69	116	71
GCALLS	69	32	12
CCALLS	1276	778	384
User Time	96.47	38.29	50.16

Table 1. Trajectory optimization & sizing problem.

For the same feasible starting point, we also ran some cases using the different options for the NLP3 algorithm. Specifically, we wished to compare the modified Newton option to the original Newton method, and the original LP factorization of variables algorithm to the new LU algorithm. The sparse linear solver was used to solve all the modified ABD linear systems discussed earlier in Section 3. We tabulate those results in Table 2. These cases demonstrate the effectiveness of the LU factorization of variables algorithm and the modified Newton method.

Newton vs. Modified Newton Constraint-solving
LP vs. LU Factorization of Variables Process
(NLP3/Sparse Linear Solver)

	Newton (original)		Modified Newton	
	LP	**LU**	**LP**	**LU**
NFE	2691	2691	1377	1332
FCALLS	69	69	123	116
GCALLS	69	69	33	32
CCALLS	1280	1276	808	778
User Time	80.21	68.14	48.22	38.29

Table 2. Trajectory optimization & sizing problem.

5. Conclusion

There remain some areas of the NLP3 algorithm to be redesigned in order to exploit further the sparsity structure as well as to improve the algorithm's convergence behavior. In practice, the initial feasible point algorithm has not performed well for these large parameter optimization problems, and should be examined and possibly redesigned or replaced. Results from a set of trajectory problems [10] consistently demonstrate that the NLP2 algorithm still outperforms NLP3 in terms of the total number of function evaluations required. Currently, during the unconstrained line search minimization process, NLP3 requests additional Jacobian matrix evaluations (i.e., a GCALL) which NLP2 does not require. In the future, we will modify NLP3 further to reduce the number of function evaluations. Different variable and constraint scaling procedures also could be employed to improve the convergence behavior of NLP3. Specifically, it may be possible to select scale weights for the optimization variables z_ξ that improve the conditioning of the reduced Hessian matrix, an idea similar to one proposed in [16] for the projected gradient algorithm NLP2.

In summary, the new version of NLP3 which exploits sparsity has already achieved a significant reduction in execution times. Future algorithm enhancements have been identified which will further reduce the run times and allow more complicated vehicles to be designed using FONSIZE.

References

[1] J. Abadie and J. Carpentier; Generalization of the Wolfe reduced-gradient

method to the case of nonlinear constraints, *Optimization* (Ed. by R. Fletcher), Academic, 1969.

[2] U. Ascher, R. Mattheij, and R. Russell; *Numerical Solution of Boundary Value Problems for Ordinary Differential Equations*, Prentice Hall, Englewood Cliffs, 1988.

[3] T. Beltracchi; A comparison of the generalized reduced gradient and generalized projected gradient methods, to appear.

[4] T. Beltracchi; Optimization software modifications to study launch vehicle sizing trajectory design problems, AIAA Paper 94-4403, *5th AIAA / NASA / USAF / ISSMO Symposium on Multidisciplinary Analysis and Optimization*, 1994.

[5] J. Betts and P. Frank; A sparse nonlinear optimization algorithm, AMS-TR-173, Boeing Computer Services, 1991.

[6] J. Betts and W. Hallman; NLP2 optimization algorithm documentation, The Aerospace Corporation, Report No. TOR-0089(4464-06)-1, 1989.

[7] J. Betts and W. Huffman; The application of sparse nonlinear programming to trajectory optimization, *J. Guid. Cont. Dynam.* **15**, 1992.

[8] K. Brenan; Differential-algebraic equations issues in the direct transcription of path constrained optimal control problems, *Ann. Num. Math.* **1**, 1994, 247–263.

[9] K. Brenan, S. Campbell, and L. Petzold; *Numerical Solution of Initial-Value Problems in Differential-Algebraic Equations*, Elsevier, New York, 1989.

[10] K. Brenan and W. Hallman; A large-scale generalized reduced gradient algorithm for flight optimization and sizing problems, The Aerospace Corporation, Report No. ATR-94(8489)-3, 1994.

[11] K. Brenan, W. Hallman, and W. Yeung; The design of a large-scale NLP code for trajectory optimization problems, The Aerospace Corporation, Report No. ATR-92(8189)-1, 1992.

[12] A. Curtis, M. Powell, and J. Reid; On the estimation of sparse Jacobian matrices, *J. Inst. Math. Appl.* **13**, 1974, 117–119.

[13] C. De Boor and R. Weiss; SOLVEBLOK: a package for solving almost block diagonal linear systems, *ACM Trans. Math. Soft.* **6**, 1980, 80–87.

[14] W. Govaerts and J. Pryce; Mixed block elimination for linear systems with wider borders, *IMA J. Num. Anal.* **13**, 1993, 161–180.

[15] W. Hallman; Optimization algorithm NLP3, The Aerospace Corporation, IOC A87-5752-45, 1987. (Aerospace internal document; not available for external distribution.)

[16] W. Hallman; Optimal scaling techniques for the nonlinear programming problem, AIAA Paper 94-4417, *5th AIAA / NASA /USAF / ISSMO Symposium on Multidisciplinary Analysis and Optimization*, 1994.

[17] C. Hargraves and S. Paris; OTIS - optimal trajectories by implicit integration, Boeing Aerospace Co., Contract No. F33615-85-c-3009, 1988.

[18] C. Hargraves and S. Paris; Direct trajectory optimization using nonlinear programming and collocation, *AIAA J. Guid. Cont. Dynam.* **10**, 1987, 338–342.

[19] N. Higham; A survey of condition number estimation for triangular matrices, *SIAM Review* **29**, 1987.

[20] C. Lawson and R. Hanson; *Solving Least Squares Problems*, Prentice-Hall, Englewood Cliffs, 1974.

[21] B. Murtagh and M. Saunders; MINOS/AUGMENTED user's manual, Report SOL 80-14, Department of Operations Research, Stanford, 1980.

[22] H. Nguyen; FONSIZE: a trajectory optimization and vehicle sizing program, The Aerospace Corporation, Report No. TOR-0091(6911-04)-1, 1991.

[23] Y. Shi, R. Nelson, D. Young, P. Gill, W. Murray, and M. Saunders; The application of nonlinear programming and collocation to optimal aeroassisted orbital transfer trajectory, AIAA Paper 92-0734, 1992.

[24] J. Tomlin; Pivoting for sparsity and size in linear programming inversion routines, *J. Inst. Maths. Applics.* **10**, 1972, 289–295.

[25] J. Tomlin; Maintaining a sparse inverse in the simplex method, *IBM J. Res. Dev.* **16**, 1972, 415–423.

The Optimal Design Problem of Céa and Malanowski Revisited

Michel C. Delfour*
Centre de Recherches Mathématiques et
Département de Mathématiques et de Statistique
Université de Montréal, C.P. 6128 Succ. Centre-ville
Montréal, Québec, Canada, H3C 3J7

Jean-Paul Zolésio
Institut Non Linéaire de Nice
CNRS
1361 Route des Lucioles
06904 Sophia Antipolis Cédex, France

In this paper we consider the optimal design problem for the compliance as studied by Céa and Malanowski in 1970. It consists in finding the optimal distribution of two materials in a fixed domain. This type of problem often leads to the occurence of a microstructure or a composite material. We complete the characterization of the solutions and establish the existence of optimal partitions into two measurable subdomains. Moreover we show that the set of solutions of the convex relaxed problem always contains measurable partitions. A numerical example is presented. The results readily extend to higher order elliptic operators and to variational inequalities over closed convex sets.

1. Introduction

We consider the *compliance problem* where the optimization variable is the distribution of two materials with different physical characteristics within a fixed domain D. It cannot a priori be assumed that the two regions are simply connected and separated by a smooth boundary. The optimal solution may lead to the design of a *microstructure* or a *composite material*. This type of solution often occurs in optimization problems over a bounded non-convex subset of a function space, and they are often *relaxed* to the convex hull of that subset. In *control theory*, this is known as *chattering control*.

In this paper we consider the optimal design problem studied by Céa and Malanowski [1] in 1970 and complete the characterization of the solutions. We also establish the existence of optimal measurable domains for the minimization of the compliance. This measurable partition is also a solution of the same problem relaxed to the closed convex hull of all measurable characteristic functions over D. Finally, any solution to the relaxed problem can be suitably modified to yield a measurable domain. A numerical example is presented to illustrate the theory. The

*Supported in part by a Killam fellowship from Canada Council, National Sciences and Engineering Research Council of Canada operating grant A-8730 and infrastructure grant INF-7939, and by a FCAR grant from the Ministère de l'Education du Québec.

results and the mathematical techniques used in this paper readily extend to higher order elliptic operators and to variational inequalities over closed convex sets.

2. The optimal design problem of Céa and Malanowski

Let $D \subset \mathbb{R}^N$ be a bounded domain with a Lipschitzian boundary ∂D. Assume that the domain D is partitioned into two subdomains Ω_1 and Ω_2 separated by a smooth boundary $\partial\Omega_1 \cap \partial\Omega_2$.

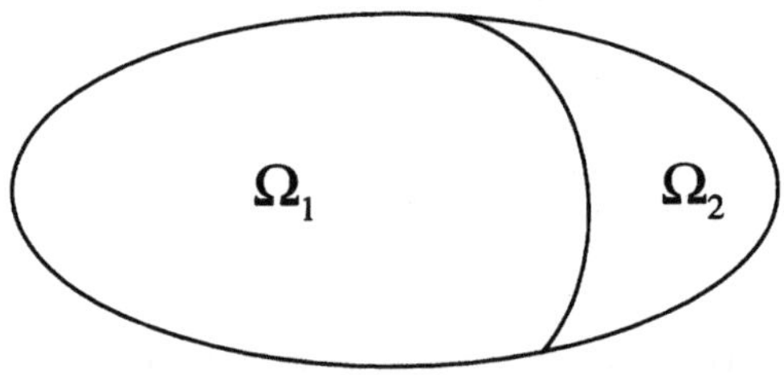

Figure 1. Domain D and its partition into Ω_1 and Ω_2

The domain Ω_1 (resp. Ω_2) is made of a material characterized by a constant $k_1 > 0$ (resp. $k_2 > 0$). Assume that $k_1 > k_2$, that is, k_1 is the *strong material*. Let y be the solution of the following *boundary boundary value problem*

$$\begin{cases} -k_1 \,\Delta y = f & \text{in } \Omega_1 \\ -k_2 \,\Delta y = f & \text{in } \Omega_2 \\ y = 0 & \text{on } \partial D \\ k_1 \dfrac{\partial y}{\partial n_1} + k_2 \dfrac{\partial y}{\partial n_2} = 0 & \text{on } \partial\Omega_1 \cap \partial\Omega_2 , \end{cases} \tag{1}$$

where n_1 (resp. n_2) is the unit outward normal to Ω_1 (resp. Ω_2) and f is a given function in $L^2(D)$. Our objective is to minimize the *compliance*

$$C(\Omega_1) = \int_D f y \, dx \tag{2}$$

over all Lebesgue measurable domains Ω_1 in D verifying the volume constraint

$$\int_{\Omega_1} dx \le \alpha \tag{3}$$

for some α, $0 < \alpha < \mathrm{m}(D)$. The case $\alpha = 0$ (resp. $\alpha = \mathrm{m}(D)$) corresponds to $\chi = 0$ (resp. $\chi = 1$). A mechanical equivalent interpretation of this problem would be the minimization of the total work of the body force f. The volume constraint means that there is an upper bound on the total volume of the strong material

k_1 available for the design. At the limit when the weak material k_2 goes to zero, the domain Ω_2 would correspond to the holes in the domain D.

For simplicity of the notation let $\Omega = \Omega_1$ and hence by complementarity $\Omega_2 = C_D\overline{\Omega}$. Let Ω be a measurable domain in $\mathbb{R}^N$ and introduce the characteristic function $\chi = \chi_\Omega$ of Ω. We want to relax the shape optimization problem (1)–(3) to an optimization problem over the set of measurable characteristic functions

$$X(D) = \left\{\chi \in L^2(D) : \chi(x) \in \{0, 1\} \text{ a.e. in } D\right\} \tag{4}$$

in $L^2(D)$.

The problem (1) can be relaxed to $\chi \in X(D)$ by introducing the following variational problem:

$$\begin{cases} \text{find } y = y(\chi) \in H_0^1(D) \text{ such that} \\ \int_D [k_1 \chi + k_2(1-\chi)]\, \nabla y \cdot \nabla\varphi \, dx = \int_D f\varphi\, dx \quad \forall \varphi \in H_0^1(D), \end{cases} \tag{5}$$

with the cost function

$$J(\chi) = -\int_D f\, y(\chi)\, dx \quad (= -C(\Omega), \text{ with } \chi = \chi_\Omega) \tag{6}$$

to be maximized over all $\chi \in X(D)$ such that

$$\int_D \chi\, dx \le \alpha\,. \tag{7}$$

Many *shape* or *geometric optimization* problems can be reformulated as optimization problems over the subset of equivalence classes of characteristic functions $X(D)$ in an appropriate function space. The set $X(D)$ is closed and bounded in $L^2(D)$ (cf. [2]) but not convex and we cannot use the weak compactness in $L^2(D)$. The weak compactness can be recovered by further *relaxing* the problem (5)-(7) from $X(D)$ to its closed convex hull

$$\overline{\text{co}}\, X(D) = \left\{\chi \in L^2(D) : \chi(x) \in [0, 1] \text{ a.e. in } D\right\}\,. \tag{8}$$

However, the elements of $\overline{\text{co}}\, X(D)$ are not necessarily characteristic functions of a domain. This new formulation allows the "mixing" of the points of Ω_1 and Ω_2 (the mixing of the materials 1 and 2) at the microscale. In general for $k_1 \neq k_2$ an optimal solution of the relaxed problem is not necessarily the characteristic function of a smooth domain Ω_1 in D, the regions Ω_1 and Ω_2 are not necessarily separated by a smooth boundary, and a new region can appear where $\chi(1-\chi) \neq 0$.

For the relaxed problem we have for all χ in $\overline{\text{co}}\, X(D)$

$$\min\{k_1, k_2\} \le k_1\, \chi(x) + k_2(1-\chi(x)) \le \max\{k_1, k_2\} \text{ a.e. in } D$$

and by the Lax-Milgram's theorem, the variational equation (5) still makes sense and has a unique solution $y(\chi)$ in $H_0^1(D)$ which coincides with the solution of the boundary value problem

$$\begin{cases} -\text{div } \{[k_2 + \chi(k_1 - k_2)] \, \nabla y\} = f \text{ in } \mathcal{D}'(D) \\ y = 0 \text{ on } \partial D \, . \end{cases}$$

To obtain the existence of maximizers χ in $\overline{\text{co}}\, X(D)$, we need a number of technical results. For instance, the property that for any sequence $\{\chi_n\}$ in $\overline{\text{co}}\, X(D)$

$$\chi_n \rightharpoonup \chi \text{ in } L^2(D)\text{–weak} \quad \Rightarrow \quad \chi_n \rightharpoonup \chi \text{ in } L^\infty(D)\text{–weak}\star \tag{9}$$

is a consequence of the following lemma (cf. [2]).

Lemma 1 *Let D be a measurable subset of* $\mathbb{R}^N$ *with finite measure, K be a bounded subset of* $\mathbb{R}$*, and*

$$\mathcal{K} = \{k : D \to \mathbb{R} \; : \; k \text{ is measurable and } k(x) \in K \quad a.e. \text{ in } D\} \, .$$

(i) For any p, $1 \le p < \infty$*, and any sequence* $\{k_n\} \subset \mathcal{K}$ *the following statements are equivalent*

a) $\{k_n\}$ converges in $L^\infty(D)$-weak $\star$,

b) $\{k_n\}$ converges in $L^p(D)$-weak,

c) $\{k_n\}$ converges in $\mathcal{D}(D)'$,

where $\mathcal{D}(D)'$ *is the space of scalar distributions on D. For c) it is assumed that D is open.*
(ii) If $\{k_n\}$ *is a convergent sequence in* $L^p(D)$*-strong for some p,* $1 \le p < \infty$*, then it is a convergent sequence in* $L^p(D)$*-strong for all p,* $1 \le p < \infty$*.*

If K is convex and compact, then the set $\mathcal{K}$ is closed for all topologies between the $L^\infty(D)$–weak$\star$ topology and the $\mathcal{D}(D)'$ topology.

It is natural to relax the problem to $\overline{\text{co}}\, X(D)$ and later look at the eventual presence of characteristic functions in the set of solutions. We first obtain the classical existence result in $\overline{\text{co}}\, X(D)$ for this specific problem. Rewrite the functional $J(\chi)$ as the minimum of the energy functional

$$E(\chi, \varphi) = \int_D ([\chi k_1 + (1 - \chi)k_2] \, \nabla\varphi \cdot \nabla\varphi - 2 \, f \, \varphi) \, dx \tag{10}$$

over $H_0^1(D)$:

$$J(\chi) = \min_{\varphi \in H_0^1(D)} E(\chi, \varphi) \, .$$

As a result we have the min max problem:

$$\max_{\chi \in M} \min_{\varphi \in H_0^1(D)} E(\chi, \varphi),$$

where

$$M = \{\chi \in \overline{\mathrm{co}}\, X(D) : \int_D \chi\, dx \leq \alpha\}.$$

For $\alpha = 0$, the maximizer is trivially equal to 0. Thus we assume that $\alpha > 0$. In the original problem considered by Céa and Malanowski an equality constraint on the volume was used rather than (7).

At this juncture it is technically advantageous to reformulate the problem by using a Lagrange multiplier $\lambda \geq 0$ to handle the volume constraint:

$$\max_{\chi \in \overline{\mathrm{co}}\, X(D)} \min_{\substack{\varphi \in H_0^1(D) \\ \lambda \geq 0}} G(\chi, \varphi, \lambda), \tag{11}$$

where

$$G(\chi, \varphi, \lambda) = E(\chi, \varphi) - \lambda \left[\int_D \chi\, dx - \alpha\right]. \tag{12}$$

The existence of saddle points follows from a general result in [3, Prop. 2.4, p. 164]. The set $\overline{\mathrm{co}}\, X(D)$ is a nonempty bounded closed convex subset of $L^2(D)$ and the set $H_0^1(D) \times [\mathbb{R}^+ \cup \{0\}]$ is trivially closed and convex. The functional G is concave–convex with the properties:

$$\left\{\begin{array}{l} \forall \chi \in \overline{\mathrm{co}}\, X(D), \\ (\varphi, \lambda) \mapsto G(\chi, \varphi, \lambda) \text{ is convex, continuous, and} \\ \exists \chi_0 \in \overline{\mathrm{co}}\, X(D) \text{ such that } \lim\limits_{\|\varphi\|_{H_0^1} + |\lambda| \to \infty} G(\chi_0, \varphi, \lambda) \to \infty \end{array}\right. \tag{13}$$

$$\left\{\begin{array}{l} \forall \varphi \in H_0^1(D),\ \forall \lambda \geq 0 \\ \chi \mapsto G(\chi, \varphi, \lambda) \text{ is affine and continuous for } L^2(D) - \text{strong}. \end{array}\right. \tag{14}$$

For the first condition recall that $\alpha > 0$ and pick

$$0 < \chi_0(x) = \frac{\alpha}{2\, \mathrm{m}(D)} \leq \frac{1}{2}.$$

To check the second condition pick any sequence $\{\chi_n\}$ in $\overline{\mathrm{co}}\, X(D)$ which converges to some χ in $L^2(D)$-strong. Then the sequence also converges in $L^2(D)$-weak and by Lemma 1 in $L^\infty(D)$-weak$\star$. Hence $G(\chi_n, \varphi, \lambda)$ converges to $G(\chi, \varphi, \lambda)$.

Moreover the set of saddle points $(\hat{\chi}, y, \hat{\lambda})$ is of the form $X \times Y \subset \overline{\mathrm{co}}\, X(D) \times \{H_0^1(D) \times [\mathbb{R}^+ \cup \{0\}]\}$ and is completely characterized by the following variational equation and inequalities (cf. [3, Proposition 1.6, p. 157]):

$$\int_D \{[k_1 \hat{\chi} + k_2(1 - \hat{\chi})] \nabla y \cdot \nabla \varphi - f \varphi\}\, dx = 0, \quad \forall \varphi \in H_0^1(D) \tag{15}$$

$$\int_D \left[(k_1 - k_2)|\nabla y|^2 - \hat{\lambda}\right] (\chi - \hat{\chi})\, dx \leq 0, \quad \forall \chi \in \overline{\text{co}}\, X(D) \tag{16}$$

$$\left[\int_D \hat{\chi}\, dx - \alpha\right] \hat{\lambda} = 0, \quad \int_D \hat{\chi}\, dx - \alpha \leq 0, \quad \hat{\lambda} \geq 0\,. \tag{17}$$

It is readily seen that for each $\chi \in \overline{\text{co}}\, X(D)$ there exists a unique $y(\chi)$ solution of (15) and then

$$\forall \hat{\chi} \in X, \quad \{\hat{\chi}\} \times Y \subset \{\hat{\chi}\} \times \left\{\{y(\hat{\chi})\} \times [\mathbb{R}^+ \cup \{0\}]\right\}\,.$$

Therefore $y(\hat{\chi})$ is independent of $\hat{\chi} \in X$, that is $Y = \{y\} \times \Lambda$ and

$$\forall \hat{\chi} \in X,\, \hat{\lambda} \in \Lambda, \qquad y(\hat{\chi}) = y\,.$$

For each $\hat{\lambda}$, inequality (16) is completely equivalent to the following characterization of the maximizer $\hat{\chi} \in X$

$$\hat{\chi}(x) = \begin{cases} 1\,, & \text{if} \quad (k_1 - k_2)|\nabla y|^2 - \hat{\lambda} > 0 \\ \in [0, 1]\,, & \text{if} \quad (k_1 - k_2)|\nabla y|^2 - \hat{\lambda} = 0 \\ 0\,, & \text{if} \quad (k_1 - k_2)|\nabla y|^2 - \hat{\lambda} < 0\,. \end{cases} \tag{18}$$

For each $\lambda \geq 0$, it is convenient to define

$$\begin{aligned} D_+(\lambda) &= \{x \in D \,:\, (k_1 - k_2)|\nabla y|^2 - \lambda > 0\}\,, \\ D_0(\lambda) &= \{x \in D \,:\, (k_1 - k_2)|\nabla y|^2 - \lambda = 0\}\,, \\ D_-(\lambda) &= \{x \in D \,:\, (k_1 - k_2)|\nabla y|^2 - \lambda < 0\}\,. \end{aligned} \tag{19}$$

The set Λ is closed and convex. Hence it has a minimal element

$$\hat{\lambda}_m = \min\{\hat{\lambda} \,:\, \hat{\lambda} \in \Lambda\} \in \Lambda\,.$$

From the characterization (18) we conclude that

$$\hat{\lambda} \geq \hat{\lambda}_m \quad \Rightarrow \quad D_+(\hat{\lambda}) \subset D_+(\hat{\lambda}_m)\,.$$

Define the charateristic function

$$\chi_m = \chi_{D_+(\hat{\lambda}_m)}\,.$$

From the characterization (18) of the maximizers with $\hat{\lambda} = \hat{\lambda}_m$

$$\forall \hat{\chi} \in X\,, \quad \hat{\chi} \geq \chi_m\,, \quad \int_D \hat{\chi}\, dx \leq \alpha$$

and necessarily

$$\int_D \chi_m\, dx \leq \int_D \hat{\chi}\, dx \leq \alpha\,.$$

Moreover it will be shown that

$$G(\hat{\chi}, y, \hat{\lambda}_m) = G(\chi_m, y, \hat{\lambda}_m)$$

and by the following lemma, $(\hat{\chi}_m, y, \hat{\lambda}_m)$ is also a solution.

Lemma 2 *Consider a functional*

$$G : X \times Y \to \mathbb{R} \tag{20}$$

for some sets X and Y. Define

$$g = \inf_{x \in X} \sup_{y \in Y} G(x, y) \tag{21}$$

$$X_0 = \left\{ x \in X \; : \; \sup_{y \in Y} G(x, y) = g \right\} \tag{22}$$

$$h = \sup_{y \in Y} \inf_{x \in X} G(x, y) \tag{23}$$

$$Y_0 = \left\{ y \in Y : \inf_{x \in X} G(x, y) = h \right\} . \tag{24}$$

When $g = h$ the set of saddle points (possibly empty) will be denoted by

$$S = \{(x, y) \in X \times Y : g = G(x, y) = h\} . \tag{25}$$

Then

(i) *In general $h \leq g$ and*

$$\forall (x_0, y_0) \in X_0 \times Y_0 , \quad h \leq G(x_0, y_0) \leq g . \tag{26}$$

(ii) *If $h = g$, then*

$$S = X_0 \times Y_0 . \tag{27}$$

Indeed, by construction

$$\hat{\chi}\{(k_1 - k_2)|\nabla y|^2 - \hat{\lambda}_m\} = \hat{\chi}_m\{(k_1 - k_2)|\nabla y|^2 - \hat{\lambda}_m\} \quad a.e. \text{ in } D$$

and

$$\begin{aligned}
&G(\hat{\chi}, y, \hat{\lambda}_m) \\
&= \int_D \{[k_1 \hat{\chi} + k_2(1 - \hat{\chi})]|\nabla y|^2 - 2fy\}\, dx - \hat{\lambda}_m \{\int_D \hat{\chi}\, dx - \alpha\} \\
&= \int_D \{k_2 |\nabla y|^2 - 2fy\}\, dx + \int_D \{(k_1 - k_2)|\nabla y|^2 - \hat{\lambda}_m\}\hat{\chi}\, dx + \hat{\lambda}_m \alpha \\
&= \int_D \{k_2 |\nabla y|^2 - 2fy\}\, dx + \int_D \{(k_1 - k_2)|\nabla y|^2 - \hat{\lambda}_m\}\chi_m\, dx + \hat{\lambda}_m \alpha \\
&= \int_D \{[k_1 \chi_m + k_2(1 - \chi_m)]|\nabla y|^2 - 2fy\}\, dx - \hat{\lambda}_m \{\int_D \chi_m\, dx - \alpha\} \\
&= G(\chi_m, y, \hat{\lambda}_m)
\end{aligned}$$

We can further characterize the elements of X. If there exists $\hat{\lambda} > 0$ in Λ, then by (17)

$$\int_D \hat{\chi}\, dx = \alpha = \int_D \chi_m\, dx .$$

Since $\hat{\chi} \geq \chi_m$, necessarily $\hat{\chi} = \chi_m$, the maximizer is unique and its integral is equal to α. If, on the contrary, $\Lambda = \{0\}$, then the characteristic function χ_m is still a maximizer but the integral is not necessarily equal to α and it is not unique. The set $D_-(0)$ is empty and

$$D_+(0) = \{ x \in D \, : \, \nabla y \neq 0 \}$$
$$D_0(0) = \{ x \in D \, : \, \nabla y = 0 \}$$

where y is the solution of

$$\int_D \{[\chi_m k_1 + (1-\chi_m)k_2]\nabla y \cdot \nabla \varphi - f\varphi\}\, dx = 0\,, \quad \forall \varphi \in H_0^1(D)$$

with $\chi_m = \chi_{D_+(0)}$. Moreover any $\hat{\chi} \in X$ is of the form

$$\chi_{D_+(0)} \leq \hat{\chi} \leq 1\,, \quad \int_D \hat{\chi}\, dx \leq \alpha\,.$$

From this we conclude that the characteristic function $\chi_m = \chi_{D_+(\hat{\lambda}_m)}$ is always a maximizer over $\overline{\text{co}}\, X(D)$. When the set of optimal multipliers $\Lambda \neq \{0\}$, this solution is unique and its integral is equal to α. When $\Lambda = \{0\}$ the maximizer is not necessarily unique and its integral is not necessarily equal to α. If $\hat{\chi}$ is not unique, then the number of maximizers which are or which are not characteristic functions is inifnite.

We have proved the following theorem.

Theorem 3 *Let D be a bounded domain in $\mathbb{R}^N$ with a Lipschitzian boundary ∂D. Let $k_1 > 0$ and $k_2 > 0$ be two real numbers such that $k_1 > k_2$.*

(i) The functional $E(\chi, \varphi)$ has saddle points of the form $(\hat{\chi}, y) \in M \times H_0^1(D)$, where y is the solution of the variational equation (5) which is independent of $\hat{\chi}$.

(ii) The functional $E(\chi, \varphi)$ has saddle points of the form $(\hat{\chi}, y) \in \overline{\text{co}}\, M \times H_0^1(D)$, where

$$\overline{\text{co}}\, M = \{\chi \in L^2(D) \, : \, \chi(x) \in [0,1] \text{ a.e.}, \int_D \chi\, dx \leq \alpha\}$$

and y is the solution of the variational equation (5) which is independent of $\hat{\chi}$. Moreover

$$\max_{\chi \in M} \min_{\varphi \in H_0^1(D)} E(\chi, \varphi) = \max_{\chi \in \overline{\text{co}}\, M} \min_{\varphi \in H_0^1(D)} E(\chi, \varphi)\,.$$

(iii) For any $\hat{\chi}$ such that $(\hat{\chi}, y) \in \overline{\text{co}}\, M \times H_0^1(D)$ is a saddle point of (ii), a minimum positive multiplier $\hat{\lambda}_m \geq 0$ can always be constructed in such a way that

$$\hat{\chi}(x) = \begin{cases} 1\,, & x \in D_+(\hat{\lambda}_m) \\ \in [0,1]\,, & x \in D_0(\hat{\lambda}_m) \\ 0\,, & x \in D_-(\hat{\lambda}_m)\,, \end{cases}$$

where the D's are given by (19). Moreover, the characteristic function $\chi_m = \chi_{D_+(\hat{\lambda}_m)}$ is also a maximizer. It is obtained by setting $\hat{\chi} = 0$ on $D_0(\hat{\lambda}_m)$. Finally, when the set of Lagrange multipliers for the volume constraint contains a non-zero element, χ_m is the unique minimizer, and, in addition,

$$\int_D \chi_m \, dx = \alpha\,.$$

3. Some numerical results

Here we considered (1) over the diamond shaped domain

$$D = \{(x, y) \,:\, |x| + |y| < 1\}$$

with $k_1 = 2$ (black in Figures 2a to 6a), $k_2 = 1$ (white on the figures) and

$$f(x, y) = 56\,(1 - |x| - |y|)^6\,. \tag{28}$$

This function has a sharp peak in (0, 0); see Figure 2. Figures 3 to 5 show the optimal partition as the grid size is reduced. Figure 6 repeats the last computation in the case where the function f is centered at (0.1, 0.2).

For this example the Lagrange multiplier associated with the problem is strictly positive. The gray zones correspond to the region $D_0(\hat{\lambda}_m)$ where $\hat{\chi} \in [0, 1]$. The presence of this gray zone in the approximated problem is due to the fact that equality for the total area where $\hat{\chi} = 1$ could not be exactly achieved with the chosen triangulation of the domain. Thus the problem had to adjust the value of $\hat{\chi}$ between 0 and 1 in a few triangles in order to achieve equality for the integral of $\hat{\chi}$.

The boundary value problem was approximated by continuous piecewise linear finite elements on each triangle and the function χ by a piecewise constant function on each triangle. The constant on each triangle was constrained to lie between 0 and 1 together with the global constraint on its integral over the whole domain D. The computations have been done on a 6-processor SGI using a standard optimization package which performs well for a large number of optimization variables.

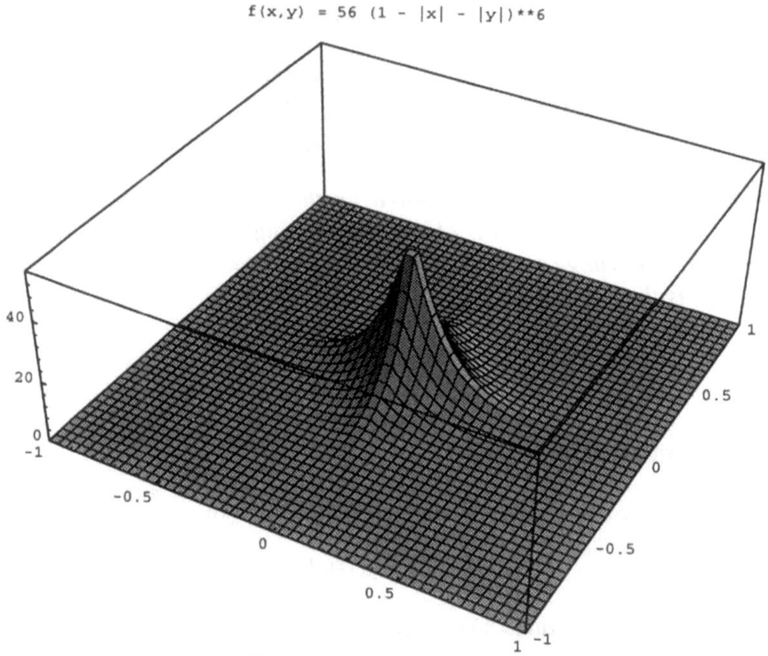

Figure 2. The function f.

4. Extensions and conclusions

In this paper we have only considered the simple elliptic equation (1). The theory readily extends without changes to more general elliptic operators of higher orders. It also naturally extends to variational inequalities over a closed convex set $C \subset V$, where V is a fixed Hilbert space of functions on D which are 0 on ∂D. For instance we can consider energy functionals of the general form

$$E(\chi, \varphi) = \int_D \{k_1\, \chi\, a_1(x, \varphi, \varphi) + k_2\, (1 - \chi)\, a_2(x, \varphi, \varphi) - 2\, f\, \varphi\}\, dx$$

where $a_1(x, \varphi, \psi)$ and $a_2(x, \varphi, \psi)$ are suitable continuous bilinear forms on V. Here the problem is of the new form

$$\max_{\chi \in M} \min_{\varphi \in C} E(\chi, \varphi),$$

but the results remain identical.

Figure 3a. The optimal distribution.

The results presented are extremely important for numerical computations. The relaxed problem can be solved and a solution $\hat{\chi} \in \overline{\mathrm{co}}\, M$ can be computed. But the theorem also says that if our computation yields a function $\hat{\chi}$ which is not a characteristic function, then it can always be modified to yield a characteristic function which does not change the state variable y and the value of $E(\hat{\chi}, y)$ at the saddle point.

References

[1] J. Céa and K. Malanowski; An example of a max-min problem in partial differential equations, *SIAM J. Control* **8**, 1970, 305–316.

[2] M. Delfour and J.-P. Zolésio; Shape analysis via distance functions, *J. Funct. Anal.* **123**, 1994, 129–201.

[3] I. Ekeland and R. Temam, *Analyse convexe et problèmes variationnels*, Dunod, Paris, 1974.

Figure 3b. Level curves.

A	=	0.357105E-02	B	=	0.357105E-01	C	=	0.714209E-01
D	=	0.107131E+00	E	=	0.142842E+00	F	=	0.178552E+00
G	=	0.214263E+00	H	=	0.249973E+00	I	=	0.285684E+00
J	=	0.321394E+00	K	=	0.357105E+00	L	=	0.392815E+00
M	=	0.428526E+00	N	=	0.464236E+00	O	=	0.499946E+00
P	=	0.535657E+00	Q	=	0.571367E+00	R	=	0.607078E+00
S	=	0.642788E+00	T	=	0.674928E+00			

Figure 4a. The optimal distribution.

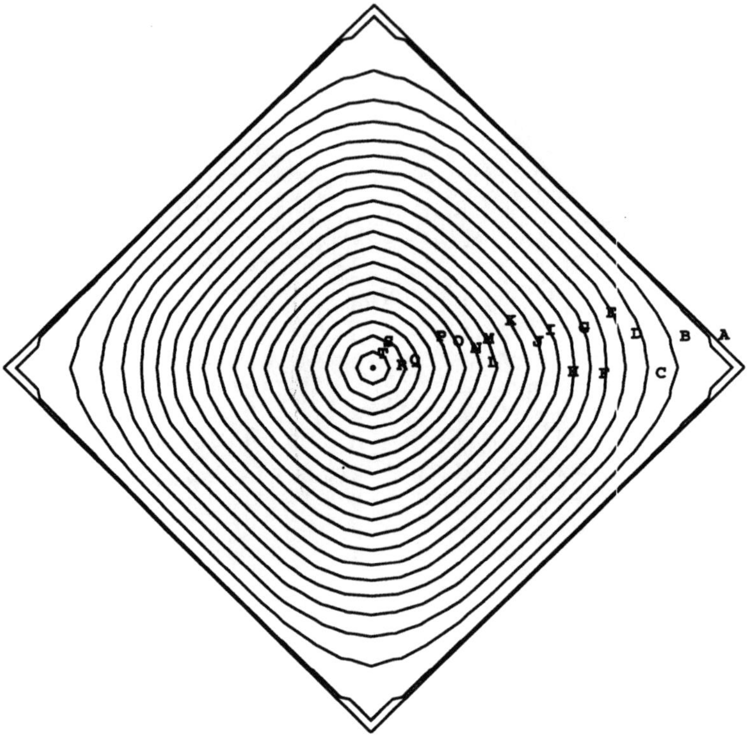

Figure 4b. Level curves.

A	=	0.364407E-02	B	=	0.364407E-01	C	=	0.728815E-01
D	=	0.109322E+00	E	=	0.145763E+00	F	=	0.182204E+00
G	=	0.218644E+00	H	=	0.255085E+00	I	=	0.291526E+00
J	=	0.327967E+00	K	=	0.364407E+00	L	=	0.400848E+00
M	=	0.437289E+00	N	=	0.473729E+00	O	=	0.510170E+00
P	=	0.546611E+00	Q	=	0.583052E+00	R	=	0.619492E+00
S	=	0.655933E+00	T	=	0.688730E+00			

Figure 5a. The optimal distribution.

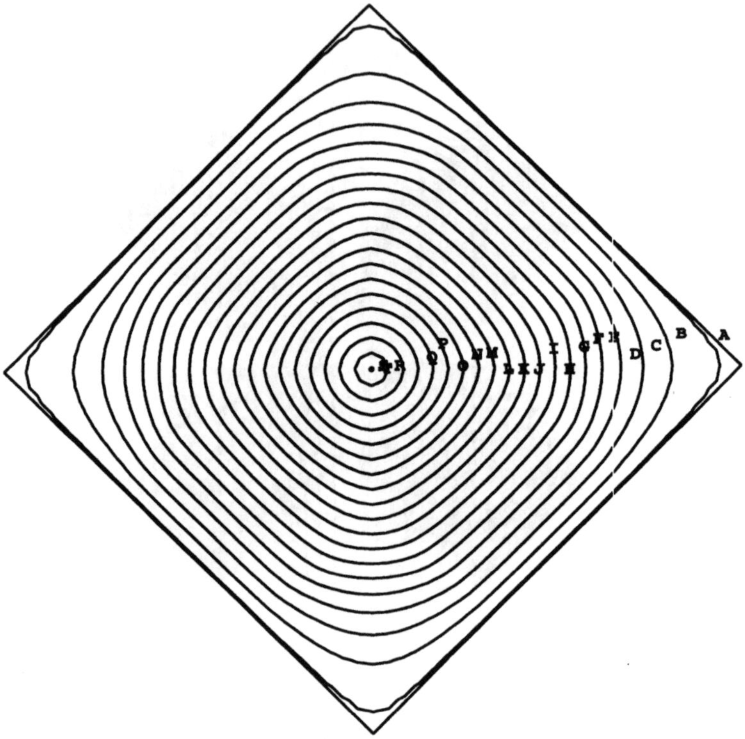

Figure 5b. Level curves.

A	=	0.365480E-02	B	=	0.365480E-01	C	=	0.730961E-01
D	=	0.109644E+00	E	=	0.146192E+00	F	=	0.182740E+00
G	=	0.219288E+00	H	=	0.255836E+00	I	=	0.292384E+00
J	=	0.328932E+00	K	=	0.365480E+00	L	=	0.402029E+00
M	=	0.438577E+00	N	=	0.475125E+00	O	=	0.511673E+00
P	=	0.548221E+00	Q	=	0.584769E+00	R	=	0.621317E+00
S	=	0.657865E+00	T	=	0.690758E+00			

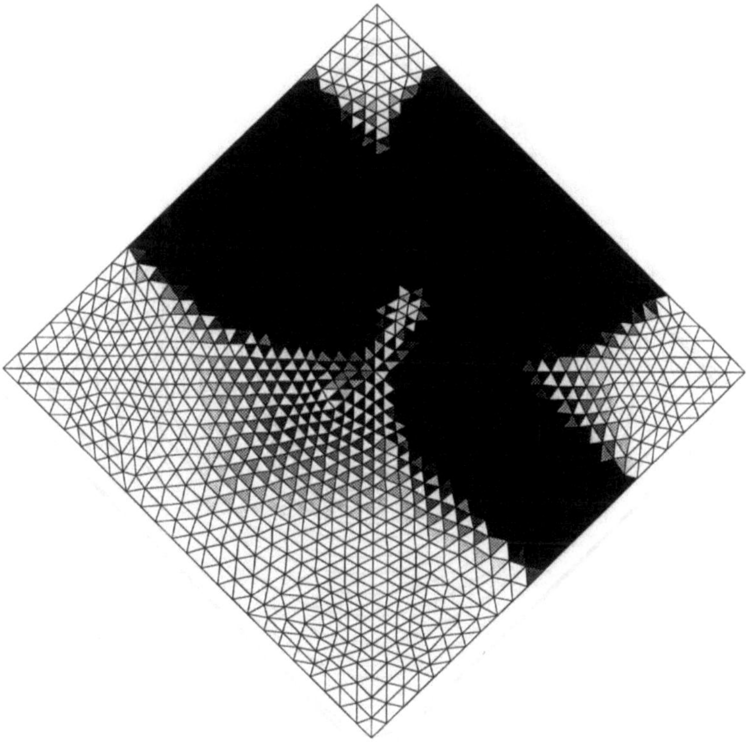

Figure 6a. The optimal distribution for the function f centered at $(0.1, 0.2)$.

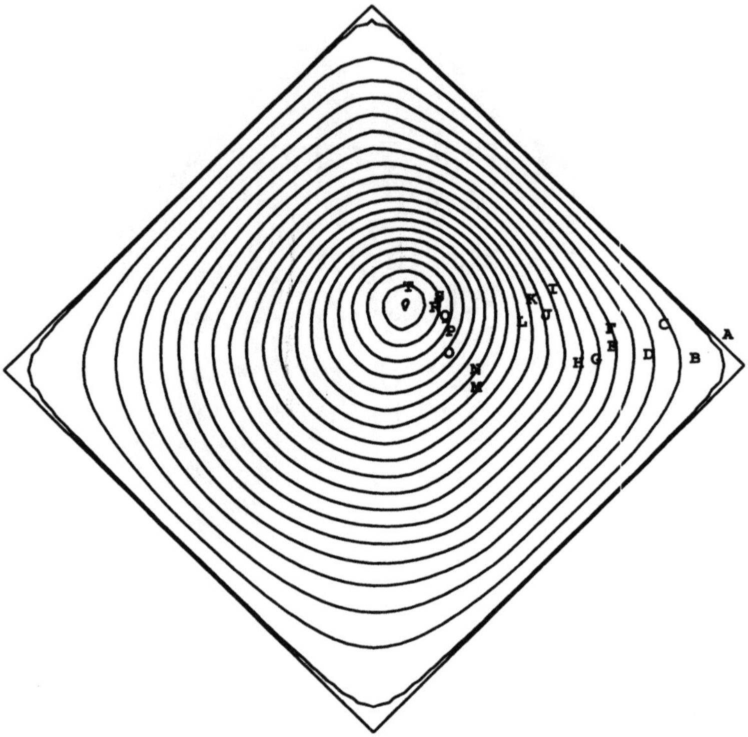

Figure 6b. Level curves for the function f centered at $(0.1, 0.2)$.

A	=	0.314670E-02	B	=	0.314670E-01	C	=	0.629340E-01
D	=	0.944010E-01	E	=	0.125868E+00	F	=	0.157335E+00
G	=	0.188802E+00	H	=	0.220269E+00	I	=	0.251736E+00
J	=	0.283203E+00	K	=	0.314670E+00	L	=	0.346137E+00
M	=	0.377604E+00	N	=	0.409071E+00	O	=	0.440538E+00
P	=	0.472005E+00	Q	=	0.503472E+00	R	=	0.534939E+00
S	=	0.566406E+00	T	=	0.594726E+00			

Uniformly Stable Spline Approximations for Scalar Delay Problems

Richard H. Fabiano
Department of Mathematics
Texas A&M University
College Station TX 77843, USA

Uniformly stable semidiscrete approximation schemes are desirable for constructing approximate solutions of certain optimal control problems for distributed parameter systems. We discuss construction of uniformly stable spline-based approximation schemes for a scalar delay equation. A method is used which employs an equivalent inner product. Convergence is established via a "hybrid" Trotter–Kato/Galerkin result.

1. Introduction

In this paper we investigate, in the context of an example in scalar delay equations, the issue of preservation of stability under approximation. To motivate ideas, consider a linear Cauchy problem

$$\begin{aligned} \dot{z}(t) &= Az(t) + f(t) \\ z(0) &= z_0 \end{aligned} \tag{1}$$

on a Hilbert space Z, and assume that A is the infinitesimal generator of a C_0-semigroup $T(t)$ on Z. Consider also a generic approximation scheme for (1) that consists of finite dimensional subspaces $Z^N \subset Z$, associated orthogonal projections $P^N : Z \to Z^N$, and operators $A^N : Z^N \to Z^N$. This defines a semidiscrete approximation of (1) given by

$$\begin{aligned} \dot{z}^N(t) &= A^N z^N(t) + P^N f(t) \\ z^N(0) &= P^N z_0 . \end{aligned} \tag{2}$$

Suppose that $T(t)$ is exponentially stable. That is, suppose there exists $M \geq 1$ and $\omega > 0$ such that $\| T(t) \| \leq Me^{-\omega t}$ for all $t \geq 0$. It is natural to ask if the sequence of finite dimensional semigroups $T^N(t) = e^{tA^N}$ is uniformly exponentially stable. That is, does there exist $\widetilde{M} \geq 1$ and $\widetilde{\omega} > 0$ such that $\| T^N(t) \| \leq \widetilde{M}e^{-\widetilde{\omega}t}$ for all $t \geq 0$ for all sufficiently large N? A related question is the following – given only the spaces $Z^N \subset Z$, can one construct $A^N : Z^N \to Z^N$ so that $T^N(t)$ is uniformly exponentially stable? These issues are important when considering the problem of constructing approximation schemes for optimal control of distributed parameter systems (such as the linear quadratic regulator problem). In particular, an important sufficient condition for convergence of feedback gains is that finite dimensional approximate problems preserve the stability properties of the original system (see [3], [6], [7], [14]).

We will consider these issues for a specific example involving a scalar delay equation. In Section 2 we analyze the delay equation and certain spline-based approximation schemes. We show how to use an equivalent inner product to construct improved stability preserving semigroup approximations. In Section 3 we state a hybrid Trotter-Kato/Galerkin convergence theorem and apply it to these schemes.

2. Semigroup formulation and approximation

Consider the scalar delay equation

$$\dot{x}(t) = ax(t) + bx(t-r) \tag{3}$$

where $a, b, r \in \mathbb{R}$ and $r > 0$. It is standard to reformulate (3), augmented with appropriate initial data, as an abstract Cauchy problem. To this end, consider the Hilbert space $Z = \mathbb{C} \times L^2(-r, 0; \mathbb{C})$ with the inner product

$$\langle(\eta, \phi), (\xi, \psi)\rangle = \eta\bar{\xi} + \int_{-r}^{0} \phi(\theta)\overline{\psi(\theta)}\, d\theta\,,$$

and a compatible norm given by

$$\| (\eta, \phi) \|^2 = |\eta|^2 + \int_{-r}^{0} |\phi(\theta)|^2\, d\theta\,.$$

Next define the operator A : dom $A \subset Z \to Z$ by dom $A := \{(\eta, \phi) \in Z : \phi \in H^1(-r, 0),\ \phi(0) = \eta\}$, and $A(\eta, \phi) = (a\eta + b\phi(-r), \frac{d}{d\theta}\phi)$. If the state at time $t \geq 0$ is defined to be $z(t) = (x(t), x_t)$, where $x_t(\theta) = x(t+\theta)$ for $-r \leq \theta \leq 0$, then (3) can be reformulated as the abstract Cauchy problem (1). It is well known (see [1], [8]) that A is the infinitesimal generator of a strongly continuous semigroup $T(t)$ of bounded linear operators on Z.

Let us now describe some spline-based Galerkin approximation schemes for (3), and for simplicity consider only linear splines. Following [2] let $\theta_j^N = -jr/N$ for $j = 0, 1, \ldots, N$ be a partition of $[-r, 0]$. Define the piecewise linear "hat" functions

$$e_0^N(\theta) = \begin{cases} \frac{N}{r}(\theta - \theta_1^N) & \text{if } \theta_1^N \leq \theta \leq 0 \\ 0 & \text{elsewhere} \end{cases}$$

$$e_N^N(\theta) = \begin{cases} -\frac{N}{r}(\theta - \theta_{N-1}^N) & \text{if } \theta_N^N \leq \theta \leq \theta_{N-1}^N \\ 0 & \text{elsewhere} \end{cases}$$

and, for $j = 1, 2, \ldots, N-1$,

$$e_j^N(\theta) = \begin{cases} -\frac{N}{r}(\theta - \theta_{j-1}^N), & \text{if } \theta_j^N \le \theta \le \theta_{j-1}^N \\ \frac{N}{r}(\theta - \theta_{j+1}^N) & \text{if } \theta_{j+1}^N \le \theta \le \theta_j^N \\ 0 & \text{elsewhere}. \end{cases}$$

Set $B_0^N = (1, e_0^N)$, $B_j^N = (0, e_j^N)$ for $j = 1, 2, \ldots, N$, and define $Z^N = \text{span}\ \{B_j^N\}_{j=0}^N$. It is useful to define the space $V = \text{dom}A$ equipped with the graph norm. Also, for $x = (\phi(0), \phi)$, $y = (\psi(0), \psi) \in V$, define the bilinear form $\sigma : V \times V \to \mathbb{C}$ by

$$\sigma(x, y) = [a\phi(0) + b\phi(-r)]\overline{\psi(0)} + \int_{-r}^{0} \phi'(\theta)\overline{\psi(\theta)}\, d\theta\,. \tag{4}$$

Observe that $\langle Ax, y\rangle = \sigma(x, y)$ for all $x \in \text{dom}A$, $y \in V$. Since $Z^N \subset V$, we can define $A^N : Z^N \to Z^N$ by

$$\langle A^N x, y\rangle = \sigma(x, y) \qquad \forall x, y \in Z^N\,. \tag{5}$$

If we let $P^N : Z \to Z^N$ denote the orthogonal projection of Z onto Z^N in the $\langle \cdot, \cdot \rangle$ inner product, then $\{Z^N, P^N, A^N\}$ is precisely the approximation scheme introduced and studied in [2].

Next consider a modified scheme devised by Kappel and Salamon ([12]). Set $\widetilde{B}_0^N = (1, 0)$, $\widetilde{B}_j^N = (0, e_{j-1}^N)$ for $j = 1, 2, \ldots, N+1$, and define $\widetilde{Z}^N = \text{span}\ \{\widetilde{B}_j^N\}_{j=0}^{N+1}$. Define the space $\widetilde{V} = \mathbb{C} \times H^1(-r, 0)$, and for $x = (\eta, \phi)$, $y = (\xi, \psi) \in \widetilde{V}$, define the bilinear form $\widetilde{\sigma} : \widetilde{V} \times \widetilde{V} \to \mathbb{C}$ by

$$\widetilde{\sigma}(x, y) = [a\eta + b\phi(-r)]\bar{\xi} + \int_{-r}^{0} \phi'(\theta)\overline{\psi(\theta)}\, d\theta + [\eta - \phi(0)]\overline{\psi(0)}\,. \tag{6}$$

Since $\widetilde{Z}^N \subset \widetilde{V}$, we can define an operator $\widetilde{A}^N$ by

$$\langle \widetilde{A}^N x, y\rangle = \widetilde{\sigma}(x, y) \qquad \text{for all } x, y \in \widetilde{Z}^N\,. \tag{7}$$

If we let $\widetilde{P}^N : Z \to \widetilde{Z}^N$ denote the orthogonal projection of Z onto $\widetilde{Z}^N$, then $\{\widetilde{Z}^N, \widetilde{P}^N, \widetilde{A}^N\}$ is exactly the scheme devised in [12]. It is possible to show semigroup convergence for each scheme ([2], [12]), but neither satisfies the uniform exponential stability condition even when the original semigroup $T(t)$ is exponentially stable ([13]). We next show how to construct "improved" spline based schemes which are uniformly exponentially stable.

Recall that the numerical range of an operator A is the subset of the complex plane defined by $\Theta(A) = \{\langle Ax, x\rangle : x \in \text{dom}A, \ \| x \| = 1\}$. It is interesting to observe that for operators A^N defined via Galerkin type relations like (5) and (7), the point spectrum of A^N is contained in the numerical range of A. It turns out that (see [4]) for the scalar delay equation under consideration, $\Theta(A)$ is the entire left half plane (even when $T(t)$ is exponentially stable). Hence it is possible for the eigenvalues of A^N and $\widetilde{A}^N$ to be near the imaginary axis. In fact, $\lim_{N\to\infty} \max_{\lambda\in\sigma(A^N)} \mathrm{Re}\,\lambda = 0$ and $\lim_{N\to\infty} \max_{\lambda\in\sigma(\widetilde{A}^N)} \mathrm{Re}\,\lambda = 0$, so that neither scheme is uniformly exponentially stable. However, the numerical range is inner product dependent. Thus, we are motivated to seek an inner product which is equivalent to the original inner product (that is, the compatible norms are equivalent) and which, for the case where $T(t)$ is exponentially stable, "pushes" the numerical range into a half-plane of the form $\mathrm{Re}\,\lambda \le -\omega$ for some $\omega > 0$. We can then use this new inner product to define new Galerkin approximations with point spectra contained in this half-plane, and hence uniformly exponentially stable.

To proceed, for any real γ we can endow Z with the following inner product

$$\langle(\eta,\phi),(\xi,\psi)\rangle_e = \eta\bar{\xi} + |b|e^{-\gamma r}\int_{-r}^{0} e^{-2\gamma\theta}\phi(\theta)\overline{\psi(\theta)}\,d\theta\,, \tag{8}$$

and a compatible norm given by

$$\|(\eta,\phi)\|_e^2 = |\eta|^2 + |b|e^{-\gamma r}\int_{-r}^{0} e^{-2\gamma\theta}|\phi(\theta)|^2\,d\theta\,.$$

It is clear that the inner products $\langle\cdot,\cdot\rangle$ and $\langle\cdot,\cdot\rangle_e$ are equivalent on Z. If we now choose γ to be the unique real root of the following equation (when $b > 0$, this is just the characteristic equation for (3))

$$a - \lambda + |b|e^{-\lambda r} = 0\,, \tag{9}$$

then it follows (see [4]) that $\mathrm{Re}\langle Ax, x\rangle_e \le \gamma|x|_e^2$ for all $x \in \mathrm{dom}A$. It turns out (see [9]) that $T(t)$ is exponentially stable if and only if the real root γ of (9) satisfies $\gamma < 0$. Thus, if $T(t)$ is exponentially stable, the numerical range of A in the inner product $\langle\cdot,\cdot\rangle_e$ is the half-plane $\mathrm{Re}\,\lambda \le -\omega$ where $\omega = -\gamma > 0$. As noted above, we can use this inner product to define new uniformly exponentially stable Galerkin approximations on the spaces Z^N and $\widetilde{Z}^N$.

To do this on the spaces Z^N, first define a bilinear form $\sigma_e : V \times V \to \mathbb{C}$ by

$$\sigma(x,y) = [a\phi(0) + b\phi(-r)]\overline{\psi(0)} + |b|e^{-\gamma r}\int_{-r}^{0} e^{-2\gamma\theta}\phi'(\theta)\overline{\psi(\theta)}\,d\theta \tag{10}$$

for $x = (\phi(0),\phi)$, $y = (\psi(0),\psi) \in V$. Now, let $P_e^N : Z \to Z^N$ denote the orthogonal projection of Z onto Z^N in the inner product $\langle\cdot,\cdot\rangle_e$, and define $A_e^N : Z^N \to Z^N$ by

$$\langle A_e^N x, y\rangle_e = \sigma_e(x,y) \qquad \forall x, y \in Z^N\,. \tag{11}$$

Then A_e^N is the infinitesimal generator of a C_0 semigroup $T_e^N(t)$ on Z^N, and $\{Z^N, P_e^N, A_e^N\}$ is a new approximation scheme.

To do this on the spaces $\widetilde{Z}^N$, define the bilinear form $\widetilde{\sigma}_e : \widetilde{V} \times \widetilde{V} \to \mathbb{C}$ by

$$\begin{aligned}\widetilde{\sigma}_e(x, y) &= [a\eta + b\phi(-r)]\bar{\xi} \\ &\quad + |b|e^{-\gamma r} \int_{-r}^{0} e^{-2\gamma\theta}\phi'\overline{\psi}\, d\theta + |b|e^{-\gamma r}[\eta - \phi(0)]\overline{\psi(0)}\end{aligned}$$

for $x = (\eta, \phi)$, $y = (\xi, \psi) \in \widetilde{V}$. Let $\widetilde{P}_e^N : Z \to Z^N$ denote the orthogonal projection of Z onto $\widetilde{Z}^N$ in the inner product $\langle\cdot,\cdot\rangle_e$, and define $\widetilde{A}_e^N : \widetilde{Z}^N \to \widetilde{Z}^N$ by

$$\langle \widetilde{A}_e^N x, y\rangle = \widetilde{\sigma}_e(x, y) \qquad \text{for all } x, y \in \widetilde{Z}^N . \tag{12}$$

Then $\widetilde{A}_e^N$ is the infinitesimal generator of a C_0 semigroup $\widetilde{T}_e^N(t)$ on $\widetilde{Z}^N$, and $\{\widetilde{Z}^N, \widetilde{P}_e^N, \widetilde{A}_e^N\}$ is a new approximation scheme. The semigroups $T_e^N(t)$ and $\widetilde{T}_e^N(t)$ are uniformly exponentially stable *by construction*, and in the next section we address the issue of convergence of these semigroups.

3. Semigroup convergence

We begin this section by stating a hybrid Trotter-Kato/Galerkin semigroup convergence result. Suppose that V and Z are Hilbert spaces, with V densely and continuously embedded in Z, and suppose that $A : \operatorname{dom} A \subset V \subset Z \to Z$ is the infinitesimal generator of a C_0-semigroup $T(t)$ on Z. In addition, assume that there is a bilinear form $\sigma : V \times V \to \mathbb{C}$ and fixed $\alpha \in \mathbb{R}$ satisfying

$$\begin{aligned}\operatorname{Re} \sigma(u, u) &\le \alpha \| u \|_Z^2 && \forall u \in V \\ \sigma(u, v) &= \langle Au, v\rangle_Z && \forall u \in \operatorname{dom} A,\ \forall v \in V .\end{aligned} \tag{13}$$

Let $\{V^N\}_{N=1}^{\infty}$ be a sequence of finite dimensional subspaces of V, and let P^N denote the orthogonal projections of Z onto V^N. For each N, define operators $A^N : V^N \to V^N$ by

$$\langle A^N u, v\rangle_Z = \sigma(u, v) \qquad \forall u, v \in V^N .$$

We have the following convergence result.

Theorem. *Suppose that the subspaces V^N satisfy the condition*

(C) there exists $s \ge 1$ and $k > 0$ such that for all $v \in \operatorname{dom} A^s$, there exists $v^N \in V^N$ satisfying

i. $|\sigma(u, v - v^N)| \le k \| u \|_Z \| v - v^N \|_V \qquad \forall u \in V$, *and*

ii. $\| v - v^N \|_V \to 0$ *as* $N \to \infty$.

Then $T^N(t)P^N u \to T(t)u$ *for all* $u \in Z$.

This theorem is closely related to, and in the spirit of, results found in [10] and [11]. It is sufficiently distinct that a proof is given in [5]. The result can be applied to the semigroup approximation schemes developed in the previous section.

In order to apply the result for the scheme $\{Z^N,\ P_e^N,\ A_e^N\}$, first observe that $\sigma_e(u,v) = \langle Au, v\rangle_e$ for all $u, v \in \mathrm{dom}A = V$. From a previous observation, it follows that Re $\sigma_e(u,u) \le \gamma \ \| u \|_e^2$, and hence (13) holds. Now, given $\psi \in H^3(-r,0)$, let $\psi_I^N(\theta)$ be the linear spline which interpolates ψ. That is, $\psi_I^N(\theta)$ is continuous, piecewise linear, and satisfies $\psi_I^N(\theta_j^N) = \psi(\theta_j^N)$ for $j = 0,\dots,N$. It is known that $\| \psi - \psi_I^N \|_{L^2(-r,0)} \le \mathcal{O}(\frac{1}{N^2})$ and $\| \frac{d}{d\theta}(\psi - \psi_I^N) \|_{L^2(-r,0)} \le \mathcal{O}(\frac{1}{N})$. Thus if $v = (\psi(0), \psi) \in \mathrm{dom}A^3$, then $\psi \in H^3(-r,0)$, and we can define $v^N = (\psi_I^N(0), \psi_I^N) = (\psi(0), \psi_I^N) \in Z^N$. Observe that $\| v - v^N \|_V^2 = \| \psi - \psi_I^N \|_{H^1(-r,0)}^2 \to 0$ as $N \to \infty$. Further, for any $u = (\phi(0), \phi) \in \mathrm{dom}A$ we have

$$
\begin{aligned}
|\sigma_e(u, v - v^N)| &= |b| e^{-\gamma r} \left| \int_{-r}^{0} e^{-2\gamma\theta} \phi'(\theta)[\psi(\theta) - \psi_I^N(\theta)] d\theta \right| \\
&\le k_1 \ \| \phi \|_{L^2(-r,0)} \| \frac{d}{d\theta}(\psi - \psi_I^N) \|_{L^2(-r,0)} \\
&\le k_2 \ \| u \|_e \| v - v^N \|_V \ .
\end{aligned}
$$

It follows from the theorem that $T_e^N(t)P_e^N u \to T(t)u$ for all $u \in Z$.

A similar argument works for the scheme $\{\widetilde{Z}^N,\ \widetilde{P}_e^N,\ \widetilde{A}_e^N\}$. Note that $\widetilde{\sigma}_e(u,v) = \langle Au, v\rangle_e$ for all $u \in \mathrm{dom}A$, $v \in \widetilde{V}$, and it is straightforward to show that Re $\widetilde{\sigma}_e(u,u) \le \gamma \ \| u \|_e^2$ for all $u \in \widetilde{V}$. As above, if $v = (\psi(0), \psi) \in \mathrm{dom}A^3$, define $v^N = (\psi_I^N(0), \psi_I^N) = (\psi(0), \psi_I^N) \in \widetilde{Z}^N$, and it follows that $\| v - v^N \|_{\widetilde{V}}^2 = \| \psi - \psi_I^N \|_{H^1(-r,0)}^2 \to 0$ as $N \to \infty$. In addition, for any $u = (\phi(0), \phi) \in \mathrm{dom}A$ we have

$$
\begin{aligned}
|\widetilde{\sigma}_e(u, v - v^N)| &= |b| e^{-\gamma r} \left| \int_{-r}^{0} e^{-2\gamma\theta} \phi'(\theta)[\psi(\theta) - \psi_I^N(\theta)] d\theta \right| \\
&\le k_1 \ \| \phi \|_{L^2(-r,0)} \| \frac{d}{d\theta}(\psi - \psi_I^N) \|_{L^2(-r,0)} \\
&\le k_2 \ \| u \|_e \| v - v^N \|_{\widetilde{V}} \ .
\end{aligned}
$$

It follows from the theorem that $\widetilde{T}_e^N(t)\widetilde{P}_e^N u \to T(t)u$ for all $u \in Z$. It is also worth noting that although $\mathrm{dom}A \ne \mathrm{dom}A^*$, it is true that $\mathrm{dom}A \subset \widetilde{V}$ *and* $\mathrm{dom}A^* \subset \widetilde{V}$. Furthermore, $\widetilde{\sigma}_e(u,v) = \langle Au, v\rangle_e$ for all $u \in \mathrm{dom}A$, $v \in \widetilde{V}$, *and* $\widetilde{\sigma}_e(u,v) = \langle u, A_e^* v\rangle_e$ for all $u \in \widetilde{V}$, $v \in \mathrm{dom}A_e^*$. Here A_e^* denotes the adjoint of A computed in the inner product $\langle\cdot,\cdot\rangle_e$. Hence the same argument used to show that $\widetilde{T}_e^N(t)\widetilde{P}_e^N u \to T(t)u$ gives adjoint semigroup convergence $\widetilde{T}_e^{N*}(t)\widetilde{P}_e^N u \to T_e^*(t)u$. This is especially useful (see [6]) for applications to optimal control such as we are interested in.

References

[1] H. Banks and J. Burns; Hereditary control problems: numerical methods based on averaging approximations, *SIAM J. Cont. Opt.* **16**, 1978, 169-208.

[2] H. Banks and F. Kappel; Spline approximations for functional differential equations, *J. Diff. Eqs.* **34**, 1979, 496-522.

[3] H. Banks and K. Kunisch; The linear regulator problem for parabolic systems, *SIAM J. Cont. Opt.* **22**, 1984, 684-698.

[4] R. Fabiano; Stability preserving spline approximations for scalar functional differential equations, *Comp. Math. Appl.*, to appear.

[5] R. Fabiano, in preparation.

[6] J. Gibson; Linear-quadratic optimal control of hereditary differential systems: infinite dimensional Riccati equations and numerical approximations, *SIAM J. Cont. Opt.* **21**, 1983, 95-139.

[7] J. Gibson and A. Adamian; Approximation theory for linear quadratic Gaussian control of flexible structures, *SIAM J. Cont. Opt.* **29**, 1991, 1-37.

[8] J. Hale; *Theory of Functional Differential Equations*, Springer, Berlin, 1977.

[9] E. Infante and J. Walker; A Lyapunov functional for a scalar differential difference equation, *Proc. Royal Soc. Edin.* **79A**, 1977, 307-316.

[10] K. Ito and F. Kappel; On variational formulations of the Trotter-Kato theorem, CAMS Report 91-7, University of Southern California, Los Angeles, 1991.

[11] K. Ito, F. Kappel, and D. Salamon; A variational approach to approximation of delay systems, *Diff. Int. Eqs.* **4**, 1991, 51-72.

[12] F. Kappel and D. Salamon; Spline approximation for retarded systems and the Riccati equation, *SIAM J. Cont. Opt.* **25**, 1987, 1082-1117.

[13] F. Kappel and D. Salamon; On the stability properties of spline approximations for retarded systems, *SIAM J. Cont. Opt.* **27**, 1989, 407-431.

[14] I. Lasiecka and R. Triggiani; Numerical approximations of algebraic Riccati equations for abstract systems modelled by analytic semigroups and applications, *Math. Comp.* **57**, 1991, 639-662.

Implicit Filtering and Optimal Design Problems

P. Gilmore
National High Magnetic Field Laboratory
Florida State University
Tallahassee FL 32306-3016, USA

C. T. Kelley[1]
Center for Research in Scientific Computation and
Department of Mathematics
North Carolina State University
Box 8205, Raleigh NC 27695-8205, USA

C. T. Miller[2] and G. A. Williams
Department of Environmental Sciences and Engineering
105 Rosenau Hall
University of North Carolina
Chapel Hill NC 27599-7400, USA

Implicit filtering is a form of the gradient projection method of Bertsekas in which the stepsize in a difference approximation of the gradient is changed as the iteration progresses. In this way the algorithm is able to avoid certain types of local minima and in some cases find accurate approximations to the global minimum. The algorithm is particularly effective in avoiding local minima that are caused by high-frequency low-amplitude terms in the objective function. In this report we will discuss the algorithm and its theoretical properties. We will also present applications for modeling of subsurface contaminant transport and high-field magnet design.

1. Introduction

In this paper we review the implicit filtering algorithm developed in [7], [9], [18], [19], [22], and [23]. We propose a method of analysis that we expect to be more generally applicable than that in [7] and [9]. We show how this approach may be applied to the implicit filtering algorithm from [7] and [9] in the one-dimensional setting as motivation. We hope that this framework will be applicable to simplex and direct search methods such as those considered in [4], [16], [20], and [21]. Finally, we report on applications of implicit filtering to partial differential equations and the design of high-field magnets.

The optimization problems we consider in this paper have objective functions that are sums of very simple, e.g., smooth and convex, functions and high frequency, often low-amplitude, perturbations that induce local minima. Often the size of

[1]This research was supported by National Science Foundation grants #DMS-9024622 and #DMS-9321938. Computing activity was also partially supported by an allocation of time from the North Carolina Supercomputing Center.

[2]This research was supported by Army Research Office grant #DAALA03-92-G-0111.

these perturbations, which we shall refer to as noise, decays near the minimum of the underlying simple function. Finite simple bound constraints are imposed.

The figure below from [23] is illustrative of this type of objective function. The plot is a graph of negative of the power added efficiency of a simulated semiconductor device against the real and imaginary parts of the second harmonic of load impedance, which are constrained to lie in the interval [0, 80]. Near the optimal point at (0, 0), the objective function is smooth, farther from (0, 0) one can see high frequency oscillations and local minima.

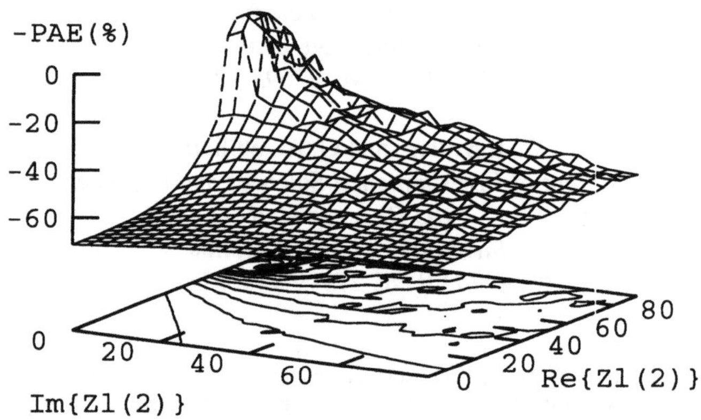

Figure 1. Power added efficiency.

The problem considered in 2.1. has large amplitude "spikes" away from the minimum. Implicit filtering successfully avoided these spikes, as it did in the work reported in [18], [19], [22], and [23], where some of the problems considered in those papers also had such features.

Implicit filtering is a variant of the gradient projection algorithm [1] in which gradients are approximated by finite differences. The stepsize in the finite difference is reduced as the iteration progresses. In this way we hope to "step over" local minima, high frequency effects, or even low amplitude discontinuities. We called the algorithm "implicit filtering" in [7] and [9] because the changing step size in the finite difference approximation of the gradient implicitly filters out high frequency terms in the objective function.

The implicit filtering algorithm as implemented in [7], [9], [18], [19], [22], and [23] was designed for general bound constrained problems. We consider the constrained minimization problem of minimizing a function f on a box in R^N

$$\Omega = \{x \in R^N \mid -l^i \leq x^i \leq u^i, \text{ for } 1 \leq i \leq N\}. \tag{1}$$

Here x^i denotes the ith component of a vector $x \in R^N$ and l and u are the vectors of lower and upper bounds. Following [9] we denote the projection onto Ω by $\mathcal{P}$

where

$$\mathcal{P}(x)^i = \begin{cases} u^i, & x^i > u^i \\ x^i, & l^i \le x^i \le u^i \\ l^i, & x^i < l^i . \end{cases}$$

We denote by $\nabla_h f$ a finite difference approximate gradient of the objective function. For example, forward or centered differences, computed with step length h, might be used. Implicit filtering begins with a difference form of the gradient projection algorithm. In the specification of Algorithm `projgrad`, we use the notation

$$x(\alpha, h, f) = \mathcal{P}(x - \alpha \nabla_h f) . \tag{2}$$

Algorithm 1 *Algorithm* `projgrad`$(f, x, h, \bar{\tau}, \bar{\alpha}, \sigma)$

1. $k = 0$
2. *Compute* $\nabla_h f$.
   (a) *If* $\|x - x(1, h, f)\| \le \bar{\tau} h$, *terminate successfully.*
3. *Set* $\alpha = 1$.
4. (a) *If* $\alpha < \bar{\alpha} = \beta^{\bar{j}}$, *terminate unsuccessfully.*
   (b) *Compute* $f(x(\alpha, h, f))$.
   (c) *If*
   $$f(x) - f(x(\alpha, h, f)) \ge \frac{\sigma \|x - x(\alpha, h, f)\|^2}{\alpha}$$
   set $x = x(\alpha, h, f)$, $k = k + 1$ *and go to step 2.*
   (d) $\alpha = \beta \alpha$

In Algorithm `projgrad` the parameters σ and β are the standard Armijo rule stepsize control parameters [1], [3], [17]. $\bar{\alpha} = \beta^{\bar{j}}$ both safeguards against the case in which $\nabla_h f$ is not a descent direction for f and keeps the number of stepsize reductions bounded by $\bar{j}$. This limit on the number of stepsize reductions means that the line search can fail and `projgrad` can terminate unsuccessfully. This does not mean that we have failed to find an acceptable solution, however, as the analysis below will indicate. We terminate `projgrad` successfully when the current point x and the projection of $x - \nabla_h f$ on the feasible region differ by a multiple of h. The choice of this multiple, $\bar{\tau}$, will be considered later.

Implicit filtering consists of repeated calls to `projgrad` with a decreasing, finite sequence of difference steps $\{h_i\}_{i=1}^m$, called *scales* in [7] and [9].

Algorithm 2 *Algorithm* `imfilter`$(f, x, \{h_i\}, \bar{\tau}, \bar{\alpha}, \sigma)$

- *for* $i = 1, \dots m$
  call `projgrad`$(f, x, h_i, \bar{\tau}, \bar{\alpha}, \sigma)$

The role of the scale, like the temperature in an annealing algorithm, the simplex size in the pattern search algorithms [16], [20], [21], the time step in the explicit filtering algorithm in [10], or the adaptivity parameter in [12], controls the resolution of the optimization. When far from the optimal point, coarse resolution enables one to avoid the nearby high frequency effects and search for a better point far from the current point. This heuristic, of course, can be defeated by a sufficiently pathological objective function. The goal in [9] was to characterize objective functions that could be effectively minimized by such an approach.

In [7] and [9], model problems of the form $f = \bar{f} + \phi$, where $\bar{f}$ was a simple function with a global minimum x^* and ϕ was a small perturbation, were considered. Conditions were given on the parameters $\bar{\tau}$, $\bar{\alpha}$, and σ for successful termination of `projgrad`. Successful termination, in turn, implies that $x - x^* = O(h)$. This is the key step, as the terminal iterate from `projgrad` for a given scale h is the initial iterate for the next (and smaller) scale. In this paper, we relax the requirement that `projgrad` terminate successfully in step 2a and allow for failure of the line search in step 4a. We then try to show that $\|x - x^*\| = O(h)$. One benefit of this approach, in a sense a "backward" form of the analysis in [7] and [9], is that one could apply it to any algorithm that evaluates f on a pattern of points arranged in some way about the current point. For instance, one could ask how a pattern search method with a fixed simplex size would perform when given an objective function of the form (3).

For this paper we consider the simple one dimensional function

$$f(x) = Mx^2/2 + \phi(x) \tag{3}$$

with $M > 0$ and the perturbation $|\phi(x)| \leq \epsilon$.
We will consider only the unconstrained problem and use forward differences.

With forward differences, for a given h and x,

$$\nabla_h f(x) = Mx + Mh/2 + \nabla_h \phi(x) .$$

If ϵ were known, then the optimal choice of h would be $h \approx \sqrt{\epsilon}$. If h were much smaller than $\sqrt{\epsilon}$, $\nabla_h f(x)$ would provide no useful information about f and algorithm `projgrad` could fail completely. We show that if $h \geq \sqrt{\epsilon}$ and the algorithmic parameter $\bar{\tau}$ is sufficiently large, then `projgrad` will terminate with $x = O(h)$, even if the termination is caused by failure in the line search.

Proposition 1 *Let $f(x)$ be given by* (3) *with $M > 0$ and $|\phi(x)| \leq \epsilon$. Assume that*

$$h \geq \sqrt{\epsilon}, 1 - 2(2\sigma + M\bar{\alpha}) > 0, \text{ and } \bar{\tau} > 4 + M . \tag{4}$$

Then `projgrad` *terminates with $x = O(h/M)$.*

Proof: We begin by remarking that if the line search always succeeds, then it is well known that the iteration will terminate successfully. Hence `projgrad` will not loop indefinitely and the only termination modes are success, step 2a, or failure of the line search, step 4a.

We let the parameters $(\bar{\tau}, \bar{\alpha}, \sigma, \beta)$ in `projgrad` be given, with $\bar{\tau}$ satisfying (4). If `projgrad` terminates on entry in step 2a then

$$M|x| \leq \bar{\tau}h + Mh/2 + 2\epsilon/h \,. \tag{5}$$

If ϵ were known, then the optimal choice of h would be $h \approx \sqrt{\epsilon}$. If $h \geq \sqrt{\epsilon}$ then (5) implies

$$|x| \leq (\bar{\tau} + M/2 + 2)h/M \,. \tag{6}$$

Hence termination of `projgrad` on entry would indicate that the minimum had been identified as accurately as possible, given the scale h and the curvature M.

Since we consider an unconstrained problem,

$$x - x(\bar{\alpha}, h, f) = \bar{\alpha}\nabla_h f(x) \,. \tag{7}$$

Unsuccessful termination of `projgrad` means that `projgrad` did not terminate in step 2a. In this case

$$\bar{\tau}h < |\nabla_h f(x)| \leq M|x| + Mh/2 + 2\epsilon/h \leq M|x| + (M/2 + 2)h \,. \tag{8}$$

By (4), $\bar{\tau} > M + 4$, and we can conclude from (8) that

$$|x| \geq C_1 h/M \geq (1/2 + 2/M)h \,, \tag{9}$$

where $C_1 = (\bar{\tau} - M/2 - 2) \geq M/2 + 2$. Therefore, by (9),

$$\begin{aligned} M|x| - Mh/2 - 2h &\leq |\nabla_h f(x)| = |M(x + h/2) + \nabla_h \phi| \\ &\leq M|x| + Mh/2 + 2h \leq 2M|x| \,. \end{aligned} \tag{10}$$

The line search will certainly fail if $x\nabla_h f(x) < 0$. This cannot happen because (10) implies that

$$x\nabla_h f(x) \geq Mx^2 - M|x|h/2 - 2h|x| \geq |x|(C_1 - M/2 - 2)h > 0 \,. \tag{11}$$

Note that (11) depends on (10) which in turn depends on the fact, expressed in (8), that `projgrad` did not terminate in step 2a.

Failure of the line search in step 4a means

$$f(x) - f(x(\bar{\alpha}, h, f)) < \sigma|x - x(\bar{\alpha}, h, f)|^2/\bar{\alpha} \,. \tag{12}$$

By (7),

$$\begin{aligned} f(x) - f(x(\bar{\alpha}, h, f)) &= \tfrac{M}{2}\{x^2 - [x - \bar{\alpha}\nabla_h f(x)]^2\} + \phi(x) \\ &\quad -\phi(x - \bar{\alpha}\nabla_h f(x)) \\ &\geq M\{x^2 - [x - \bar{\alpha}\nabla_h f(x)]^2\}/2 - 2\epsilon \\ &= M[x\bar{\alpha}\nabla_h f(x) - \bar{\alpha}^2\nabla_h f(x)^2/2] - 2\epsilon \,. \end{aligned}$$

Hence, the line search fails only if

$$\begin{aligned} M[x\bar{\alpha}\nabla_h f(x) \quad & -\bar{\alpha}^2\nabla_h f(x)^2/2] - 2\epsilon \\ & \leq f(x) - f(x(\bar{\alpha}, h, f)) < \sigma|x - x(\bar{\alpha}, h, f)|^2/\bar{\alpha} \qquad (13) \\ & = \sigma\bar{\alpha}\nabla_h f(x)^2 . \end{aligned}$$

Recalling our assumption that $\bar{\tau} > 4 + M$ and hence $x\nabla_h f(x) \geq 0$, and using (10), we see that equation (13) implies

$$\begin{aligned} M|x|(M|x| - Mh/2 - 2h) \quad & \leq M|x||\nabla_h f(x)| \\ & \leq (\sigma + M\bar{\alpha}/2)\nabla_h f(x)^2 + 2\epsilon/\bar{\alpha} . \end{aligned} \qquad (14)$$

By (10), (14) implies

$$M^2|x|^2 - M(M/2 + 2)|x|h \leq (2\sigma + M\bar{\alpha})2M^2x^2 + 2\epsilon/\bar{\alpha} .$$

Therefore the line search fails only if

$$x^2 \leq \frac{(M/2 + 2)|x|h + 2\epsilon/(M\bar{\alpha})}{M[1 - 2(2\sigma + M\bar{\alpha})]} . \qquad (15)$$

Since $h \geq \sqrt{\epsilon}$, we may write (15) as

$$x^2 \leq C_2 \max\left(\frac{h|x|}{M}, \frac{h^2}{\bar{\alpha}M^2}\right) \qquad (16)$$

where

$$C_2 = \frac{M/2 + 4}{1 - 2(2\sigma + M\bar{\alpha})} .$$

$C_2 > 0$ by (4). By (9),

$$h/(M|x|) \leq 1/C_1$$

and hence

$$|x| \leq C_2 \max[1, 1/(\bar{\alpha}C_1)]h/M \qquad (17)$$

which completes the proof. □

In this analysis of the failure modes of `projgrad` we assumed that $h \geq \sqrt{\epsilon}$, *i. e.* h is large enough to avoid numerical differentiation of the noise term ϕ. Repeated failures of `projgrad` at a sequence of scales may well indicate that one has reduced h too much.

Note that for both successful and unsuccessful termination the curvature M plays an important role and that for small curvature h and ϵ must be small if algorithm `projgrad` will find a useful solution. Unsuccessful termination indicates that no further progress is possible with the current value of the scale h and that a reduction of the scale is needed for further progress.

In [7] and [9] we proposed restarting `imfilter` until each call to `projgrad` left x unchanged, because if each call to `projgrad` terminates successfully with a solution x, there is no guarantee that a second call to `imfilter` would leave x invariant since ϕ could change the output from the early calls to `projgrad`. In practice, we have never found that restarting after convergence changed x, but restarting was necessary for the theory in [7] and [9]. The analysis here for problems of the form (3), however, shows that $x = O(h_m + \epsilon/h_m)$ after `imfilter` terminates. Hence restarts are not necessary in this case.

The idea of the analysis here is that the essential properties of any objective function f for algorithms like implicit filtering are curvature M and the size of the perturbation ϵ. We believe that the analysis above can be generalized in several ways. While here we considered a one-dimensional model problem and did not take constraints into account, such an extension can be done directly. Dependence of ϵ on $x - x^*$ would add some complexity, but could be incorporated into the analysis in a way similar to that in [9]. With h playing the role of simplex size, a similar analysis could be done for pattern search methods. We will explore these extensions in future work.

2. Applications

In this section we discuss two applications to illustrate that the structure we discussed in 1. does indeed arise in practice. The full details of the applications have been or will be presented elsewhere. We also refer the reader to [18], [19], [22], and [23] for more discussion of applications.

2.1. Optimization of Petrov-Galerkin coefficients

Petrov-Galerkin methods are finite element methods that use test functions of a higher order than the basis functions. We consider some recent results from [13] and [14] in which Petrov-Galerkin methods are applied to advective-dominated transport problems. The goal of the method, in which piecewise linear basis functions are used with piecewise cubic test functions, is to reduce the distortion of fronts associated with typical upstream weighting formulations.

We consider the advective-dispersive equation in one space dimension

$$\frac{\partial C}{\partial t} = D_x \frac{\partial^2 C}{\partial x^2} - v_x \frac{\partial C}{\partial x} \tag{18}$$

where C is the solute concentration, D_x is the longitudinal hydrodynamic dispersion coefficient, and v_x is the mean solute pore velocity in the x direction. Dirichlet-Neumann boundary conditions are imposed.

Letting $\{\phi_i\}$ denote the standard finite element basis of "hat functions" centered at node x_i, consider test functions of the form

$$\psi_i = \phi_i + \alpha M_{2i} + \beta M_{3i}$$

where M_{2i} and M_{3i} are appropriate scalings and translates of the functions

$$M_2(x) = \begin{cases} -3x(x+1), & -1 \le x \le 0 \\ 3x(x-1), & 0 \le x \le 1 \end{cases}$$

and

$$M_3(x) = \begin{cases} 5x(2x+1)(x+1)/2, & -1 \le x \le 0 \\ 5x(2x-1)(x-1)/2, & 0 \le x \le 1 \end{cases}$$

which are defined on $[-1, 1]$. The nonsymmetry of the test functions makes this scheme an upstream method.

We considered test problems with initial conditions corresponding to a Gaussian source having standard deviation SD. These problems have known analytic solutions. For given spatial mesh width Δx and time step Δt we seek to minimize the error as a function of the Courant number,$Cr = v_x \Delta t/\Delta x$; the Peclet number, $Pe = v_x \Delta x / D_x$; and the dimensionless standard deviation of the source, $\bar{\sigma}_x = SD/\Delta x$. As reported in [13], the use of Taylor series analysis to eliminate the lowest order terms in the expansion of truncation error in powers of Δx did not lead to the minimal error. Moreover, the effects of the higher order terms increased with $\bar{\sigma}_x$. It was also found that the optimal α and β were time dependent. In [13] this time dependence was incorporated into α and β by allowing them to depend on C in the spatial interval $[x_i, x_{i+1}]$ through the relations

$$\alpha = \alpha_0 + \alpha_1 \frac{C_{i+1} - C_i}{\max\{|C_{i+1}|, |C_i|\}} \text{ and } \beta = \beta_0 + \beta_1 \frac{C_{i+1} - C_i}{\max\{|C_{i+1}|, |C_i|\}}.$$

α and β are time dependent because C is.

Hence, the minimization problem to be solved is to find the values of α_0, α_1, β_0, and β_1 that minimize the error for fixed values of Cr, Pe, and $\bar{\sigma}_x$. In [13] a Levenberg-Marquardt code was used. In this report we compare the results of implicit filtering with the results of the Levenberg-Marquardt code. In many cases, the Levenberg-Marquardt iteration found local minima that were dependent on the initial iterate and far from optimal. We report on one such case here. In all cases, the initial iterate was the zero vector. In Figure 2 we compare the analytic solution and the solutions found with the Levenberg-Marquardt code and implicit filtering (using a C-language implementation of the IFFCO code from [7] and [8]). The Levenberg-Marquardt result is poor and corresponds to a local minimum. The optimal point, as computed with Levenberg-Marquardt, was

$$(\alpha_0, \beta_0, \alpha_1, \beta_1) = (0.43, -0.265, 1.03, 1.16)$$

with an objective function value of 77.7. The optimal point, computed with implicit filtering, was

$$(\alpha_0, \beta_0, \alpha_1, \beta_1) = (0.0249, 0.230, 1.36, 0.00191)$$

with an objective function value of 0.962.

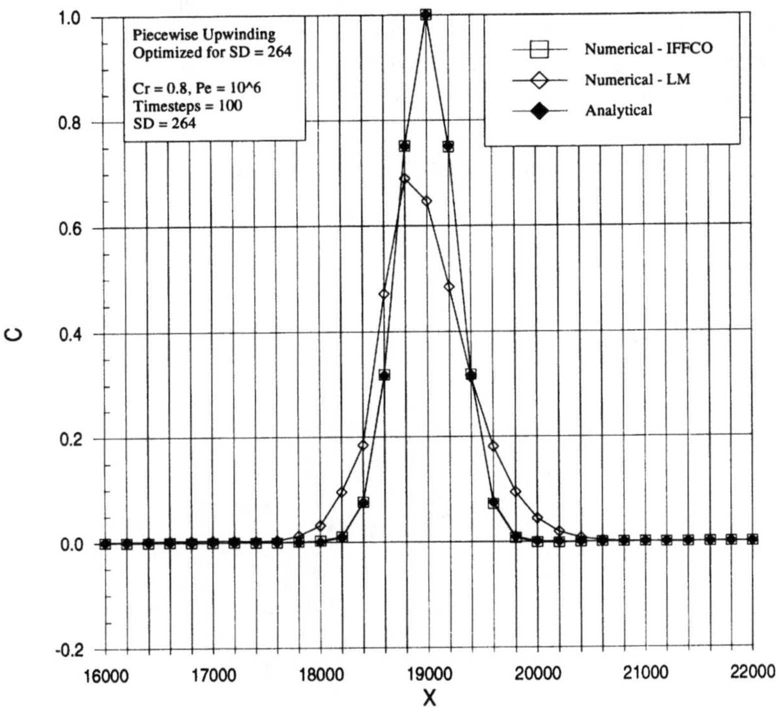

Figure 2. Comparison of errors.

We can examine the graph of the objective function in a neighborhood of the terminal points for both the implicit filtering and Levenberg-Marquardt iterations. In Figures 3 and 4, we plot the error as a function of α_0 and β_0, varying them about the terminal point. In Figures 5 and 6 we do the same for α_1 and β_1. The smooth graph near the implicit filtering solution and the spikes near the Levenberg-Marquardt solution are indicative of the type of problem that implicit filtering is intended to solve.

2.2. Design of high-field magnets

This subsection describes how IFFCO, the FORTAN implementation of implicit filtering, [7], [8] is being used for the design of high-field pulsed magnets at the National High Magnetic Field Laboratory (NHMFL). Pulsed magnets are multi-layer solenoids. Each layer of the magnet consists of a sublayer of conductor and a sublayer of reinforcement. The sublayer of conductor is made up of tightly wound insulated conducting wire. The internal reinforcement is made by winding glass or carbon fibers around the conductor sublayer. The magnet is usually surrounded by a shell of some strong stiff material, such as steel or carbon fiber. The pulse of magnetic field is created by discharging a capacitor in series with the magnet. The magnet is often in parallel with a crowbar resistor, which helps absorb energy in the circuit after the magnet reaches peak field.

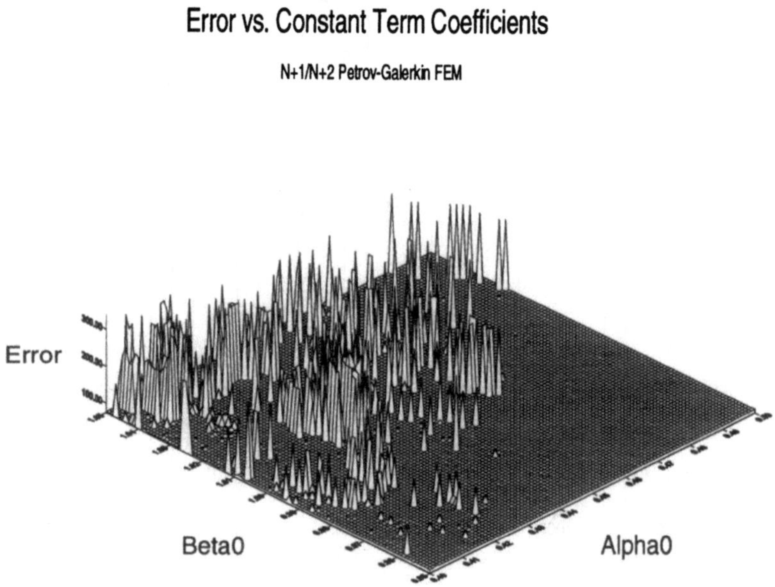

Figure 3. Error surface near Levenberg-Marquardt solution.

The object of the pulsed magnet design optimization project at NHMFL is to find a design that obtains the highest possible field for a given energy and capacitance. The most common cause of magnet destruction is mechanical failure due to stresses within the coil created by the Lorentz forces F

$$F = \hat{I} \times \hat{B}\,. \tag{19}$$

In (19), $\hat{I}$ represents the current and $\hat{B}$ represents the field. The axial component of the field interacts with the current to create a radial body force in the conductor. This radial force tends to displace the conductor outward against the following layers of the magnet. The radial component of the field interacts with the current to create an axially directed force, which tends to displace the windings towards the midplane of the magnet [15]. Mechanical failure occurs when the total stress in the reinforcement exceeds the ultimate stress of the reinforcement material, or when the strain in the conductor exceeds the maximum strain of the conductor material.

Even if the conductor itself does not fail, excessive conductor strain may damage the insulation, leading to electrical failure of the magnet. Much of the energy in the magnet is dissipated as Joule heat. Excessive heat is a frequent cause of insulation failure.

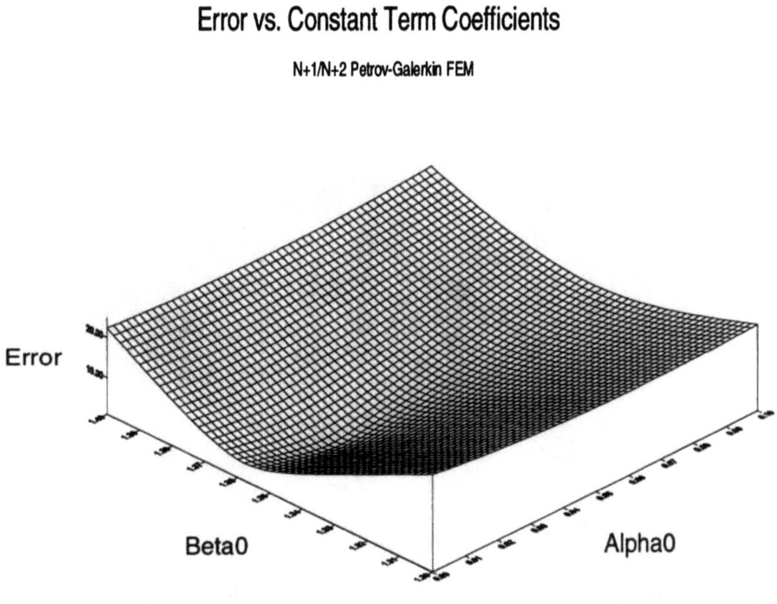

Figure 4. Error surface near implicit filtering solution.

2.2.1. Pulsed magnet modeling

The NHMFL pulsed magnet design optimization package consists of three codes: IFFCO; LHEAT, a code for analyzing the temperature and electro-magnetic behavior of the magnet; and CYCLE, a code for calculating the stress and strain in the magnet at peak field. LHEAT and CYCLE use closed form expressions for the equations that they solve. LHEAT uses the general solution to the circuit equation (20)

$$\frac{d^2 I}{dt^2} + \left(\frac{L + RR_cC}{R_cCL}\right)\frac{dI}{dt} + \frac{R + R_c}{R_cCL} I = 0\,, \tag{20}$$

along with temperature and resistance calculations at a sequence of time steps up through the current pulse, to calculate magnet performance. In (20), I represents the current, L the magnet inductance, R_c the crowbar resistance, and R the AC magnet resistance taking into consideration eddy currents and magneto resistance [5].

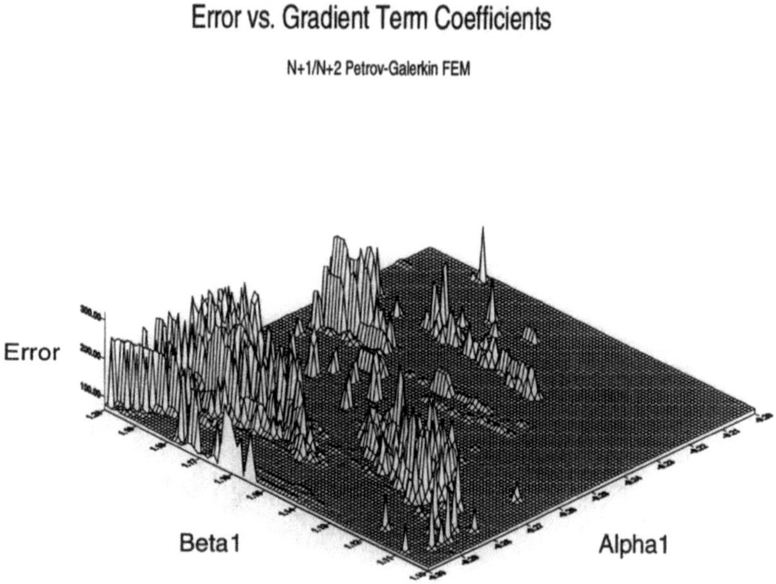

Figure 5. Error surface near Levenberg-Marquardt solution.

CYCLE calculates the stress and strain in the magnet by solving the coupled algebraic differential equations given by equations (21), (22), and (23) at a sequence of load increments until the peak load is reached.

$$\begin{pmatrix} \epsilon_\theta \\ \epsilon_r \\ \epsilon_z \end{pmatrix} = \begin{pmatrix} \frac{\sigma_\theta}{E_\theta} & -\nu_{r\theta}\frac{\sigma_r}{E_r} & \nu_{z\theta}\frac{\sigma_z}{E_z} \\ -\nu_{\theta r}\frac{\sigma_\theta}{E_\theta} & \frac{\sigma_r}{E_r} & -\nu_{zr}\frac{\sigma_z}{E_z} \\ -\nu_{\theta z}\frac{\sigma_\theta}{E_\theta} & -\nu_{rz}\frac{\sigma_r}{E_r} & \frac{\sigma_z}{E_z} \end{pmatrix} \begin{pmatrix} \sigma_\theta \\ \sigma_r \\ \sigma_z \end{pmatrix} . \tag{21}$$

In (21), each ϵ represents a component of strain, each σ a component of stress, each E a component of the modulus of elasticity, and each ν a component of Poisson's ratio [11]

$$\epsilon_\theta = \frac{u}{r} \quad \epsilon_r = \frac{du}{dr} . \tag{22}$$

In (22), u indicates the radial displacement of the material.

$$\frac{d}{dr}(r\sigma_r) - \sigma_\theta + rJB_z = 0 . \tag{23}$$

In (23), J indicates the current density, and B_z indicates the component of the magnetic field in the axial direction.

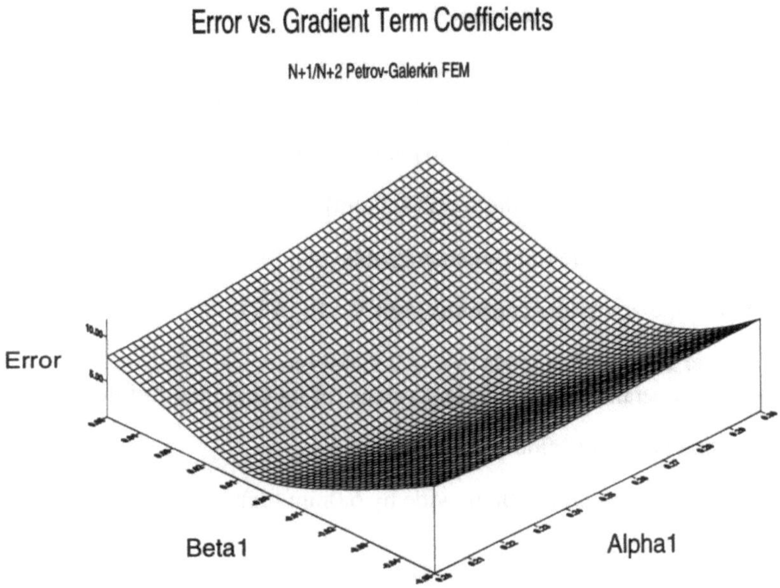

Figure 6. Error surface near implicit filtering solution.

2.2.2. Pulsed magnet optimization

As mentioned earlier, the goal in pulsed magnet optimization is to maximize the magnetic field at the center of the magnet. Because IFFCO is designed for minimization problems, the objective function for the pulsed magnet problem is $-B(r = 0, z = 0)$. The physical constraints imposed by the temperature and the Lorentz forces give rise to mathematical constraints in the formulation of the minimization problem. The physical constraints and the number of nonlinear constraints each causes are listed below; we also provide typical upper and lower bounds for the constraints.

- The conductor cannot strain beyond a certain percentage, depending on the material. Values for this percentage vary between 30.0% for copper to 3.0% for copper silver. For an n-layer magnet this leads to n nonlinear constraints.
- The von Mises stress in the reinforcement cannot exceed the ultimate stress for the material. Values range from 1.0GPa for steel to 4.5GPa for glass fiber. This leads to n nonlinear constraints.
- The radial stress in the conductor cannot exceed the ultimate stress for the material. Values range from 0.3GPa for copper to 1.0GPa for copper silver. This leads to n nonlinear constraints.

- The temperature in the conductor cannot exceed 650.0K. This requirement adds one nonlinear constraint.
- The time to peak field must exceed some designer-supplied minimum, ranging from 0.0ms to 20.0ms. This requirement adds one nonlinear constraint.

Currently the variables and their typical bounds are the following:

- the magnet height, [50.0mm , 200.0mm]
- the thickness of the reinforcement sublayers, [0.1mm , 10.0mm]
- the voltage across the capacitor [5.0kV, 10.0kV].

For an n-layer magnet, this leads to $n + 2$ variables.

Some fixed parameters and typical values for them are the following:

- the capacitance (30.0mF)
- the wire dimensions (4.0mm wide by 6.0mm high)
- conductor materials
  - copper
  - glidcop (copper doped with aluminum oxide)
  - copper silver
- reinforcement materials
  - glass fiber
  - carbon fiber
  - steel (for the shell)
- resistance of the crowbar resistor (0.25Ω).

The mathematical formulation of the pulsed magnet optimization problem is given in (24).

$$\begin{aligned} \text{Minimize} \quad & -B(x), \quad x \in Q = \{x \mid l_i \leq x_i \leq u_i\} \\ \text{Subject to} \quad & g_i(x) \leq b_i, \quad i = 1, \ldots, m\,. \end{aligned} \tag{24}$$

In (24), Q represents the hyper-box defined by the variable constraints and the g_i are the various nonlinear constraints. Using quadratic penalty functions [6], we reformulate (24) into a problem that IFFCO can solve. This formulation is given in (25).

$$\begin{aligned} &\text{Minimize } f_k(x),\ k = 1, \ldots K,\ \ x \in Q = \{x \mid l_i \le x_i \le u_i\} \\ &f_k(x) = -B(x) + r_k \textstyle\sum_{i=1}^{m} \alpha_i \left(\frac{g_i(x)-b_i+|g_i(x)-b_i|}{2}\right)^2 . \end{aligned} \tag{25}$$

The r_k in (25) satisfy

$$r_k \longrightarrow \infty \text{ as } k \longrightarrow \infty.$$

For small k, r_k is small and the effect of the nonlinear constraints is small near the feasible region. For large k, r_k is also large and the constraints dominate the calculation outside of the feasible region. Note that for a point in the feasible region, the constraints have no effect on the calculation of the merit function. A typical sequence $\{r_k\}$ is $\{1, 2, 5, 10, 17, 26, 37, 50, 65, 82, 100\}$.

2.2.3. Conclusions

The problems that arise in pulsed magnet design optimization are discontinuous, have many local minima, and are also expensive to solve. An optimization code designed for smooth problems, LANCELOT, [2], was applied to an earlier version of the pulsed magnet design problem. In general, LANCELOT was unable to converge to an acceptable solution. This is not surprising given the discontinuities and local minima of the merit function. We also tested IFFCO in "noise-free" mode, setting the scale to a small value, and also failed to find an acceptable solution. In this mode IFFCO is a a finite difference projected SR1 code [7], [8]. The failure of LANCELOT and IFFCO when used in noise-free mode indicates that the current formulation of the pulsed magnet problem is one of global optimization. Figure 7 shows the merit function with $r_k = 5.0$ of a four-layer magnet, plotted over a two-dimensional subplane of the hyper-box. The heavy black square indicates the final iterate of the optimization using IFFCO.

As mentioned earlier, the function evaluations are fairly expensive. A typical evaluation takes approximately 10 seconds of CPU time on a dedicated IBM RISC 6000 580. Because of the expensive function evaluations, a fast global optimization code is critical. IFFCO has been very successful in solving these problems and usually converges to a point within the feasible region that has a high field value, relative to both existing magnet designs and typical field values within the hyper-box. A typical optimization takes about 6000 function evaluations (overnight on a RISC 6000 580).

The optimization software has produced a number of high-field pulsed magnet designs which are currently undergoing experimental analysis.

Acknowledgment

The authors thank Julie N. Straight for her editorial assitance with the preparation of this paper.

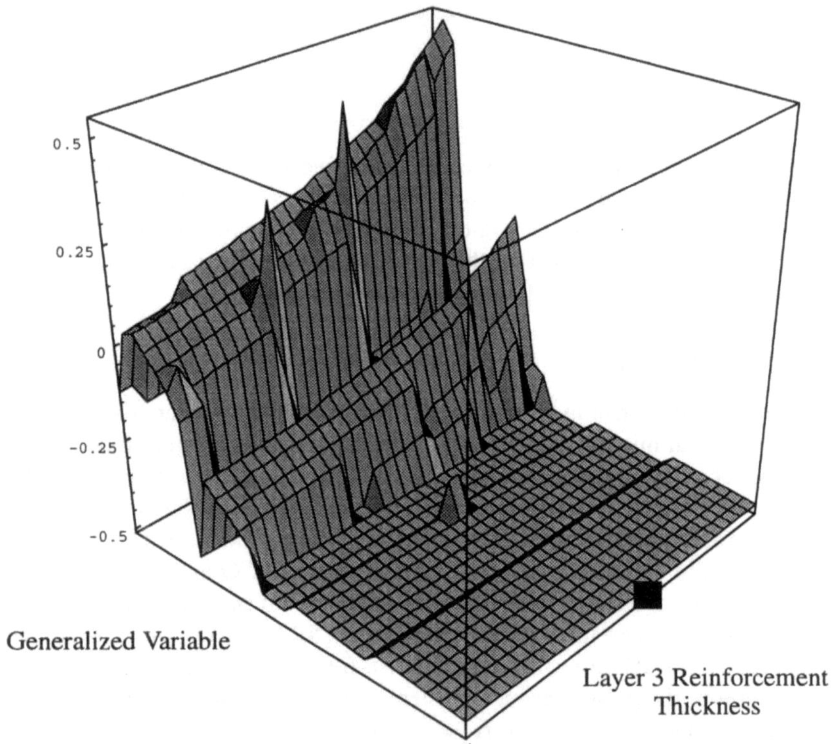

Figure 7. Merit function plotted over a 2D subplane of the hyper-box.

References

[1] D. Bertsekas; On the Goldstein-Levitin-Polyak gradient projection method, *IEEE Trans. Autom. Cont.* **21**, 1976, 174–184.

[2] A. Conn, N. Gould, and P. Toint; LANCELOT: a fortran package for large-scale nonlinear optimization (Release A), in Springer Series in Computational Mathematics No. 17, Springer, Berlin, 1992.

[3] J. Dennis and R. Schnabel; *Numerical Methods for Nonlinear Equations and Unconstrained Optimization*, Prentice-Hall, Englewood Cliffs, 1983.

[4] J. Dennis and V. Torczon; Direct search methods on parallel machines, *SIAM J. Optim* **1**, 1991, 448 – 474.

[5] Y. Eyssa; Electro-magnetic and temperature analysis of pulsed magnets, to appear in *Proc. 14th International Conference on Magnet Technology*, 1995.

[6] A. Fiacco and G.. McCormick; *Nonlinear Programming*, Wiley, New York, 1968.

[7] P. Gilmore; *An Algorithm for Optimizing Functions with Multiple Minima*, PhD thesis, North Carolina State University, Raleigh, 1993.

[8] P. Gilmore; IFFCO: implicit filtering for constrained optimization, Tech. Report CRSC-TR93-7, Center for Research in Scientific Computation, North Carolina State University, Raleigh, 1993. (Available by anonymous ftp from math.ncsu.edu in pub/kelley/iffco/ug.ps.)

[9] P. Gilmore and C. Kelley; An implicit filtering algorithm for optimization of functions with many local minima, *SIAM J. Optim.*, to appear.

[10] J. Kostrowicki and L. Piela; Diffusion equation method of global minimization: Performance for standard test functions, *J. Optim. Theo. Appl.* **50**, 1991, 269–284.

[11] W. Markiewicz, M. Vaghar, I. Dixon, H. Garmestani, and J. Jimeian, Generalized plane strain analysis of solenoid magnets: Formulation and examples, Tech. Report NHMFLIR 93-788, National High Magnetic Field Laboratory, 1993.

[12] D. Mayne and E. Polak; Nondifferential optimization via adaptive smoothing, *J. Optim. Theo. Appl.* **43**, 1984, 601–613.

[13] C. Miller and F. Cornew; A Petrov-Galerkin method for resolving advective-dominated transport, in *Proceedings of Computational Methods in Water Resources IX, Denver, Colorado, Vol. 1 Numerical Methods in Water Resources*, Ed. by T. Russell, R. Ewing, C. Brebbia, W. Gray, and G. Pinder, Computational Mechanics Publications, Southhampton and Boston and Elsevier Applied Science, London and New York, 1992, 157–164.

[14] C. Miller and A. Rabideau; Development of split-operator, Petrov-Galerkin methods to simulate transport and diffusion problems, *Water Resour. Res.* **29**, 1993, 2227–2240.

[15] D. Montgomery, *Solenoid Magnet Design*, Krieger, Malabar, 1980.

[16] J. Nelder and R. Mead; A simplex method for function minimization, *Comput. J.* **7**, 1965, 308–313.

[17] J. Ortega and W. Rheinboldt; *Iterative Solution of Nonlinear Equations in Several Variables*, Academic, New York, 1970.

[18] D. Stoneking, G. Bilbro, R. Trew, P. Gilmore, and C. Kelley; Yield optimization using a GaAs process simulator coupled to a physical device model, *IEEE Trans. Microwave Theo. Tech.* **40**, 1992, 1353–1363.

[19] D. Stoneking, G.. Bilbro, R. Trew, P. Gilmore, and C. Kelley; Yield optimization using a GaAs process simulator coupled to a physical device model, in *Proc. IEEE/Cornell Conference on Advanced Concepts in High Speed Devices and Circuits*, IEEE, 1991, 374–383.

[20] V. Torczon; On the convergence of the multidimensional direct search, *SIAM J. Optim.* **1**, 1991, 123–145.

[21] V. Torczon; On the convergence of pattern search methods, Tech. Report TR93-10, Department of Computational and Applied Mathematics, Rice University, Houston, 1993.

[22] T. Winslow, R. Trew, P. Gilmore, and C. Kelley; Doping profiles for optimum class B performance of GaAs mesfet amplifiers, in *Proc. IEEE/Cornell Conference on Advanced Concepts in High Speed Devices and Circuits*, IEEE, 1991, 188–197.

[23] T. Winslow, R. Trew, P. Gilmore, and C. Kelley; Simulated performance optimization of GaAs MESFET amplifiers, in *Proc. IEEE/Cornell Conference on Advanced Concepts in High Speed Devices and Circuits*, IEEE, 1991, 393–402.

Generating Conjugate Directions Using Limited Second Derivatives

Andreas Griewank and Manfred Fruth
Institute of Scientific Computing
Technical University Dresden
and
Mathematics and Computer Science Division
Argonne National Laboratory

Gradients of objective functions in very many variables can be evaluated cheaply by the reverse, or adjoint, mode of automatic differentiation. One can simultaneously evaluate arbitrary Hessian–vector products with only a constant increase in complexity. These limited second derivatives can be used within a truncated Newton code or to improve nonlinear conjugate gradient codes with respect to the aspects: stepsize prediction, search direction conjugacy, restart criteria. We report preliminary results with an experimental conjugate gradient implementation.

1. Introduction

Virtually all matrix-free optimization methods are variations of the conjugate gradient method for unconstrained minimization. Especially in applications with functional parameters or 'unknowns' that need to be discretized for practical calculations, the optimization of designs and/or identification of parameters requires the minimization of scalar valued functions in very many variables. We will assume here that, at least in some neighborhood of their local minimizers, these functionals $f : \mathbf{R}^n \mapsto \mathbf{R}$ are sufficiently smooth that quadratic approximations based on the gradient ∇f and Hessian $\nabla^2 f$ make sense. It is known that $\nabla^2 f$ can only be sparse if f is the sum of several element functions, whose Hessians form structurally dense submatrices of $\nabla^2 f$. In stationary finite element calculations these dense sub–blocks may be quite small, in which case the evaluation, storage and manipulation of $\nabla^2 f$ in this partitioned element–by–element form may be feasible and algorithmically useful. We will assume here that the number of nonzero entries in $\nabla^2 f$ is so large and the sparsity pattern so irregular that forming and factoring the overall Hessian in any way is out of the question. Formally we may assume that

$$\frac{\partial^n f}{\partial x_1 \partial x_2 \ldots \partial x_n} \not\equiv 0 \quad \text{with} \quad n \gg 1,$$

which ensures that f is not partially separable [9].

This scenario applies for example in the 4D data assimilation problem [8], in weather forecasting, and in other time–dependent inverse problems. The number of variables in these problems is always in the thousands and can easily exceed a million, so that the number of minimization steps cannot even come close to the dimension of the search space. On the other hand, one is satisfied with a low accuracy solution and can assume some eigenvalue clustering in the Hessian. Typically, no more than 30 conjugate gradient type iterations are performed.

Derivative complexities

Under the conditions sketched above, gradient values are indispensable for a successful minimization. As has long been realized in the 4D assimilation case, with hand coded adjoints one can always calculate $\nabla f(x)$ in the so–called reverse mode of automatic differentiation with an operations count no greater than five times that of $f(x)$. Due to various overheads, this theoretical factor may be somewhat larger for actual implementations and it is also enlarged if repeated forward calculations from suitable checkpoints are used to reduce the potentially very large memory requirement of the reverse mode [4]. On the other hand, carefully crafted adjoint codes are quite likely to yield gradients at a complexity no greater than twice that of the underlying scalar function.

In any case we may assume that evaluating $\nabla f(x)$ costs only a little more than computing $f(x)$. The same is true for the Hessian–vector products

$$\nabla^2 f(x)\, v \quad = \quad \nabla_x \left(v^T \nabla f(x)\right) \quad = \quad \frac{\partial}{\partial t} \nabla f(x + tv)\bigg|_0$$

for any given vector $v \in \mathbf{R}^n$. This gradient of the directional derivative $\nabla f^T v$, sometimes called a it second order adjoint, can easily be added to the tangent linear code that evaluates $\nabla f(x)^T v$. Alternatively, $\nabla^2 f(x)\, v$ can be calculated in the forward mode as a directional derivative of the gradient $\nabla f(x)$, provided a code for the evaluation of that vector function is available. This was the case for our test calculations reported in Section 4, where Fortran programs for evaluating both function and gradients where directionally differentiated using the preprocessor ADIFOR [1]. Identical numerical results were obtained by differentiating the pair f, $\nabla f(x)^T v$ in the reverse mode as implemented in the package ADOL-C [5] and applied to C versions of the same evaluation codes. This approach comes closer to our target scenario but resulted in longer run–times due to various redundancies and overheads.

Evaluating n of the vectors $\nabla^2 f(x)\, v$ with v ranging over the Cartesian basis vectors yields the full Hessian with, in general, minimal complexity (up to a constant). Within our scenario, this effort far exceeds the total amount of computation we can perform throughout the whole minimization calculation, even if we had a way of storing the resulting huge array of numbers. Therefore we have to make do with second order information in just a few selected directions v. In truncated Newton methods a sequence of directions v which need to be multiplied by the local Hessian is generated recursively. Since the various v are not known *a priori*, this requires repeated forward and reverse sweeps, i.e., executions of the tangent linear and second order adjoint codes. Since each of these double sweeps may entail very significant data transfer and other overheads, we will aim at an algorithm for which only one of them must be executed per major iteration. This rules out Newton iterative methods but allows for nonlinear generalizations of block solvers, which iterate on subspaces rather than individual directions. In the latter case a family of directions v is known at the beginning of the double sweep evaluation at each new major iterate.

What can be improved in CG?

In this paper we will only consider the unblocked case, which restricts our choices to modifications of the classical conjugate gradient scheme. We believe that the following three aspects of nonlinear CG algorithms can be improved on the basis of selected second derivative information:

1. Prediction of Step–Size
2. Conjugacy of Directions
3. Criteria for Restarting

The simplest formulations of nonlinear CG are based on unit trial steps, whose actual length is strongly dependent of the scaling of the objective function. More sophisticated strategies, such as the Phua–Shanno[7] step selection, incorporate curvature information from previous steps, which tends to be a conservative choice as the eigenvalues are successively eliminated from the top. Consequently, one usually will need at least one extra function evaluation since conjugacy is thought to require a reasonably accurate line–search. In some implementations, at least one readjustment of the step size α_k based on parabolic interpolation is always performed to ensure finite termination in up to n steps on quadratics. This rather pleasing theoretical property of CG is of limited practical importance when n is large. The key properties of conjugate direction methods for linear problems, namely orthogonality of gradients and conjugacy of search directions, cannot be achieved or are not even well–defined on nonlinear problems. Ideally, one would probably want the search directions to be conjugate with respect to the Hessian $\nabla^2 f(x_*)$ at the (local) minimum x_*. For obvious reasons, one must settle for conjugacy of two successive search–directions d_k, d_{k+1} with respect to the Hessian $\nabla^2 f(x_{k+1})$ at the iterate x_{k+1} between them. Without exact second derivatives, approximate conjugacy can be achieved by estimating

$$\nabla^2 f(x_{k+1})d_k \approx \left[\nabla f(x_{k+1}) - \nabla f(x_k)\right] / \alpha_k \, , \tag{1}$$

which leads to the classical Polak–Ribière formula [6].

Descent and curvature

One advantage of this difference formula is that the positive curvature condition $y_k^T d_k > 0$ can always be achieved by a sufficiently accurate line–search. It serves the dual purpose of making $|g_{k+1}^T d_k|$ with $g_{k+1} \equiv \nabla f(x_{k+1})$ much smaller than $-g_k^T d_k \approx \| g_k \|^2$ and ensuring that the latter relationship applies also at the new iterate x_{k+1}. Strictly speaking, the Wolfe condition[10]

$$|g_{k+1}^T d_k| < \bar{\sigma} |g_k^T d_k| \quad \text{with} \quad \bar{\sigma} < 1/2$$

only ensures $g_{k+1}^T d_{k+1} < 0$ in case of the Fletcher–Reeves [2] formula

$$d_{k+1} = -g_{k+1} + \beta_k d_k \tag{2}$$

with

$$\beta_k = \| g_{k+1} \|^2 / \| g_k \|^2 \ . \tag{3}$$

In practice a reasonably small $\bar{\sigma} \approx 0.01$ also yields descent for the numerically more successful Polak–Ribière formula

$$\beta_k \equiv y_k^T g_{k+1} / y_k^T d_k \tag{4}$$

which is only equivalent to (3) when f is quadratic and $d_0 = -g_0$. In the calculation to be reported here we used the Polak–Ribière formula with

$$y_k \equiv \nabla^2 f(x_{k+1}) d_k \tag{5}$$

so that the two successive search directions d_k and d_{k+1} are exactly conjugate with respect to the Hessian at x_{k+1}. Inadvertently, we once ran the program with $y_k = \nabla^2 f(x_k) d_k$ so that conjugacy was enforced with respect to the Hessian at the previous iterate. This modification made no difference on quadratics, but led to a serious deterioration in the performance on general problems. Compared to the customary approximation our exact conjugacy led to some notable but certainly not dramatic improvements.

There have been extensive theoretical and experimental investigations into the if's and how's of restarting, but apparently no consensus approach has emerged. We believe that the availability of second and higher derivative information should help with the design of rational restart procedures, but we have not yet been able to find a convincing answer.

The paper is organized as follows: in Section 2 we develop a quadratic predictor, which yields initial guesses for the stepsizes that are asymptotically of second order; in Section 3 we utilize a "hypothetical point" to ensure descent for virtually arbitrary y_k and extremely weak line–searches; Section 4 reports the results of numerical experiments on the test set used by Gilbert and Nocedal [3] in their seminal paper on the global convergence of conjugate gradient methods.

2. Quadratic prediction

Suppose that at the k–th iteration we have available the information

$$x_k, \quad \tilde{f}_k \approx f(x_k), \quad \tilde{g}_k \approx \nabla f(x_k), \quad d_k, \quad \tilde{y}_k \approx \nabla^2 f(x_k) d_k, \quad \tilde{z}_k \approx \nabla^2 f(x_k) g_k .$$

The key element, which is not found in classical CG implementations, is the curvature vector $\tilde{y}_k$. The tildes indicate that on most iterations x_k will be a *hypothetical* point in the sense that the attributed values of f and its derivatives are obtained by extrapolation from a nearby trial evaluation point. If it subsequently transpires that the discrepancies between the actual and extrapolated values are too large, the hypothetical point may be abandoned and the iteration continues from the nearby *established* point. Some explanation of the hypothetical point mechanism can be found in Section 3.

Assuming that $\tilde{g}_k$ and $\tilde{y}_k$ are exact and that the objective f is in fact quadratic, we conclude that it attains its minimum along d_k at

$$x_{k+1} = x_k + \alpha_k d_k \tag{6}$$

where

$$\alpha_k \equiv -\tilde{g}_k^T d_k / \tilde{y}_k^T d_k \tag{7}$$

is positive, provided

$$\tilde{g}_k^T d_k < 0 \tag{8}$$

and

$$0 < \tilde{y}_k^T d_k . \tag{9}$$

The "descent" condition (8) will always be met by our definition of d_k in the following section. The "curvature" condition (9) may not be satisfied if $\nabla^2 f(x_k)$ is actually indefinite or $\tilde{y}_k$ is a poor approximation to $\nabla^2 f(x_k) d_k$. Neither contingency is likely to occur during the final approach to a local minimum and when it does we simply reset the search direction to that of steepest descent. In view of the desirable possibility that x_{k+1} is in fact an acceptable new iterate we estimate the next gradient as

$$\tilde{g}_{k+1} = \tilde{g}_k + \alpha_k \tilde{y}_k \tag{10}$$

and evaluate using a single forward sweep

$$f_{k+1} = f(x_{k+1}), \quad \delta_k \equiv \nabla f(x_{k+1})^T d_k, \quad \tilde{\gamma}_k \equiv \nabla f(x_{k+1})^T \tilde{g}_{k+1} . \tag{11}$$

The second directional derivative is of no use in itself; its calculation is solely in preparation for a subsequent reverse sweep, which is performed only if x_{k+1} is acceptable. To answer that question we employ the classical Goldstein–Armijo test by requiring that

$$\psi_k \equiv (f_{k+1} - \tilde{f}_k)/(\alpha_k \tilde{g}_k^T d_k) \geq \sigma \in (0, \tfrac{1}{2}) . \tag{12}$$

This condition enforces a reduction in the function that is significant relative to the linearly extrapolated gain $\alpha_k \tilde{g}_k^T d_k$. Even when x_k is 'only' a hypothetical point, $\tilde{g}_k^T d_k$ is negative and $\tilde{f}_k$ always represents the smallest function value attained so far. Hence f_{k+1} must represent an improvement if the trial value α_k meets this test. Otherwise, we reset x_k to the last *established* point (where actually $f = \tilde{f}_k$) and determine the search direction so that the just computed trial point x_{k+1} lies on the search ray.

A very simple line–search procedure based on parabolic interpolation yields the recursive formula

$$\alpha_k \leftarrow \tfrac{1}{2}\alpha_k/(1-\psi_k)\,, \tag{13}$$

which is stationary exactly when $\psi_k = \frac{1}{2}$. While it is quite likely that one or two such adjustments will yield an α_k satisfying $\psi_k \geq \sigma$, one might prefer to utilize a higher order interpolant based on extra information in the form of the first directional derivative δ_k at $\alpha = \alpha_k$ and the approximate second derivative $\tilde{y}_k^T d_k$ at $\alpha = 0$. Thus we have altogether five pieces of data, which one might use to fit a quartic or a rational function, e. g. a cubic divided by a linear factor. We have chosen the latter approach with the hope that the pole facilitates better approximations to exponential terms or barrier functions. The question of how to best perform a higher order line search warrants further investigation but does not seem critical for this study since an evaluation–to–iteration ratio close to 1 could be achieved on the Gilbert and Nocedal test set even with the simple parabolic adjustment. The crucial ingredient to that achievement is the good initial stepsize prediction based on $\tilde{y}_k$, which is updated from iteration to iteration in a manner which we will now describe.

Once the step–multiplier α_k has been accepted, one can follow up on the forward evaluation (11) at the corresponding point x_{k+1} with a reverse sweep yielding the corresponding "gradients"

$$g_{k+1} = \nabla f(x_{k+1}), \quad y_k = \nabla^2 f(x_{k+1})d_k, \quad \tilde{z}_{k+1} \equiv \nabla^2 f(x_{k+1})\tilde{g}_{k+1} \tag{14}$$

at a "small" multiple of the cost for evaluating f by itself. Motivated by the large–scale scenario described in the introduction, we will count the combined double sweep to calculate the quantities (11) and (14) as a single evaluation, whose cost is more or less the same as the inevitable effort of evaluating $f(x_{k-1})$ and $\nabla f(x_{k+1})$ themselves. For any multiplier β_k in the search direction update (2) we may finally compute the new quadratic estimate

$$\tilde{y}_{k+1} = \beta_k y_k - \tilde{z}_{k+1} \approx \nabla^2 f(x_{k+1})d_{k+1}\,. \tag{15}$$

Since g_{k+1} is by now exactly known it might seem strange that we use the $\tilde{z}_{k+1}$ based on the extrapolated value $\tilde{g}_{k+1} \approx g_{k+1}$. The motivation for this is that "multiplying" g_{k+1} or d_{k+1} by the new Hessian $\nabla^2 f(x_{k+1})$ would require an extra double sweep. However, this simplification has a price, since it was found in our numerical experiments that the discrepancies between the updated estimates $\tilde{y}_k$ and the actual curvature vectors $\nabla^2 f(x_k)d_k$ tend to grow linearly. This is not surprising since for fixed $\tilde{g}_k$ and x_{k+1} the error propagation is controlled by the Jacobian matrix

$$\frac{\partial \tilde{y}_{k+1}}{\partial \tilde{y}_k} = -\frac{\partial \tilde{z}_{k+1}}{\partial \tilde{y}_k} \tag{16}$$

$$= -\nabla^2 f(x_{k+1}) \frac{\partial \tilde{g}_{k+1}}{\partial \tilde{y}_k} \tag{17}$$

$$= -\alpha_k \nabla^2 f(x_{k+1}), \tag{18}$$

where the last equation follows from (10). It is well known that in the quadratic case the reciprocal α_k^{-1} always lies between the smallest and largest eigenvalue of the Hessian $\nabla^2 f$, which implies that the eigenvalues of the Jacobian are spread about 1. In the worst case, when α_k^{-1} equals the smallest eigenvalue, the spectral radius of the Jacobian equals the condition number of the Hessian. To control the error growth we reevaluate $y_k = \nabla^2 f(x_k)d_k$ whenever the relative error between g_{k+1} and $\tilde{g}_{k+1}$ is greater than some threshold, say 0.1. In our numerical experience this typically happens periodically after every 5–10 steps and the resulting step prediction was entirely satisfactory. Finally, we used a modification $\hat{g}_{k+1}$ of g_{k+1} in order to achieve descent under all circumstances, as discussed in the next section.

3. Hypothetical points

Especially during the final approach to a local minimizer, which can take many iterations on a large–dimensional problem, the predicted step sizes will be quite good in that the ratios

$$\Delta\alpha_k \equiv -g_{k+1}^T d_k / y_k^T d_k \tag{19}$$

are very small. Here the vectors y_k and g_{k+1} are exact since they were evaluated at the trial point x_{k+1}. The increment $\Delta\alpha_k$ represents a Newton step from the current estimate $\alpha = \alpha_k$ for solving the scalar equation $\nabla f(x_k + \alpha_k d_k)^T d_k = 0$. If one actually took this step everything would have to be reevaluated, which seems wasteful, especially if $x_{k+1} + \Delta\alpha_k d_k$ is close to x_{k+1}. Assuming that f is more or less quadratic we may compute the extrapolated values

$$\hat{x}_{k+1} = x_{k+1} + \Delta\alpha_k d_k = x_k + (\alpha_k + \Delta\alpha_k)d_k \tag{20}$$

and

$$\hat{g}_{k+1} = g_{k+1} + \Delta\alpha_k y_k = \left(I - \frac{y_k d_k^T}{y_k^T d_k}\right) g_{k+1} \tag{21}$$

where the last equation follows directly from the definition of $\Delta\alpha_k$ in (16). Hence we see immediately that $\hat{g}_{k+1}^T d_k = 0$ and $\hat{x}_{k+1}$ is the minimum of f along d_k if the function f is indeed quadratic.

Now applying the Polak–Ribière formula with y_k still defined as in (14), we obtain at the hypothetical point $\hat{x}_{k+1}$ the hypothetical search direction

$$\hat{d}_{k+1} = -\hat{g}_{k+1} + \hat{\beta}_k d_k$$

with

$$\hat{\beta}_k = y_k^T \hat{g}_{k+1} / y_k^T d_k = (y_k^T g_{k+1} + \Delta\alpha_k y_k^T y_k) / y_k^T d_k \, .$$

Now one can easily check that

$$\hat{g}_{k+1}^T \hat{d}_{k+1} = -\|\hat{g}_{k+1}\|^2 = g_{k+1}^T \hat{d}_{k+1} \, ,$$

which means that $\hat{d}_{k+1}$ is a descent direction not only at the hypothetical point $\hat{x}_{k+1}$ but also at the established point x_{k+1}.

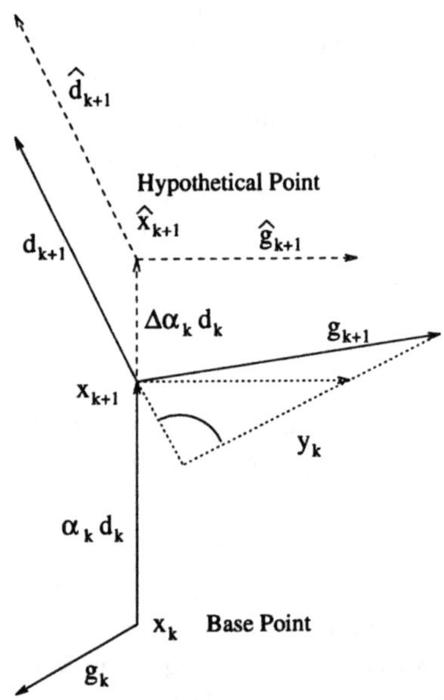

Figure 1. Guaranteed descent based on hypothetical point.

The geometry of the situation is displayed in Figure 1. One should note in particular that the difference between g_{k+1} and $\hat{g}_{k+1}$ is a multiple of y_k, which is orthogonal to $\hat{d}_{k+1}$. Hence we may set

$$d_{k+1} = \hat{d}_{k+1} = -g_{k+1} - \Delta\alpha_k y_k + \hat{\beta}_k d_k$$

and obtain for any $\varepsilon < 1$

$$|\|g_{k+1}\| / \|\hat{g}_{k+1}\| - 1| \leq \varepsilon$$

provided we can ensure that

$$|\Delta\alpha_k| \equiv \left|\frac{g_{k+1}^T d_k}{y_k^T d_k}\right| \leq \varepsilon \frac{\|g_{k+1}\|}{\|y_k\|}$$

This last condition can be enforced by the line–search and would probably be needed for proofs of global convergence to a stationary point. We have not yet performed this analysis but did not encounter any convergence difficulties. There are several variations in the hypothetical point idea worth exploring. Currently our experimental code computes

$$\hat{z}_{k+1} \equiv \tilde{z}_{k+1}(1 + \frac{\Delta\alpha_k}{\alpha_k}) - \frac{\Delta\alpha_k}{\alpha_k}\hat{z}_k$$

$$\hat{y}_{k+1} \equiv \hat{\beta}_k y_k - \hat{z}_{k+1}$$

and

$$\hat{\alpha}_{k+1} \equiv -\|\hat{g}_{k+1}\|^2/(\hat{y}_{k+1}^T \hat{d}_{k+1}) .$$

This hypothetical step multiplier easily leads to the trial point

$$\tilde{x}_{k+2} = \hat{x}_{k+1} + \hat{\alpha}_{k+1}\hat{d}_{k+1} .$$

If this point is found to be acceptable relative to x_{k+1} one may ignore its hypothetical origins and proceed from there. Otherwise one can easily see that $\tilde{x}_{k+2} - x_{k+1}$ is always a descent direction at the established point x_{k+1} so that a conventional line–search must yield a reduction. One might also perform a line–search along a parabola or other curve connecting x_{k+1} and $\tilde{x}_{k+2}$ in the plane spanned by d_{k+1} and $\hat{d}_{k+1}$. Probably such modifications would not significantly alter the results of the current study because we found $\Delta\alpha_k/\alpha_k$ to be very small on almost all iterations.

4. Numerical results

In this final section we compare preliminary results for the new method with the results obtained by Gilbert and Nocedal with their CG variants. They provided us with their Fortran sources for methods and problems. These evaluation routines compute not only the function f but also the gradient ∇f with very little extra storage. None of the test problems has the *time–dependent* characteristics we used as motivation for our method in the introduction. Hence, only the resulting iteration and evaluation counts will be considered significant. Even though these counts were typically halved compared to the best runs of Gilbert and Nocedal, the run–times grew roughly two- or threefold when the second derivatives required were obtained by applying the preprocessor ADIFOR [1] directly to the Fortran sources. The numerical results were identical when the Fortran codes were translated to C using the language processor **f2c** and subsequently slightly modified for use in combination with the automatic differentiation package ADOL-C [5]. This conversion was done in a very simple–minded fashion, so that the resulting run–times were not at all competitive with the original times obtained by Gilbert and Nocedal using hand–coded functions and gradients. For most test functions the gradient evaluation procedures themselves were (unnecessarily) differentiated again since there was no easy way to separate them out from the function evaluation. Anyway, the algebraic simplicity of these test functions means that their evaluations have spatial and temporal complexities that are not representative for the kinds of problems we eventually want to solve. Consequently, we have only compared function evaluation and iteration counts. Following the development in Sections 2 and 3, we use the algorithm outlined in Figure 2.

Set $\hat{x}_0 = x_0,\ f_0 = f(x_0),\ \hat{g}_0 = g_0 = \nabla f(x_0),\ d_0 = -g_0,$
$\tilde{y}_0 = \nabla^2 f(x_0) d_0,\ \hat{z}_0 = \tilde{z}_0 = \nabla^2 f(x_0) g_0$

For $k = 0, 1, \ldots$

$\alpha_k \equiv -\hat{g}_k^T d_k / \tilde{y}_k^T d_k$

• set $x_{k+1} = \hat{x}_k + \alpha_k d_k\ ,\quad \tilde{g}_{k+1} = \hat{g}_k + \alpha_k \tilde{y}_k$

evaluate $f_{k+1} = f(x_{k+1})\ ,\quad g_{k+1} = \nabla f(x_{k+1})\ ,\quad y_k = \nabla^2 f(x_{k+1}) d_k,$

if $(\|g_{k+1} - \tilde{g}_{k+1}\| \leq \|g_{k+1}\|/10)$

then $\tilde{z}_{k+1} = \nabla^2 f(x_{k+1}) \tilde{g}_{k+1}$

else $\tilde{z}_{k+1} = \nabla^2 f(x_{k+1}) g_{k+1}$

if $\psi_k \equiv (f_{k+1} - f_k)/(\alpha_k \hat{g}_k^T d_k) \geq \sigma$

then $\hat{x}_{k+1} = x_{k+1} + \Delta\alpha_k d_k$ with $\Delta\alpha_k \equiv -g_{k+1}^T d_k / y_k^T d_k$

$\hat{g}_{k+1} = g_{k+1} + \Delta\alpha_k y_k\ ,\quad \hat{z}_{k+1} = \tilde{z}_{k+1}(1 + \frac{\Delta\alpha_k}{\alpha_k}) - \frac{\Delta\alpha_k}{\alpha_k} \hat{z}_k\ ,$

Set $d_{k+1} = -\hat{g}_{k+1} + \hat{\beta}_k d_k$ and $\tilde{y}_{k+1} = \hat{\beta}_k y_k - \hat{z}_{k+1}$

with $\hat{\beta}_k \equiv y_k^T \hat{g}_{k+1} / y_k d_k$

else $d_k = (x_{k+1} - x_k)/\alpha_k\ ,\quad \hat{x}_k = x_k\ ,\quad \hat{g}_k = g_k\ ,$

$\alpha_k = \frac{1}{2}\alpha_k/(1 - \psi_k)\ ,$ goto •

Figure 2. The limited second derivative conjugate gradient algorithm.

The overall stopping criteria were left unchanged and the Goldstein–Armijo condition (12) was imposed with $\sigma = 10^{-4}$ in all our calculations. As one can see in Figure 3 for the smaller as well as the larger problems, the average number of function evaluations per step was reduced almost to 1, yet the number of iterations hardly ever went up, and was sometimes significantly reduced. The main ingredient for this success was the quadratic stepsize prediction developed in Section 2. A slight further improvement was achieved by basing the first stepsize correction on a rational interpolation function formed by the quotient of a cubic polynomial and a linear factor.

Smaller Problems

P	N	Best of G&N	New Method
2	100	400/812	353/383
3	100	1299/2599	1377/1544
6	100	254/525	245/309
9	100	7/20	10/17
10	100	93/191	82/112
28	100	120/280	103/124
31	100	2/3	1/3
38	100	70/142	70/75
39	121	59/122	59/94
40	100	132/266	265/285
41	100	24/50	23/31
42	1000	9/25	10/16
43	100	13/37	18/25
45	100	37/90	51/70
47	100	59/122	62/78
48	1000	23/70	19/46
49	100	117/281	52/67
50	100	72/146	73/77
51	100	42/94	46/75
52	1000	3/10	2/5
		it/fev	it/fev

Larger Problems

P	N	Best of G&N	New Method
2	200	701/1420	699/751
3	200	2631/5263	2742/3043
6	500	1067/2149	1012/1222
		/	/
		/	/
28	1000	97/229	181/211
31	200	2/4	1/3
38	1000	262/527	264/278
39	961	142/287	184/284
		/	/
		/	/
42	1000	6/26	8/13
43	1000	10/27	9/15
		/	/
47	1000	113/231	110/138
48	10000	19/62	19/46
49	1000	97/229	181/211
50	1000	273/549	274/285
51	1000	40/91	42/64
		/	/
		it/fev	it/fev

Figure 3. Numerical results on test set of Gilbert and Nocedal.

Obviously the results are encouraging but by no means conclusive. Apart from speeding up the differentiation process and applying it to truly complex problems we need to run comparisons with truncated Newton and limited memory BFGS methods. A particular promising approach is the generalization of block CG methods to nonlinear problems, which would hardly be possible without the analytic evaluation of Hessian–vector products utilized here.

References

[1] C. Bischof, A. Carle, G. Corliss, A. Griewank, and P. Hovland; ADIFOR: Generating derivative codes from Fortran programs, *Scient. Prog.* **1**, 1992, 11–29.

[2] R. Fletcher and C. Reeves; Function minimization by conjugate gradients, *Comp. J.* **7**, 1964, 149–154.

[3] J. Gilbert and J. Nocedal; Global convergence properties of conjugate gradient methods for optimization, INRIA Rapport de Recherche n°1268, Le Chesnay, 1991.

[4] A. Griewank; Achieving logarithmic growth of temporal and spatial complexity in reverse automatic differentiation, *Optim. Meth. Soft.* **1**, 1992), 35–54.

[5] A. Griewank, D. Juedes, and J. Utke, ADOL-C, a package for the automatic differentiation of algorithms written in C/C++, *ACM Trans. Math. Soft.*, 1994, to appear.

[6] E. Polak and G. Ribière; Note sur la convergence de méthods de directions conjuguées, *Rev. Franç. d'Inform. Rech. Opér.* **16**, 1969, 35–43.

[7] D. Shanno and K. Phua; Remark on algorithm 500: minimization of unconstrained multivariate functions, *ACM Trans. Math. Soft.* **6**, 1980, 618–622.

[8] O. Talagrand and P. Courtier; Variational assimilation of meteorological observations with the adjoint vorticity equation. I: Theory, *Q.J.R. Meteor. Soc.* **113**, 1987, 1311–1328.

[9] A. Griewank and P. Toint; On the unconstrained optimization of partially separable objective functions, in *Nonlinear Optimization 1981* Ed. by M.Powell), Academic, London, 1981, 301–312.

[10] P. Wolfe; Convergence conditions for ascent methods, *SIAM Review* **11**, 1969, 226–235.

[11] P. Wolfe; Convergence conditions for ascent methods II: some corrections, *SIAM Review* **13**, 1971, 185–188.

Contact Shape Optimization

Jaroslav Haslinger
Faculty of Mathematics and Physics
Charles University
Prague, The Czech Republic

1. Introduction

The goal of optimal shape design problems is to find a shape of the structure, belonging to an apriori defined class U_{ad} and minimizing a cost functional J on U_{ad}. Shape optimization of systems, governed by equations is now more or less standard. On the other hand there exist a lot of problems where the mathematical model leads to the so called *variational inequality*. It is the aim of this contribution to extend optimal shape design problems to systems described by variational inequalities. It is well known that the optimal control of such systems has one specific feature, namely the problem is *non-smooth*. Here we deal with one of the most typical applications of variational inequalities in solid mechanics, namely with the so called *contact problems* for linearly elastic bodies. A detailed mathematical analysis can be found in [3].

2. Setting of the problem

We begin with the definition of the state problem.

Let us assume an elastic body Ω, subjected to body forces F and surface tractions P on a part Γ_p of the boundary $\partial\Omega$. The body is supported by a rigid, frictionless foundation $\mathcal{F}$. The aim is to determine the equilibrium state. We restrict ourselves to the planar problem only, i.e., $\Omega \subset \mathbb{R}^2$ will be represented by a bounded domain, with Lipschitz boundary $\partial\Omega$. Moreover, the rigid foundation $\mathcal{F}$ will be given by the half plane $\mathbb{R}^2_- = \{[x, y], x \in \mathbb{R}^1, y \leq 0\}$. This simple shape of the rigid foundation makes possible to formulate the non-penetration conditions exactly.

We start with the classical formulation of our problem. The boundary $\partial\Omega$ will be decomposed as follows:

$$\partial\Omega = \bar{\Gamma}_u \cup \bar{\Gamma}_p \cup \bar{\Gamma}_c,$$

where Γ_u , Γ_p and Γ_c are non-empty, open parts in $\partial\Omega$. On each part different boundary conditions will be prescribed.

By a *classical solution* of our problem we call any displacement field $u = (u_1, u_2)$, which is in the equilibrium state with the applied forces $F = (F_1, F_2)$ and $P = (P_1, P_2)$, i.e., satisfies the equilibrium equations

$$\tau_{ij,j}(u) + F_i = 0 \qquad i = 1, 2 \text{ in } \Omega \tag{1}$$

subject to

$$T_i \equiv \tau_{ij}(u)n_j = P_i \qquad i = 1, 2 \text{ on } \Gamma_p. \tag{2}$$

Here the symbol τ_{ij} stands for the i, j-th component of the stress tensor τ, related to the displacement field u by means of the linear Hooke's law

$$\tau_{ij}(u) = C_{ijkl}\epsilon_{kl}(u), \tag{3}$$

where $\epsilon_{kl}(u) = \frac{1}{2}\left(\frac{\partial u_k}{\partial x_l} + \frac{\partial u_l}{\partial x_k}\right)$ is the k, l-th component of the linearized strain tensor. We suppose, that the elasticity coefficients C_{ijkl} are given by bounded, measurable functions, satisfying the usual symmetry and ellipticity conditions. Moreover, $n = (n_1, n_2)$ denotes the unit normal vector to $\partial\Omega$. The conditions (1) - (3) have to be completed by the classical boundary conditions of the Dirichlet type on Γ_u:

$$u_i = 0, \qquad i = 1, 2 \text{ on } \Gamma_u \tag{4}$$

and the unilateral conditions on Γ_c:

$$\begin{aligned} u_2(x_1, \alpha(x_1)) &\geq -\alpha(x_1) \quad \forall x_1 \in (a, b) \\ T_2(u) &\equiv \tau_{2j}(u)n_j \geq 0, \quad T_2(u)(u_2 + \alpha) = 0 \\ T_1(u) &\equiv \tau_{1j}(u)n_j = 0 \end{aligned} \tag{5}$$

on Γ_c. Here we assume that the part Γ_c is described by the graph of a non-negative function $\alpha : [a, b] \to \mathbb{R}^1_+$, i.e.,

$$\Gamma_c = \{[x_1, x_2], x_2 = \alpha(x_1), x_1 \in (a, b)\}$$

(see Figure 1).

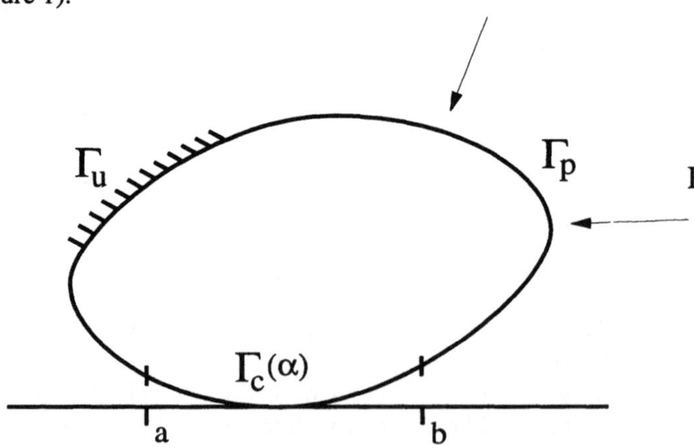

Figure 1.

The inequality $(5)_1$ expresses the fact that Ω cannot penetrate into the rigid foundation $\mathcal{F}$. The next conditions $(5)_2$ state that the contact forces cannot be tensile and moreover at points where no contact occurs, i.e., $u_2(x_1, \alpha(x_1)) > -\alpha(x_1)$, no contact forces can arise, i.e., $T_2 = 0$. As only the frictionless case is assumed here, the first component T_1 vanishes along Γ_c (condition $(5)_3$).

A displacement field $u = (u_1, u_2)$ satisfying (1)–(5) will be called the *classical solution* of the contact problem.

In order to give the definition of the variational problem under consideration some functional sets will be necessary. Define

$$V = \{v = (v_1, v_2) \in (H^1(\Omega))^2 | \; v_i = 0, \; i = 1, 2 \text{ on } \Gamma_u\} \tag{6}$$

as the space of virtual displacements and

$$K = \{v \in V | \; v(x_1, \alpha(x_1)) \geq -\alpha(x_1) \;\; \forall x_1 \in (a, b)\} \tag{7}$$

as the set of kinematically admissible displacements. The symbol $H^1(\Omega)$ denotes the classical Sobolev space of functions which together with their first derivatives are square integrable in Ω.

Furthermore, let

$$a(y, v) = \int_\Omega \tau_{ij}(y)\epsilon_{ij}(v)dx \equiv (\tau_{ij}(y), \epsilon_{ij}(v))_{0,\Omega}$$

and

$$L(v) = \int_\Omega F_i v_i dx + \int_{\Gamma_p} P_i v_i ds \equiv (F_i, v_i)_{0,\Omega} + (P_i, v_i)_{0,\Gamma_p}$$

be the inner energy and the work of external forces, respectively.

We call any function $u \in K$ the *variational solution* of our problem if it satisfies

$$a(u, v - u) \geq L(v - u) \qquad \forall v \in K, \tag{8}$$

or equivalently,

$$u \in K \text{ such that } J(u) \leq J(v) \qquad \forall v \in K, \tag{9}$$

where $J(v) = \frac{1}{2}a(v, v) - L(v)$ is the total potential energy functional. It can be shown that under our assumptions, the problem (8) (or (9)) has exactly one solution. A detailed analysis of such problems, including their discretization by finite elements, can be found in [1], [2].

Until now, the shape of Ω has been given. Next, we shall suppose that the contact part Γ_c will be the object of the shape optimization, i.e., the shape of Ω will change through the variation of Γ_c. To this end, let the function α, describing the contact part Γ_c, belong to the admissible set U_{ad}, where

$$U_{ad} = \{\alpha \in C^{0,1}([a, b]) | \; 0 \leq \alpha \leq \gamma, \; |\alpha'(x_1)| < C_0, \; \text{meas } \Omega(\alpha) = C_1\}.$$

In other words, U_{ad} contains all Lipschitz continuous functions α defined on $[a, b]$ that are uniformly bounded, uniformly continuous, and preserve the area of $\Omega(\alpha)$. The constants γ, C_0 and C_1 are chosen in such a way that $U_{ad} \neq \emptyset$. Here $\Omega(\alpha)$ denotes the body, bounded by Γ_u, Γ_p and $\Gamma_c(\alpha)$. To emphasize the dependence of all data on the *design variable* α, we shall write α as the argument. Thus, we shall use the following notations: $V(\alpha)$, $K(\alpha)$, $a_\alpha(y, v)$, $L_\alpha(v)$, (8), (9), etc. All these symbols are defined in the same way as before, only instead of Ω, the domain $\Omega(\alpha)$ for $\alpha \in U_{ad}$ will be used.

Remark. Restricting the first derivatives of α appearing in the definition of U_{ad} avoids undesidered oscilations of the contact part $\Gamma_c(\alpha)$.

Finally, let

$$I : (\alpha, y) \to \mathbb{R}^1, \alpha \in U_{ad}, y \in V(\alpha)$$

be an *objective functional* which will be used in the optimization. The *optimal shape design* problem reads as follows:

Find $\alpha^* \in U_{ad}$ such that

$$I(\alpha^*, u(\alpha^*)) \leq I(\alpha, u(\alpha)) \qquad \forall \alpha \in U_{ad}, \tag{10}$$

where $u(\alpha)$ solves the contact problem (8) on $\Omega(\alpha)$, $\alpha \in U_{ad}$. A domain $\Omega^* = \Omega(\alpha^*)$, with α^* solving (10), will be called *optimal* with respect to U_{ad} and I.

In order to guarantee the existence of at least one solution of (10), additional continuity assumptions on I will be necessary. Next, assume that

$$\begin{aligned} \alpha_n &\Rightarrow \alpha \text{ (uniformly) in } [a, b], \alpha, \alpha_n \in U_{ad} \\ \hat{y}_n &\rightharpoonup \hat{y} \text{ (weakly) in } (H^1(\hat{\Omega}))^2 \\ &\Rightarrow \liminf I(\alpha_n, \hat{y}_n|_{\Omega_n}) \geq I(\alpha, \hat{y}|_{\Omega(\alpha)}), \end{aligned} \tag{11}$$

where $\Omega_n \equiv \Omega(\alpha_n)$ and $\hat{\Omega} \supseteq \Omega(\alpha)\ \forall \alpha \in U_{ad}$.

If the condition (11) is satisfied, the problem (10) has at least one solution α^*. The detailed mathematical analysis of this problem can be found in [3].

3. Approximation of the variational solution

The aim of this section is to describe the discretization of (10), which will be based on finite element approximation of the state problem (8).

Let $a = a_0 < a_1 < \ldots < a_{D(h)} = b$ be a partition of the interval $[a, b]$ and define

$$U_{ad}^h = \{\alpha_h \in C([a, b]) |\ \alpha_h|_{a_{i-1}a_i} \in P_1,\ i = 1, \ldots, D(h)\} \cap U_{ad}$$

i.e., U_{ad}^h contains those functions from U_{ad}, which are piecewise linear over the given partition. Let us assume that all $\Omega(\alpha_h)$, $\alpha_h \in U_{ad}^h$ are already *polygonal domains*. Let $\mathcal{T}(h, \alpha_h)$ be a triangulation of $\Omega(\alpha_h)$, satisfying the usual requirements on the mutual position of triangles $T_i \in \mathcal{T}(h, \alpha_h)$. The domain $\Omega(\alpha_h)$ with the triangulation $\mathcal{T}(h, \alpha_h)$ will be denoted by Ω_h.

Let $\alpha_h \in U^h_{ad}$ be given. With any $\mathcal{T}(h, \alpha_h)$ we associate the following finite element sets:

$$\begin{aligned} V_h(\alpha_h) &= \{v_h = (v_{h1}, v_{h2}) \in (C(\bar{\Omega}_h))^2 \mid v_h|T_i \in (P_1(T_i))^2 \\ &\quad \forall T_i \in \mathcal{T}(h, \alpha_h), v_{hi} = 0 \text{ on } \Gamma_u, i = 1, 2\} \qquad (12) \\ K_h(\alpha_h) &= \{v_h \in V_h(\alpha_h) | v_{h2}(a_i, \alpha_h(a_i)) \geq -\alpha_h(a_i) \\ &\quad \forall a_i \in \bar{\Gamma}_c(\alpha_h)/\bar{\Gamma}_u\} \qquad (13) \end{aligned}$$

The sets $V_h(\alpha_h)$, $K_h(\alpha_h)$ are finite element approximations of: $V(\alpha)$, $K(\alpha)$, respectively, using piecewise linear functions over the triangulation $\mathcal{T}(h, \alpha_h)$.

Instead of (8), its finite-element approximation will be assumed:

Find $u_h \equiv u_h(\alpha_h) \in K_h(\alpha_h)$ such that

$$a_{\alpha_h}(u_h, v_h - u_h) \geq L_{\alpha_h}(v_h - u_h) \qquad \forall v_h \in K_h(\alpha_h) \qquad (14)$$

or equivalently:

Find $u_h \in K_h(\alpha_h)$ such that

$$J_{\alpha_h}(u_h) \leq J_{\alpha_h}(v_h) \qquad \forall v_h \in K_h(\alpha_h), \qquad (15)$$

where

$$J_{\alpha_h}(v_h) = \frac{1}{2} a_{\alpha_h}(v_h, v_h) - L_{\alpha_h}(v_h).$$

Finally, let

$$I_h : (\alpha_h, y_h) \to \mathbb{R}^1, \ \alpha_h \in U^h_{ad}, \ y_h \in V_h(\alpha_h)$$

be an *approximation* of the cost functional I.

The *approximation* of (10) now reads as follows:

Find $\alpha_h^* \in U^h_{ad}$ such that

$$I_h(\alpha_h^*, u_h(\alpha_h^*)) \leq I_h(\alpha_h, u_h(\alpha_h)) \qquad \forall \alpha_h \in U^h_{ad}, \qquad (16)$$

where $u_h \in K_h(\alpha_h)$ is the solution of (14) .

It is possible to formulate assumptions, under which the problem (16) has at least one solution α_h^* and to establish the mutual relation between (10) and (16) , when $h \to 0^+$. A detailed analysis can be found in [3]. Below we present the matrix formulation of (16) provided $h > 0$ is fixed.

The position of nodes belonging to $\mathcal{T}(h, \alpha_h)$ is usually determined from the position of the boundary nodes lying on $\bar{\Gamma}_c(\alpha_h)$, which are called *principal moving nodes*. Let $\vec{\alpha}$ be a vector of discrete design variables, related to the principal moving nodes, i.e., there exists one to one mapping $\Phi : U^h_{ad} \to \mathbb{R}^q$

$$\alpha_h \leftrightarrow \vec{\alpha}, \qquad \alpha_h \in U^h_{ad}, \qquad \vec{\alpha} \in \mathbb{R}^q.$$

Define $\mathcal{U} = \Phi(U_{ad}^h)$. The matrix from of (14) , $\alpha_h \in U_{ad}^h$, reads as follows: Find $x(\vec{\alpha}) \in \mathcal{K}(\vec{\alpha})$ such that

$$\mathcal{J}(\vec{\alpha}, x(\vec{\alpha})) \leq \mathcal{J}(\vec{\alpha}, y) \qquad \forall y \in \mathcal{K}(\vec{\alpha}) \tag{17}$$

where

$$\mathcal{J}(\vec{\alpha}, y) = \frac{1}{2}(y, A(\vec{\alpha})y) - (L(\vec{\alpha}), y)$$

and

$$\mathcal{K}(\vec{\alpha}) = \{x \in \mathbb{R}^n | \ x_{j_i} \geq -\alpha_h(a_i), \ \ a_i \in \bar{\Gamma}_c(\alpha_h)/\bar{\Gamma}_u\}.$$

The symbol $(\cdot, \cdot)$ denotes the scalar product in $\mathbb{R}^n$, $A(\vec{\alpha})$ is the stiffness matrix of the problem and $L(\vec{\alpha})$ is a vector arising from the discretization of the linear term L_{α_h}. It is readily seen that (17) is a quadratic programming problem, which can be numerically solved by using various minimization methods (the SOR method with additional projection, conjugate gradient methods and methods of the Uzawa type among others). The whole optimal shape design problem (16) now reads as follows:

Find $\vec{\alpha}^* \in \mathcal{U}$ such that

$$\mathcal{I}(\vec{\alpha}^*, x(\vec{\alpha}^*)) \leq \mathcal{I}(\vec{\alpha}, x(\vec{\alpha})) \qquad \forall \vec{\alpha} \in \mathcal{U}, \tag{18}$$

where $x(\vec{\alpha}) \in \mathcal{K}(\vec{\alpha})$ is a unique solution of (17). The problem (18) is a *nonlinear mathematical programming problem*. The mapping $\vec{\alpha} \to \mathcal{I}(\vec{\alpha}, x(\vec{\alpha}))$ is *non-convex* and *non-smooth*, in general. By non-smoothness we mean that the mapping is *Lipschitz continuous* but not *continuously differentiable* everywhere. The non-differentiability of the cost functional $\mathcal{I}$ is due to the fact that the mapping $\vec{\alpha} \to x(\vec{\alpha})$ is only *directionally differentiable* and consequently only directional derivatives of $\mathcal{I}$ are available. This fact complicates the practical realization of (18) and explains why the numerical results are not satisfactory when classical gradient type minimization methods are used. There are several ways to overcome this difficulty. One is to use regularization of the state problem by a penalty approach, for example. Another way is to use non-smooth optimization methods developed for the minimization of non-differentiable functions [4,5]. Nevertheless it is possible that for a special choice of the cost functional I, the function $\alpha \to I(\alpha, u(\alpha))$ will be continuously differentiable regardless of the fact that the inner mapping $\alpha \to u(\alpha)$ is not. In the next Section, we shall discuss two choices of cost functionals, for which the above mapping is *one times continuously differentiable*. Therefore, classical gradient type minimization methods can be applied.

4. Total potential energy and reciprocal energy as objective fuctionals

In order to obtain a constant distribution of contact forces T_2 along $\Gamma_c(\alpha)$, the authors in [6] proposed the total potential energy functional evaluated at the equilibrium state as a possible cost functional, i.e.,

$$I_{1\alpha}(u(\alpha)) \equiv \frac{1}{2}a_\alpha(u(\alpha), u(\alpha)) - L_\alpha(u(\alpha)), \tag{19}$$

with $u(\alpha) \in K(\alpha)$ being the solution of (8), $\alpha \in U_{ad}$. It is possible to prove that the functional (19) satisfies the assumption (11) from Section 3. Thus the problem (10) has at least one solution. In order to derive the corresponding sensitivity of $I_{1\alpha}$, we use the method of mappings.

Let $\mathcal{V} \in (H^{1,\infty}(\Omega(\alpha)))^2$ be a vector field, defining the direction of the domain variation. We shall define the mapping $F_t : \mathbb{R}^2 \to \mathbb{R}^2$ by means of the relation

$$F_t(x) = id + t\mathcal{V}(x), \quad t \geq 0.$$

By Ω_t we denote the image of Ω with respect to $F_t : \Omega_t = F_t(\Omega)$. We shall suppose that $\mathcal{V} = 0$ on $\bar{\Gamma}_u \cup \bar{\Gamma}_p$, i.e., the body Ω changes along $\Gamma_c(\alpha)$ only. Denote by $\Gamma_t(\alpha_t)$ the image of $\Gamma_c(\alpha) : \Gamma_t(\alpha_t) = F_t(\Gamma(\alpha))$. Finally, let $u_t(\alpha_t)$ be the solution of $(8)_{\alpha_t}$ on Ω_t. Our aim will be to compute the *material derivative* of $I_{1\alpha}$ at configuration $\Omega(\alpha)$ with respect to $\mathcal{V}$:

$$\dot{I}_1 = \lim_{t \to 0^+} \frac{I_{1\alpha_t}(u_t(\alpha_t)) - I_{1\alpha}(u(\alpha))}{t}$$

After some calculation it is possible to show that

$$\begin{aligned} \dot{I}_1 &= \frac{1}{2}\int_{\Gamma_c(\alpha)} \tau_{ij}(u(\alpha))\epsilon_{ij}(u(\alpha))\mathcal{V}_k n_k ds - \int_{\Gamma_c(\alpha)} F_i u_i(\alpha)\mathcal{V}_k n_k ds \\ &\quad - \int_{\Gamma_c(\alpha)} \tau_{2i}(u(\alpha))n_i \mathcal{V}_2 ds - \int_{\Gamma_c(\alpha)} \tau_{2i}(u(\alpha))\frac{\partial u_2(\alpha)}{\partial x_2}\mathcal{V}_2 ds, \end{aligned}$$

assuming the vector field $\mathcal{V}$ of the form $\mathcal{V} = (0, \mathcal{V}_2)$, $\mathcal{V}_2 = 0$ on $\bar{\Gamma}_u \cup \bar{\Gamma}_p$ (for the proof see [7]). From the expression (20) we see that the cost functional $I_{1\alpha}$ is once continuously differentiable because no derivative of the state variable $u(\alpha)$ appears. The expression of $\dot{I}_1$ in terms of integrals along $\Gamma_c(\alpha)$ is useful, when the corresponding optimality conditions are interpreted. In [7], the result concerning constant distribution of normal forces along an optimal $\Gamma_c(\alpha^*)$ was justified. An equivalent expression of $\dot{I}_1$ in terms of volume integrals (suitable for practical applications) is presented in [8].

The fact that shape optimization with the total potential energy (19) as an objective functional leads to an even distribution of contact forces along $\Gamma_c(\alpha^*)$ is not readily seen. That is why we shall introduce another cost functional, expressed in terms of stresses on the boundary.

A natural idea arises, namely to define a new cost functional I as follows:

$$I_{2\alpha}(u(\alpha)) \equiv \frac{1}{2}\int_{\Gamma_c(\alpha)} \|T(u) - z_d\|^2 ds, \tag{20}$$

where z_d is a given element and $\| \cdot \|$ is the norm of the element "·". The functional (20) is nothing else but the $L^2(\Gamma_c(\alpha))$ least squares method. This definition however, is meaningless from the mathematical point of view. The reason is very simple: the solution u of (8) is not smooth enough in general, so that the stress vector is not a square integrable function. Instead of the $L^2(\Gamma_c(\alpha))$ norm, a dual norm over the space of traces has to be introduced (for details see [9]).

Let $M(\alpha) \subseteq \partial\Omega(\alpha)$ be an open part of $\partial\Omega(\alpha)$ such that

$$M(\alpha) \cap \Gamma(\alpha) = \emptyset, \qquad M(\alpha) \cap \Gamma_c(\alpha) \neq \emptyset,$$

and define

$$V_M(\alpha) = \{v \in (H^1(\Omega(\alpha)))^2 \mid v_i = 0,\ i = 1, 2 \text{ on } \partial\Omega(\alpha)/\bar{M}(\alpha)\}.$$

By $W_M(\alpha)$ we denote the space of traces of functions from $V_M(\alpha)$, i.e., $w \in W_M(\alpha)$ if and only if $\exists v \in V_M(\alpha)$ such that $w = v$ on $\partial\Omega(\alpha)$. Let $W'_M(\alpha)$ be a space of linear continuous functionals over $W_M(\alpha)$. Using Green's theorem, one can show that $T(u) \in W'_M(u)$. Denote by $\| \cdot \|_{*,M(\alpha)}$ the corresponding dual norm in $W'_M(\alpha)$. Instead of (20) the following cost functional will be used:

$$I_{2\alpha}(u(\alpha)) \equiv \frac{1}{2}\|T(u) - z_d\|^2_{*,M(\alpha)}. \tag{21}$$

The problem is now correct from the mathematical point of view; unfortunately the form of $I_{2\alpha}$ is not suitable for the computation. An alternative (and equivalent) expression has to be introduced. It is possible to show (see [9]) that

$$\|T(u) - z_d\|^2_{*,M(\alpha)} = |z(u)|^2_{1,\Omega(\alpha)},$$

where $z(u) \in V_M(\alpha)$ is the unique solution of an auxiliary elasticity problem

$$(\tau_{ij}(z(u)), \epsilon_{ij}(v))_{0,\Omega(\alpha)} = \langle T(u) - z_d, v\rangle \qquad \forall v \in V_M(\alpha), \tag{22}$$

where $| \cdot |^2_{1,\Omega(\alpha)} = (\tau_{ij}(\cdot), \epsilon_{ij}(\cdot))_{0,\Omega(\alpha)}$, $< \cdot, \cdot >$ denotes the corresponding duality between $W'_M(\alpha)$ and $W_M(\alpha)$ and $u(\alpha) \in K(\alpha)$ is the solution of the state problem $(8)_\alpha$. Finally, we use the following form of the cost functional

$$I_{2\alpha}(u(\alpha)) \equiv \frac{1}{2}|z(u(\alpha))|^2_{1,\Omega(\alpha)}, \tag{23}$$

with $z \in V_M(\alpha)$ being the solution of (22).

It has been shown, [9] that if $M(\alpha) = \partial\Omega(\alpha)/\bar{\Gamma}_u$ (i.e., $V_M(\alpha) = V(\alpha)$, $z_d = P$ on Γ_p and $z_d \equiv 0$ on $\Gamma_c(\alpha)$), then the function $\alpha \to I_{2\alpha}(u(\alpha))$ is once continuously differentiable, as well. Moreover, its relation to the so called *reciprocal energy* was established.

These two cost functionals ((19) and (23)) were used in [10] and the numerical results were compared. Here we present one example from [10].

The body in its undeformed state is shown in Figure 2 along with the rigid foundation $\mathcal{F} = \mathbb{R}^2_-$ and the contact part $\Gamma_c = \{[x, y] \in \mathbb{R}^2 | 0 < x < 4, y = 0.01 \cdot x\}$. The body is submitted to an internal pressure $f = 10.0$ and has material constants $E = 3000$ and $\mu = 0.3$. Results are shown in Figure 3 (final optimal shapes for each cost functionals are very close). Finally, in Figure 4 the distribution of contact stresses along the optimal contact part is shown.

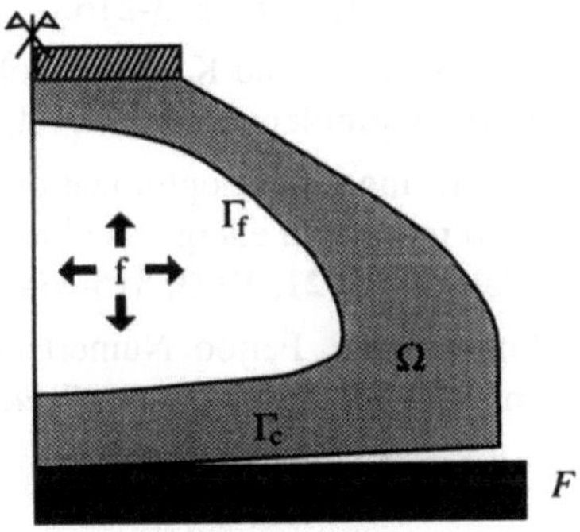

Figure 2.

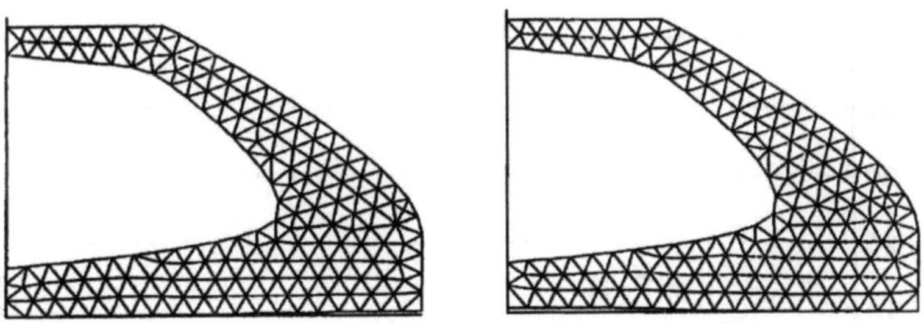

Figure 3.

References

[1] N. Kikuchi and J. Oden; *Contact Problems in Elasticity. Study of Variational Inequalities and Finite Element Methods*, SIAM, Philadelphia, 1987.

[2] I. Hlaváček, J. Haslinger, J. Nečas, and J. Lovíšek; *Numerical Solution of Variational Inequalities*, Springer, 1988.

[3] J. Haslinger and P. Neittaanmäki; *Finite Element Approximation for Optimal Shape Design: Theory and Applications*, Wiley, New York, 1988.

[4] K. Kiwiel; *Methods of Descent for Nondifferentiable Optimization*, Springer, New York, 1985.

[5] C. Lemarechal and R. Miffeir; *Non-smooth Optimization*, Pergamon, Oxford, 1978.

[6] R. Benedict and J. Taylor; Optimal design for elastic bodies in contact, in *Optimization of Distributed Parameter Structures* (Edited by E. Haugh and J. Cea), Sijhoff and Nordhoff, Holland, 1981, 917-930.

[7] A. Klarbring and J. Haslinger; On almost contact stress distribution by shape optimization, *Struct. Optim.* **3**, 1993, 213-216.

[8] J. Haslinger, P. Neittaanmäki P., and K. Salmenjoki; Sensitivity analysis for discretized plane elasticity problem, *Fin. Elem. Anal. Des* **12**, 1992, 13-25.

[9] J. Haslinger and A. Klarbring; Shape optimization in unilateral contact problems using generalized reciprocal energy as objective functional, *J. Nonlinear Anal.: Theory Meth. Appl.* **21**, 1993, 815-834.

[10] E. Fancello, J. Haslinger, and R. Feijóo; Numerical comparison between two cost functions in contact shape optimization, *Struct. Optim.*, to appear.

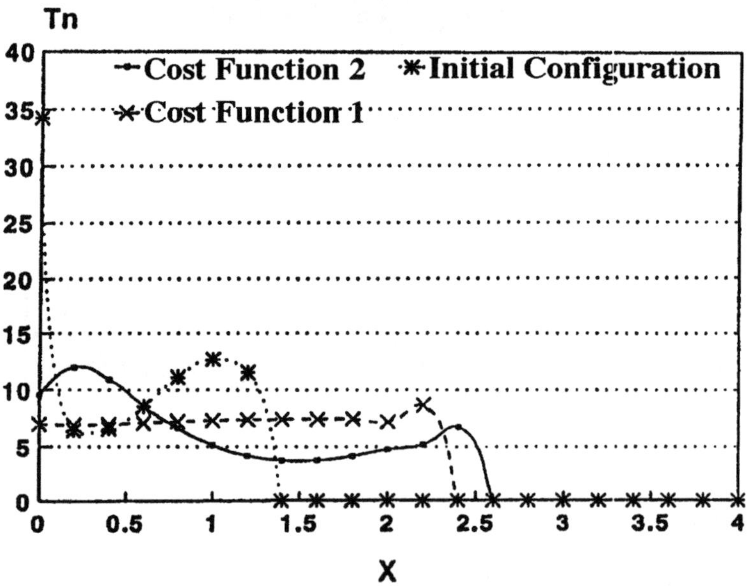

Figure 4. Stress along contact boundary.

Optimal Control of Thermally Coupled Navier Stokes Equations

Kazufumi Ito, Jeffrey S. Scroggs, and Hien T. Tran
Center for Research in Scientific Computation
North Carolina State University
Raleigh NC 27695-8205, USA

The optimal boundary temperature control of the stationary thermally coupled incompressible Navier-Stokes equation is considered. Well-posedness and existence of the optimal control and a necessary optimality condition are obtained. Optimization algorithms based on the augmented Lagrangian method with second order update are discussed. A test example motivated by control of transport process in the high pressure vapor transport (HPVT) reactor is presented to demonstrate the applicability of our theoretical results and proposed algorithm.

1. Introduction

In this paper we discuss the optimal control problem of the stationary thermally coupled incompressible Navier-Stokes equations. Consider the following optimal control problem

$$\text{minimize} \quad J(g) = \varphi(u, T - T_0) + \frac{\beta}{2}\,|g - T_0|^2_{L^2(\Gamma_1)}, \quad g \in \mathcal{C} \tag{1}$$

subject to

$$\begin{aligned} u \cdot \nabla u + \nabla p &= \nu\,\Delta u + \gamma\,(T - T_0)\,e_d + f\,, \\ \nabla \cdot u &= 0, \quad u|_\Gamma = 0\,, \\ u \cdot \nabla T &= \nabla \cdot (\kappa\,\nabla T)\,, \\ T = T_0 \ \text{ on } \Gamma_0 \ \text{ and } \ n \cdot \nabla T &= H\,(g - T) \ \text{ on } \Gamma_1\,, \end{aligned} \tag{2}$$

where $f \in L^2(\Omega)^d$ is a source field, $u,\ p,\ T$ stand for the nondimensionalized velocity vector in R^d with $d = 2,\ 3$, pressure, and temperature, respectively and $\mathcal{C}$ is the closed convex set in $L^2(\Gamma_1)$ such that

$$\bar{T}_1 \le g \le \bar{T}_2 \quad \text{on } \Gamma_1 \tag{3}$$

Here, $\bar{T}_1 \le T_0 \le \bar{T}_2$ and $\Gamma_i,\ i = 0, 1$ are disjoint open sets in Γ such that $\Gamma = \overline{\Gamma_0} \cup \overline{\Gamma_1}$. The constant γ is given by $\gamma = \dfrac{\bar{g}}{T_0}$ where $T_0 > 0$ is a constant reference temperature and $\bar{g}$ is the gravitational constant, and e_d denotes the d−th unit vector of R^d. Throughout this paper we assume that Ω is sufficiently smooth and $\nu,\ \kappa$ and H are positive constants.

This control problem is motivated by control of transport and growth processes in the high pressure vapor transport (HPVT) reactor [13]. For example, we may consider the Scholz geometry (Figure 1 in [13]). The source material and the growing crystal are sealed in a fused silica ampoule that is heated by a furnace liner at its outer cylindrical surface. The substrate Γ_0 (the single crystal) is located on a fused silica window (the bottom of the ampoule) which is cooled by a jet of helium gas from the outer surface. HPVT processes are based on physical vapor transport and can be described very roughly as proceeding via evaporation at the polycrystalline source and condensation at the surface of the cooler substrate. The system (1)-(2) is called the Boussinesq equations, where we assume that the flow is incompressible and the transport phenomena of a single (carrier) gas is modeled. At the wall we assume Newton's law of cooling holds.

The cost-functional can be of tracking type

$$\varphi(u, T - T_0) = \int_\Omega |u - u_d|^2 + |T - T_d|^2 \, dx$$

where (u_d, T_d) is the desired state, or minimization of friction force of flow in a subregion Ω_1 of Ω

$$\varphi(u, T - T_0) = \int_{\Omega_1} |\nabla \times u|^2 \, dx \, .$$

2. Well-posedness

In this section, we discuss existence and uniqueness of solutions to (2). Let $U = L^2(\Gamma_1)$ be the control space, $V = V_0 \times V_1$ where V_0 is the divergence-free subspace of $(H_0^1(\Omega))^d$ [7] and

$$V_1 = H^1_{\Gamma_0} = \{\phi \in H^1(\Omega) : \phi|_{\Gamma_0} = 0\}$$

and set $H = H_0 \times L^2(\Omega)$. H_0 is the closure of V_0 with respect to the $L^2(\Omega)^d$-norm, and is defined by

$$H_0 = \{\phi \in L^2(\Omega)^d \, : \, \nabla \cdot \phi = 0 \text{ and } n \cdot \phi = 0 \text{ on } \Gamma\}.$$

H is equipped with the natural L^2-norm and V is equipped with the norm

$$|(\phi, \chi)|_V^2 = |\phi|_{V_0}^2 + |\chi|_{V_1}^2$$

where

$$|\phi|_{V_0}^2 = |\nabla\phi|^2_{L^2(\Omega)} \quad \text{and} \quad |\chi|^2_{V_1} = |\nabla\chi|^2_{L^2(\Omega)} + H\,|\chi|^2_{L^2(\Gamma_1)}$$

for $\psi = (\phi, \chi) \in V$. Define the trilinear form b on $H^1(\Omega)^d$ by

$$b(\phi_1, \phi_2, \phi_3) = \langle \phi_1 \cdot \nabla\phi_2, \, \phi_3 \rangle \tag{4}$$

for $\phi_i \in H^1(\Omega)^d$, $i = 1, 2, 3$. Then we have

Lemma 1 *The trilinear form b satisfies*

$$\begin{aligned}
&(a) \quad |b(\phi_1, \phi_2, \phi_3)| \le |\phi_1|_{L^4} |\nabla \phi_2|_{L^2} |\phi_3|_{L^4} \le M_1\, |\phi_1|_{H^1} |\phi_2|_{H^1} |\phi_3|_{H^1} \\
&(b) \quad b(\phi_1, \phi_2, \phi_3) + b(\phi_1, \phi_3, \phi_2) = 0 \\
&\qquad\quad \textit{provided that } \nabla \cdot \phi_1 = 0 \textit{ and } (n \cdot \phi_1)(\phi_2 \cdot \phi_3) = 0 \textit{ on } \Gamma \\
&(c) \quad |b(\phi_1, \phi_2, \phi_3)| \le M_2 |\phi_1|_{L^2}^{1/2} |\phi_1|_{H^1}^{1/2} |\phi_3|_{L^2}^{1/2} |\phi_3|_{H^1}^{1/2} |\phi_2|_{H^1} \textit{ for } d = 2 \\
&(d) \quad |b(\phi_1, \phi_2, \phi_3)| \le M_2 |\phi_1|_{L^2}^{1/4} |\phi_1|_{H^1}^{3/4} |\phi_3|_{L^2}^{1/4} |\phi_3|_{H^1}^{3/4} |\phi_2|_{H^1} \textit{ for } d = 3
\end{aligned}$$

for $\phi_i \in H^1(\Omega)^d,\ i = 1, 2, 3$

Proof: By Green's formula

$$b(\phi_1, \phi_2, \phi_3) + b(\phi_1, \phi_3, \phi_2) = (\phi_1, \nabla(\phi_2 \cdot \phi_3)) = (n \cdot \phi_1, \phi_1 \cdot \phi_3)_\Gamma$$

for $\phi_2 \in C^1(\Omega)$ and $\nabla \cdot \phi_1 = 0$. Hence (b) follows from the continuity of b. The last two assertions follow from the fact that $|\psi|_{L^4} \le c\, |\psi|_{L^2}^{1/2} |\psi|_{H^1}^{1/2}$ for $d = 2$ and $|\psi|_{L^4} \le c\, |\psi|_{L^2}^{3/4} |\psi|_{H^1}^{1/4}$ for $\psi \in H^1(\Omega)$ and some constant c. □

In particular, Lemma 1 implies that

$$b(u, \phi, \phi) = 0 \quad \text{for } u,\ \phi \in V_0\,. \tag{5}$$

The weak or variational form of (2) is given by

$$\nu\, (\nabla u, \nabla \phi) + b(u, u, \phi) = (f + \gamma\, (T - T_0) e_d, \phi) \tag{6}$$

for all $\phi \in V_0$ and

$$\begin{aligned}
\kappa\, (\nabla(T - T_0), \nabla \chi) \quad &+ \quad \langle u \cdot \nabla(T - T_0),\ \chi \rangle_{V_1^* \times V_1} \\
&+ \quad \kappa\, (H\, (T - T_0), \chi)_{\Gamma_1} = \kappa\, (H\, (g - T_0),\ \chi)_{\Gamma_1},
\end{aligned} \tag{7}$$

for all $\chi \in V_1$. The pair $(u, T - T_0) \in V$ is said to be a weak solution of (2) if (6), (7) holds for all $\psi = (\phi, \chi) \in V$. Then we have the following theorem.

Theorem 2 *Given* $g \in L^2(\Gamma_1)$ *there exists a weak solution* $(u, T - T_0) \in V$ *to (2) and*

$$|(u, T - T_0)|_V \le \textit{const}\, (|f|_{L^2} + |g|_{L^2(\Gamma_1)})\,.$$

Moreover, if $\bar{T}_1 \le g(x) \le \bar{T}_2$ *a.e. in* $x \in \Gamma_1$ *then* $\bar{T}_1 \le T(x) \le \bar{T}_2$ *a.e in* Ω *for every solution* $(u, T - T_0) \in V$.

Step 1: (Existence) We show that (6), (7) has a solution $z = (u, T - T_0) \in V$. Given $\bar{u} \in V_0$, we consider the linear equation

$$\nu\, (\nabla u, \nabla \phi) + b(\bar{u}, u, \phi) = (\gamma\, (T - T_0) e_d + f, \phi) \quad \text{for } \phi \in V_0 \tag{8}$$

$$\kappa\,(\nabla(T-T_0),\ \nabla\chi)+\langle\bar{u}\cdot\nabla(T-T_0),\ \chi\rangle \ + \ \kappa\,(H\,(T-T_0),\chi) \\ = \ \kappa\,(H\,(g-T_0),\ \chi)_{\Gamma_1}, \quad (9)$$

for $\chi \in V_1$. First, we show that (8),(9) has the unique solution $(u, T - T_0) \in V$. Then, we show that the solution map S on V_0 defined by $S(\bar{u}) = u$ where $(u, T-T_0) \in V$ is the unique solution to (8),(9) has a fixed point by Schauder fixed point theorem. The fixed point $u \in V_0$ and the corresponding solution $T-T_0 \in V_1$ define a solution to (6), (7). By Green's formula

$$\langle\bar{u}\cdot\nabla\theta,\chi\rangle+\langle\bar{u}\cdot\nabla\chi,\theta\rangle=0 \quad \text{and in particular} \quad \langle\bar{u}\cdot\chi,\chi\rangle=0 \qquad (10)$$

for $\bar{u} \in V_0$, and $\theta,\ \chi \in H^1(\Omega)$. Hence, from Lemma 1 the sesquilinear form

$$\kappa\,(\nabla\chi_1,\ \nabla\chi_2)+\langle\bar{u}\cdot\nabla\chi_1,\ \chi_2\rangle+(\kappa H\,\chi_1,\ \chi_2)_{\Gamma_1}=0$$

on $V_1 \times V_1$ is bounded and V_1-coercive. It thus follows from the Lax-Milgram theorem that equation (9) has a unique solution $T-T_0 \in V_1$. Choosing $\chi = T-T_0$ in (9), we have (independent of $\bar{u} \in V_0$)

$$|T-T_0|_{V_1}^2 \le H\,|g-T_0|_{L^2(\Gamma_1)}^2. \qquad (11)$$

Next, the sesquilinear form on $V_0 \times V_0$ defined by

$$\nu\,(\nabla\phi_1,\ \nabla\phi_2)+b(\bar{u},\phi_1,\ \phi_2)$$

is bounded and V_0-coercive from Lemma 1 and (5). Thus, by the Lax-Milgram theorem, equation (8) has a unique solution $u \in V_0$, and we have

$$|u|_{V_0} \le \frac{M_3}{\nu}\,(|f|_{L^2}+\gamma\,|T-T_0|_{L^2}) \qquad (12)$$

where $|\phi|_{H_0} \le M_3\,|\phi|_{V_0},\ \phi \in V_0$. Let C be a closed convex subspace of V_0, defined by

$$C=\{\phi\in V_0 : |\phi|_{V_0} \le \frac{M_3}{\nu}\,(|f|_{L^2}+\gamma M_4 H\,|g-T_0|_{L^2(\Gamma_1)}\},$$

where $|\chi|_{L^2} \le M_4\,|\chi|_{V_1}$ for $\chi \in V_1$. Then it follows from (11)-(12) that S maps from C into C. Moreover, the solution map S is compact. In fact, if $\bar{u}_k$ converges weakly to $\bar{u}$ in V_0 then $|\bar{u}_k-\bar{u}|_{L^4} \to 0$ since $H^1(\Omega)$ is compactly embedded into $L^4(\Omega)$. Let $(u_k, T_k-T_0) \in V$ and $(u, T-T_0) \in V$ be the corresponding solution of (8),(9), respectively to $\bar{u}_k \in V_0$ and $\bar{u} \in V_0$. Then we have

$$\kappa\,(T_k-T,\ \chi)_{V_1}+\langle(\bar{u}_k-\bar{u})\cdot\nabla(T-T_0)+\bar{u}_k\cdot\nabla(T_k-T),\ \chi\rangle=0$$

for $\chi \in V_1$ from Lemma 1 and (10)

$$\kappa\,|T_k-T|_{V_1} \le M_1\,|\bar{u}_k-\bar{u}|_{L^4}|T-T_0|_{V_1}$$

which implies $|T_k - T|_{V_1} \to 0$. Similarly, we have

$$\nu\,|u_k - u|_{V_0} \le M_1\,|\bar{u}_k - \bar{u}|_{L^4}|u|_{V_0} + \gamma M_3\,|T_k - T|_{L^2}$$

and thus $|u_k - u|_{V_0} \to 0$. Now, by the Schauder fixed point theorem (e.g, see [19]) there exists at least one solution to (6), (7).

Step 2: (L^∞ estimate) We show that if $\bar{T}_1 \le g \le \bar{T}_2$ then

$$\bar{T}_1 \le T \le \bar{T}_2 \quad \text{a.e. } x \in \Omega\,.$$

for all solutions $(u, T - T_0) \in V$ to (6), (7). In fact, let $\chi = \inf(T, \bar{T}_1)$. Then $\chi \in V_1$ [19] and we have from (6), (7)

$$\kappa\,(\nabla T,\ \nabla\chi) + \langle u\cdot\nabla T,\ \chi\rangle + (\kappa H\,(T - g),\ \chi)_{\Gamma_1} = 0\,.$$

Since from (10) $\langle u\cdot\nabla T,\ \chi\rangle = 0$ we have

$$\kappa\,(\nabla\chi,\ \nabla\chi) + (\kappa H\,(T - g),\ \chi)_{\Gamma_1} = 0$$

where

$$(T - g)\chi = (T - \bar{T}_1 - (g - \bar{T}_1))\,\chi \ge |\chi|^2 \ \text{ on } \ \Gamma_1\,.$$

Thus, we obtain $|\chi|^2_{\underline{V_1}} = 0$ which implies $\chi = 0$ and hence $T \ge \bar{T}_1$. Similarly, one can prove that $T \le \bar{T}_2$, choosing the test function $\chi = \sup(T, \bar{T}_2)$.

We have also the uniqueness of solutions under the smallness assumption on f and $g - T_0$.

Theorem 3 *If* $|f|_{L^2}$ *and* $\frac{1}{T_0}|g - T_0|_{L^2(\Gamma_1)}$ *are sufficiently small then (6), (7) has a unique solution in V.*

Proof: Suppose $(u_i, T_i - T_0) \in V$, $i = 1, 2$ are two solutions to (6), (7). Then we have

$$\begin{aligned}
&\nu\,(\nabla\tilde{u},\ \nabla\phi) + \langle u_1\cdot\nabla\tilde{u} + \tilde{u}\cdot\nabla u_2,\ \phi\rangle = \gamma\,(\tilde{T}\,e_d,\ \phi)\\
&\kappa\,(\nabla\tilde{T},\ \nabla\chi) + \langle u_1\cdot\nabla\tilde{T} + \tilde{u}\cdot\nabla(T_2 - T_0),\ \chi\rangle + (\kappa H\,\tilde{T},\ \chi)_{\Gamma_1} = 0
\end{aligned}$$

for $\phi \in V_0$ and $\chi \in V_1$, where $\tilde{u} = u_1 - u_2 \in V_0$ and $\tilde{T} = T_1 - T_2 \in V_1$. Setting $\phi = \tilde{u}$ and $\chi = \tilde{T}$ we obtain from (5) and (10)

$$\begin{aligned}
&\nu\,|\tilde{u}|^2_{V_0} \le M_1\,|u_2|_{V_0}|\tilde{u}|^2_{V_0} + \gamma M_3 M_4\,|\tilde{T}|_{V_1}|\tilde{u}|_{V_0}\\
&\kappa\,(|\nabla\tilde{T}|^2 + H\,|\tilde{T}|^2_{\Gamma_1}) \le M_1|T_2 - T_0|_{V_1}|\tilde{u}|_{V_0}|\tilde{T}|_{V_1}\,.
\end{aligned}$$

If we set $X = |\tilde{u}|_{V_0}$ and $Y = |\tilde{T}|_{V_1}$ then this implies

$$(\nu - M_1\,|u_2|_{V_0})\,X^2 \le \gamma M_3 M_4\,XY \quad \text{and} \quad \kappa\,Y^2 \le M_1|T_2 - T_0|_{V_1}XY\,.$$

Hence if $\kappa(\nu - M_1\,|u_2|_{V_0}) - \gamma M_1 M_3 M_4\,|T_2 - T_0|_{V_1} > 0$ then $X = Y = 0$ and thus $(u_1, T_1) = (u_2, T_2)$. From (11)-(12) we have

$$\begin{aligned} &|T_2 - T_0|_{V_1} \le H\,|g - T_0|_{L^2(\Gamma_1)} \\ &|u_2|_{V_0} \le \frac{M_3}{\nu}\,(|f|_{L_2} + \gamma M_4 H\,|g - T_0|_{L^2(\Gamma_1)}). \end{aligned} \tag{13}$$

Thus, if $|f|_{L^2}$ and $\frac{1}{T_0}|g - T_0|_{L^2(\Gamma_1)}$ are sufficiently small then (6), (7) has a unique solution. □

Moreover, we can make the following demonstration of the regularity of the solution $(u, T - T_0) \in V$. Define the Stokes operator A on H_0 by

$$(Au,\ \phi)_{H_0} = (\nabla u,\ \nabla\phi) \quad \text{for } \phi \in V_0 \tag{14}$$

with domain

$$\text{dom}(A) = \{u \in V_0 :\ |(\nabla u,\ \nabla\phi)| \le c\,|\phi|_{H_0} \text{ for all } \phi \in V_0\}. \tag{15}$$

Then it is known [18] that A is a positive self-adjoint operator on H_0, $\text{dom}(A) \subset H^2(\Omega)^3$ and $V_0 = \text{dom}(A^{1/2}) = [H_0, \text{dom}(A)]_{1/2}$. Let $d = 3$. Since

$$|\langle u \cdot \nabla u,\ \phi\rangle| \le |u|_{L^6}|\nabla u|_{L^2}|\phi|_{L^3}$$

for $u,\ \phi \in V_0$ and

$$H^1(\Omega) \subset L^6(\Omega) \quad \text{and} \quad V_{1/2} \subset L^3(\Omega),$$

where $V_{1/2} = [V_0, H_0]_{1/2}$, $u\nabla u \in V_{-1/2} = \text{dom}(A^{-1/4})$ for $u \in V_0$. Thus,

$$u = A^{-1}(\rho(T - T_0)\,e_3 + f - u\cdot\nabla u) \in \text{dom}(A^{3/4}) = [V, \text{dom}(A)]_{1/2} \subset H^{3/2}(\Omega)^3.$$

Hence, $u \cdot \nabla u \in L^2(\Omega)^d$ and $u \in \text{dom}(A) \subset H^2(\Omega)^3$.

3. Necessary optimality condition

We now show the existence of solutions to the optimal control problem (1)-(2). Let us denote by $S(g)$, the solution set of (6), (7) for $g \in L^2(\Gamma_1)$.

Theorem 4 *Consider the minimization problem (1)-(2):*

$$\begin{aligned} \textit{minimize} \quad & J(g) = \varphi(u, T - T_0) + \tfrac{\beta}{2}\,|g - T_0|^2_{L^2(\Gamma_1)} \\ & \textit{over } (u, T - T_0) \in S(g) \textit{ and } g \in \mathcal{C}. \end{aligned} \tag{16}$$

Assume that φ satisfies

$$\begin{aligned} \varphi(z) :\ z = (u, T - T_0) \in V \to R^+ \ & \textit{is convex and lower semicontinous} \\ & \textit{and } \varphi(z) \le b_1\,|z|^2_V + b_2 \textit{ for } b_1,\ b_2 \in R^+. \end{aligned}$$

Then Problem (1)-(2) has a solution.

Proof: Let $((u_k, T_k - T_0), g_k) \in S(g_k) \times \mathcal{C}$ be a minimizing sequence. Since $\beta > 0$, $|g_k - T_0|_{L^2(\Gamma_1)}$ is uniformly bounded in k and thus from (11)-(12) so is $|(u_k, T_k - T_0)|_V$. Hence there exists a subsequence of $\{k\}$, which will be denoted by the same index, such that $(u_k, T_k - T_0, g_k)$ converges weakly to $(u, T - T_0, g) \in V \times \mathcal{C}$ since $V \times L^2(\Gamma_1)$ is a Hilbert space and $\mathcal{C}$ is closed and convex. Define the trilinear form $\tilde{b}$ on V^3 by

$$\tilde{b}(z_1, z_2, \psi) = b(u_1, u_2, \phi) + (u_1 \cdot \nabla\theta_2, \chi)$$

for $z_i = (u_i, \theta_i)$, $i = 1, 2$, $\psi = (\phi, \chi) \in V$. Since $H^1(\Omega)$ is compactly embedded into $L^4(\Omega)$ it follows from Lemma 1 that $\tilde{b}(z_k, z_k, \psi) \to \tilde{b}(z, z, \psi)$ for $\psi \in V$, where $z_k = (u_k, T_k - T_0)$ and $z = (u, T - T_0)$. Hence, for $\psi = (\phi, \chi) \in V$ the limit (z, g) satisfies (6), (7) and thus $z \in S(g)$. Now, since φ is convex and lower semicontinuous, it follows from [5] that (z, g) minimizes (16). □

Problem (16) is equivalently written as a constrained minimization on $x = ((u, T - T_0), g) \in X = V \times L^2(\Gamma_1)$ with

$$\begin{aligned} &\text{minimize} \quad J(x) = \varphi(u, T - T_0) + \tfrac{\beta}{2}\,|g - T_0|^2_{L^2(\Gamma_1)} \quad \text{over } x \in X \\ &\qquad\text{subject to } e(x) = 0 \text{ and } g \in \mathcal{C} \end{aligned} \tag{17}$$

where the equality constraint $e : X \to Y = V^*$ is defined by (6), (7); i.e.,

$$\langle e(x), \psi\rangle = a(z, \psi) + \tilde{b}(z, z, \psi) - \kappa\,(H(g - T_0), \chi)_{\Gamma_1} - (f, \phi) \tag{18}$$

for $\psi = (\phi, \chi) \in V$, where

$$a(z, \psi) = \nu\,(\nabla u, \nabla\phi) - \gamma\,((T - T_0)\,e_d, \phi) + \kappa\,(\nabla(T - T_0), \nabla\chi) + \kappa H\,(T - T_0, \chi)_{\Gamma_1}$$

for $z = (u, T - T_0)$, $\psi = (\phi, \chi) \in V$. Assume that $x^* = (z^* = (u^*, T^* - T_0), g^*)$ denotes the optimal pair of (16). Then we have

Theorem 5 *Assume that x^* is a regular point in the sense [14] that*

$$0 \in \mathit{int}\,\{e'(x^*)(v, h - g^*) : v \in V \text{ and } h \in \mathcal{C}\}. \tag{19}$$

Then there exists a Lagrange multiplier $\lambda^* \in V$ such that

$$a(\lambda^*, \psi) + \tilde{b}(\psi, z^*, \lambda^*) + \tilde{b}(z^*, \psi, \lambda^*) + \langle\varphi'(z^*), \psi\rangle = 0 \tag{20}$$

for $\psi \in V$ and

$$(\beta\,g^* - \kappa H\,\lambda_2^*,\ h - g^*)_{\Gamma_1} \geq 0 \quad \text{for all } h \in \mathcal{C}. \quad □ \tag{21}$$

Proof: It follows from [14] that if (19) is satisfied, then there exists a Lagrange multiplier $\lambda^* = (\lambda_1^*, \lambda_2^*) \in V$ such that

$$\langle\varphi'(z^*), \psi\rangle + \beta\,(g^*, h - g^*)_{\Gamma_1} + e'(x^*)(\psi, h - g^*) \geq 0$$

for all $\psi \in V$ and $h \in \mathcal{C}$, that is

$$\begin{aligned}\langle \varphi'(z^*), \psi\rangle + \beta\,(g^*, h - g^*)_{\Gamma_1} + a(\lambda^*, \psi) \\ + \tilde{b}(\psi, z^*, \lambda^*) + \tilde{b}(z^*, \psi, \lambda^*) - \kappa H\,(h - g^*, \lambda^*)_{\Gamma_1} \geq 0\end{aligned} \tag{22}$$

for all $\psi \in V$ and $h \in K$. Setting $\psi = 0$, we obtain (21). Next, setting $h = g^*$ in (22), we obtain (20). □

Concerning the regular point condition (19), we have

Lemma 6 *If $g^* \in int\,(\mathcal{C})$ then the regular point condition (19) is equivalent to the condition that for $v = (v_1, v_2) \in V$*

$$\begin{aligned} a(v, \psi) + \tilde{b}(\psi, z^*, v) + \tilde{b}(z^*, \psi, v) = 0 \\ \text{for all } \psi \in V \quad \text{and} \quad v_2 = 0 \;\; \text{on} \;\; \Gamma_1\end{aligned} \tag{23}$$

implies $v = 0$.

Proof: If $g^* \in \text{int}\,(\mathcal{C})$ then (19) is equivalent to the condition that $G = e'(x^*)$ is surjective. Define the linear map $C \in \mathcal{L}(X, V)$ by $C(v, h) = \xi$ where $\xi \in V$ is the unique solution to

$$a(\xi, \psi) + \tilde{b}(v, z^*, \psi) + \tilde{b}(z^*, v, \psi) - \kappa H\,(h, \chi)_{\Gamma_1} = 0 \quad \text{for } \psi = (\phi, \chi) \in V.$$

Then, since $H^1(\Omega)$ is embedded compactly to $L^4(\Omega)$, Lemma 1 implies that C is compact. Thus, by the Banach closed range theorem and the Riesz-Schauder theorem, $e'(x^*)(v, h)$ is surjective if and only if $ker(G^*) = \{0\}$ [4], which is equivalent to (21). □

4. Augmented Lagrangian method

In this section we discuss applications of the augmented Lagrangian method for the constrained minimization problem (17). The augmented Lagrangian method [8], [15] is based on an equivalent formulation of (17):

$$\text{minimize} \quad J(x) + \frac{c}{2}\,|e(x)|_Y^2 \quad \text{over } x \in X \text{ and } g \in \mathcal{C}\,. \tag{24}$$

subject to $e(x) = 0$, where $x = ((u, T - T_0), g) \in X = V \times L^2(\Gamma_1)$ and $c > 0$ is the penalty parameter. The augmented Lagrangian algorithm [8], [15] is the multiplier method applied to (17), i.e., it involves a sequence of minimizations of the functional

$$\begin{aligned} L_{c_k}(x, \lambda^k) = J(x) + \langle \lambda^k, e(x)\rangle + \frac{c}{2}\,|e(x)|_Y^2 \\ \text{subject to } g \in \mathcal{C}\,,\end{aligned} \tag{25}$$

where the multiplier sequence $\{\lambda^k\}$ in Y^* is generated by the first order update

$$\lambda^{k+1} = \lambda^k + c_k\, e(x_k)\,, \tag{26}$$

for $k \geq 1$, where x_k is a minimizer of $L_{c_k}(\cdot, \lambda^k)$ and we assumed that $Y^* = Y$, otherwise each element in Y^* has its Riesz representation. To carry out this iteration, a sequence of monotonically nondecreasing, positive real numbers $\{c_k\}$, $c_1 > c_0 \geq 0$ and a start up value λ^1 for the Lagrange multiplier for the equality constraint $e(x) = 0$ need to be chosen. The convergence results of the augmented Lagrangian method for the infinite dimensional optimization problem are established, for example, in [9], [16]. The augmented Lagrangian method is a hybrid of the penalty method (i.e., $\lambda^k = 0$) and the Lagrange multiplier method (i.e., $c_k = 0$) and combines good properties of both. It overcomes the difficulty of the penalty method which requires a large value of c_k. The augmented functional $L_{c_k}(x, \lambda^k)$ is locally strictly convex provided that λ^k is sufficiently close to λ^* and the second order optimality condition

$$\begin{aligned} &L_0''(x^*, \lambda^*)((v, h), (v, h)) \geq \sigma\, (|v|_V^2 + |h|_U^2) \\ &\qquad \text{for all } (v, h) \in X \text{ satisfying } e'(x^*)(v, h) = 0\,, \end{aligned} \tag{27}$$

for some $\sigma > 0$, is satisfied. Here, $L_0''(x^*, \lambda^*)$ denotes the bilinear form that characterizes the second derivative of $L_0(x, \lambda) = J(x) + \langle \lambda,\ e(x) \rangle$ with respect to x at x^*. That is, it is not necessary that the cost functional J be (locally) convex, which is required for convergence of the multiplier method. The algorithm (25)-(26) has been successfully applied to parameter estimation problems in elliptic PDEs [10], [12] and optimal control problems for 2-D incompressible Navier-Stokes [4]. The first order update (26) provides Q-linear convergence of the iterates x_k in X. In [11] we have investigated a second order update scheme for the augmented Lagrangian method. In what follows we assume that $g^* \in \mathrm{int}\,(\mathcal{C})$. Thus, (19) reduces to $e'(x^*)$ is surjective (see Lemma 6). Hence the necessary condition implies that

$$L_c'(x^*, \lambda^*) = 0 \quad \text{and} \quad e(x^*) = 0, \tag{28}$$

for all $c \geq 0$. An algorithm proposed in [11] applies Newton's method to (28). The resulting algorithm is stated as: given a current iterate (x, λ) the next iterate (x_+, λ_+) satisfies

$$\begin{pmatrix} L_c''(x, \lambda) & e'(x)^* \\ & \\ e'(x) & 0 \end{pmatrix} \begin{pmatrix} x_+ - x \\ \\ \lambda_+ - \lambda \end{pmatrix} = - \begin{pmatrix} L_c'(x, \lambda) \\ \\ e(x) \end{pmatrix}. \tag{29}$$

Note that

$$L_c'(x, \lambda) = L_0'(x, \lambda + c\, e(x))$$

and

$$L_c''(x, \lambda) = L_0''(x, \lambda + c\, e(x)) + c\, \langle e'(x)(\cdot),\ e'(x)(\cdot) \rangle\,. \tag{30}$$

Consequently, suppose $|(x, \lambda) - (x^*, \lambda^*)|$ is sufficiently small. Then it follows from (27) [11] that $L_c''(x, \lambda)$ is coercive on $X \times Y$. Thus equation (29) can be regarded as a general Stokes equation. Following an argument due to Bertsekes [2], we can avoid forming L_c'' during the iteration. From the second equation of (29) we have $e'(x)(x_+ - x) = -e(x)$. Thus the first equation can be written as

$$L_0''(x, \lambda + c\,e(x))(x_+ - x) + e'(x)^*(\lambda_+ - (\lambda + c\,e(x))) = -L_0'(x, \lambda + c\,e(x))$$

and hence (29) is equivalent to

$$\begin{pmatrix} L_0''(x, \hat{\lambda}) & e'(x)^* \\ e'(x) & 0 \end{pmatrix} \begin{pmatrix} x_+ - x \\ \lambda_+ - \hat{\lambda} \end{pmatrix} = - \begin{pmatrix} L_0'(x, \hat{\lambda}) \\ e(x) \end{pmatrix} \tag{31}$$

$$\text{where } \hat{\lambda} = \lambda + c\,e(x).$$

Note that $\hat{\lambda}$ is nothing but the first order update of the Lagrange multiplier if the current iterate x minimizes $L_c(x, \lambda)$. Equation (31) is more advantageous than (29) since the squaring term $c\,e'(x)^*e'(x)$ is absorbed and less calculation is involved. If we define a matrix operator S on $X \times Y$ by

$$S(x, \lambda) = \begin{pmatrix} L_0''(x, \lambda) & e'(x)^* \\ e'(x) & 0 \end{pmatrix},$$

then it follows from (27) that $S(x^*, \lambda^*)$ is boundedly invertible. Thus, if (x, λ) is sufficiently close to (x^*, λ^*), then equation (31) has a unique solution. We summarize our discussions as

Algorithm 1

(1) Choose $\lambda^1 \in Y,\ c > \bar{c} \geq 0$, and set $\hat{c} = c - \bar{c},\ k = 1$.

(2) Determine $x = ((u, T - T_0), g) \in V \times C$ such that

$$L_c(x, \lambda^k) \leq L_c(x^*, \lambda^k) = f(x^*).$$

(3) Set $\hat{\lambda} = \lambda^k + \hat{c}\,e(x)$.

(4) Solve for $(x_+, \lambda_+) \in X \times Y$:

$$S(x, \hat{\lambda}) \begin{pmatrix} x_+ - x \\ \lambda_+ - \hat{\lambda} \end{pmatrix} = - \begin{pmatrix} L_0'(x, \hat{\lambda}) \\ e(x) \end{pmatrix}.$$

(5) Set $x_{k+1} = x_+$ and $\lambda^{k+1} = \lambda_+$. If the convergence criterion is not satisfied then set $k = k + 1$ and go to (2).

Remark. A variant of Algorithm 1 is obtained by skipping step (2). Then it is reduced to the Newton method applied to equation (28). Step (2) implies a sufficient reduction of the merit functional (the augmented Lagrange functional). For example, if $x = (z, g)$ minimizes $L_c(\cdot, \lambda^k)$ over $V \times C$ then step (2) is completed. Assume that (19) and (27) hold. It is proved in [11] that if $|\lambda^1 - \lambda^*|_Y$ is sufficiently small, then Algorithm 1 is well-posed and (x^k, λ^k) converges to (x^*, λ^*) Q-quadratically.

5. Hybrid method and test example

In this section we present an example to demonstrate the applicability of our theoretical results and proposed algorithm. We consider the optimal control of the two dimensional stationary thermally driven cavity flow

$$\begin{aligned} \text{minimize} \quad & J(g) = \tfrac{1}{2} \int_\Omega |u - u_d|^2 + |T - T_d|^2 \, dx + \tfrac{\beta}{2} |g - T_0|^2_{L^2(\Gamma_1)} \\ & \text{over } g \in L^2(\Gamma_1) \end{aligned} \tag{32}$$

subject to (2), where $\Omega = (0, L)^2$, $\Gamma_1 = (x_1, 1),\ 0 < x_1 < L$ and Γ_0 is the relative interior of $\Gamma - \Gamma_1$. On Γ_0 we have the Dirichlet boundary condition:

$$\begin{aligned} & T(0, x_2) = T(L, x_2) = T_0, \quad 0 < x_2 < L \\ & T(x_1, 0) = T_0 - 50 \min(1, \tfrac{2}{3}(2.5 - |\tfrac{5}{L}x_1 - 2.5|), \quad 0 < x_1 < L\,. \end{aligned} \tag{33}$$

with a reference temperature $T_0 = 1350^o K$, and the temperature on the substrate $(x_1, 0)$, $\frac{1.5}{5}L \le x_1 \le \frac{3.5}{5}L$ is $1300^o K$. The desired state (u_d, T_d) appearing in (32) is chosen as follows. $u_d = L\, u^0$ where (u^0, T^0) is the solution pair for the flow with $L = 1$ and $g = T_0$ and $T_d = T^1$ where (u^1, T^1) is the solution pair for the flow with $L = 5$ and $g = T_0$. Here, our numerical calculation strongly indicates that the solution to (2) is unique and (dynamically) stable for the range of physical parameters that we chosen for our calculation.

The problem is scaled so that the velocity field u has dimension cm/sec and T has dimension K^o. The constants ν, κ are chosen for P_2 at $3atm$ pressure and are given by $\nu = .155$ and $\kappa = .110$. H, β and L are set to be $H = 100$, $\beta = \frac{1000}{1350^2} = 5.5 \times 10^{-5}$ and $L = 5$ respectively. Figures 1 and 2 show the vector field of u^0 and u^1, respectively. It can be observed that u^0 has more vertical transport than u^1 does and u^1 is confined in the two bottom corners. Hence the cost-functional (24) is formulated so that the thermal control g on the top Γ_1 increases the vertical transport of flow with $L = 5$ while retaining the temperature distribution T^1 at the reference temperature $g = T_0$.

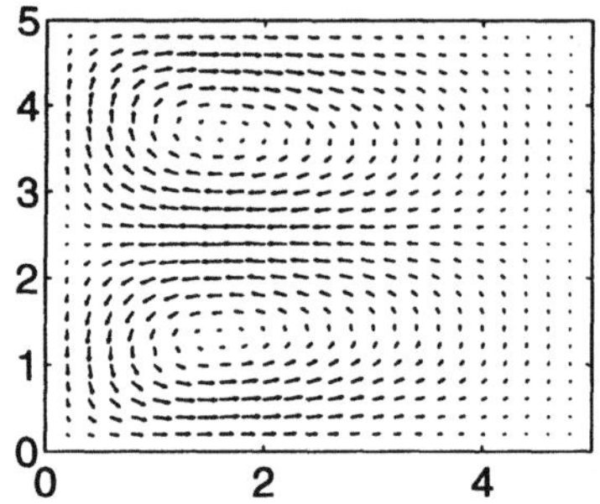

Figure 1. The desired flow.

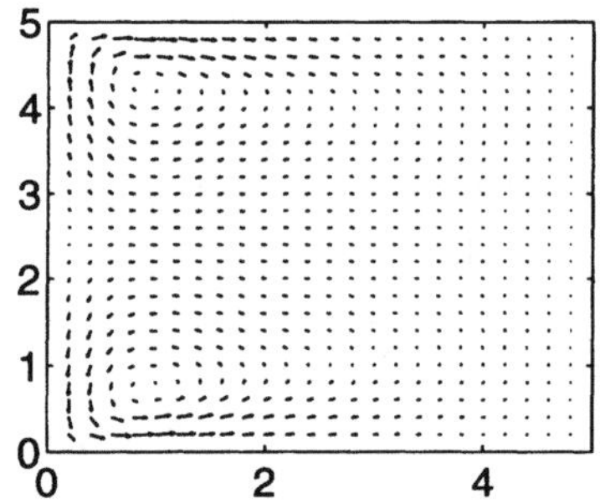

Figure 2. The uncontrolled flow.

The second order augmented Lagrangian method (Algorithm 1) described in Section 4 is used to solve problem (32). To obtain a good starting value λ^1 for the Lagrange multiplier in Algorithm 1, we employed a few steps of the gradient method. The gradient of the cost functional can be calculated by the adjoint equation as follows. Consider the cost functional $J(z, g)$ subject to $E(z, g) = 0$ where $z \in V$ and $g \in U$. Assume that V and U are Hilbert spaces, $J(z, g) : X = V \times U \to R$ is Fréchet differentiable and $E(z, g) : X \to Y$ is continuously Fréchet differentiable in a neighborhood of $x_0 = (z_0, g_0)$. Suppose that $E_z(x_0) \in \mathcal{L}(V, Y)$, the $F-$derivative of E with respect to z at x_0, has a bounded inverse. Then by the implicit function theorem, there exists a unique C^1 mapping Ψ defined in a neighborhood N of g_0 in U, $\Psi : U \to V$ such that $\Psi(g_0) = z_0$ and $E(\Psi(g), g) = 0$ for $g \in N$. Then the $F-$derivative d of $J(\Psi(g), g)$ with respect to g in N exists and is given by

$$d = J_g + E_g^* \psi \tag{34}$$

where $\psi \in Y$ satisfies the adjoint equation

$$E_z^* \psi + J_z = 0\,. \tag{35}$$

Here we assume that V, U and Y are identified with their dual spaces. In fact, if $v = \Psi'(h)$, $h \in U$ then

$$(d, h)_U = (J_g, h)_U + (J_z, v)_V$$

and $v \in V$ satisfies

$$E_z v + E_g h = 0\,.$$

Hence, for $h \in U$

$$(d, h)_U = (J_g, h)_U - (E_z^* \psi, v)_V = (J_g, h)_U + (\lambda, E_z h)_V = (J_g + E_z^* \psi, h)$$

which implies (34). The projected gradient method can be written as:

Algorithm 2

(1) Choose $g_1 \in U$ and set $k = 1$.

(2) Let z_k be a solution of $E(z, g_k) = 0$, ψ_k be the solution of $E_z(z_k, g_k)^* \psi_k + J_z(z_k, g_k) = 0$ and set $d_k = J_g(z_k, g_k) + E_g(z_k, g_k)^* \psi_k$.

(3) Set $d_k = J_g(z_k, g_k)$ and determine $\alpha_k \geq 0$ such that $J(\Psi(g_\alpha), g_\alpha)$ is minimized where $g_\alpha = Proj_C(g_k - \alpha\, d_k)$.

(4) Set $g_{k+1} = g_k - \alpha_k\, d_k$. If the convergence criterion is not satisfied, then set $k = k + 1$ and go to Step (2).

In our specific example equation (35) is written as

$$\begin{gathered}
-\nu\,\Delta\lambda - u \cdot \nabla\lambda + \lambda^t \nabla u - (T - T_0)\,\nabla\mu + \nabla q + u - u_d = 0 \\
\nabla \cdot \lambda = 0 \quad \text{and} \quad \lambda|_\Gamma = 0 \\
-\kappa\,\Delta\mu - u \cdot \nabla\mu - \frac{g}{T_0}\,\lambda_2 + T - T_d = 0 \\
\mu = 0 \text{ on } \Gamma_0 \quad \text{and} \quad n \cdot \nabla\mu + \kappa H\,\mu = 0 \text{ on } \Gamma_1\,,
\end{gathered} \tag{36}$$

where $\psi = (\lambda, \mu) \in V = V_0 \times V_1$, $q \in L^2(\Omega)$ and

$$E_z^* \psi = -\kappa H\,\mu \quad \text{on } \Gamma_1\,. \tag{37}$$

We used the mixed-finite element method [7] based on the Legendre polynomials to approximate problem (1)-(2) numerically. Detailed discussions about the method are given in [13]. In our implementation we calculated the adjoint system for the approximated problem and solved equations (2) and (36) using GMRES [3]. The specific implementation of GMRES applied to our example is described in [13] in detail. Concerning the divergence-free constraint $\nabla \cdot u = 0$, we employed the feasible method, projecting the first equation onto the divergence–free space V_0 as in [4], [6], [13]. The line search in Step 3 of Algorithm 2 was performed by the linearization of the constraint $E(z, g) = 0$ at (z_k, g_k) since the cost functional is quadratic. That is, if v_k is the solution to $E_z(z_k, g_k)v_k + E_g(z_k, g_k)d_k = 0$ then $\alpha > 0$ is chosen so that $J(z_k - \alpha\, v_k, g_k - \alpha\, d_k)$ is minimized.

For this specific example, three steps of Algorithm 2 were performed. We then set x^1 and λ^1 as $x^1 = (z_4, g_4)$ and $\lambda^1 = \psi_3$ for Algorithm 1 without Step 2. The matrix operator $S(x, \lambda)$ was calculated for the approximated system and the resulting linear equation (31) was again solved by GMRES.

The calculations were performed using a 20×20 Cartesian product of Legendre polynomials, choosing $c = 1$ and $g_1 = 0$. Algorithm 1 was terminated after three iterates, since the necessary and sufficient optimality condition (28) was satisfied within a residual norm of 1×10^{-7}.

We may compare this rapid convergence of the hybrid method with the results of using either algorithm by itself. Algorithm 2 did not fully converge after 50 iterates. Algorithm 1, with the start-up $x^1 = ((u^1, T^1), 0) \in X$ and $\lambda^1 = 0$, also failed to converge. Thus, the use of the hybrid method combining the gradient method and the second-order augmented Lagrangian was essential for the success of our numerical calculations. Figure 3 shows the iterates g_k (the first three curves from top to middle) for Algorithm 1 and the (calculated) optimal control g^* (the lowest curve). The iterates for Algorithm 1 are not shown because they coincide with g^* within the accuracy of the plotting. Figure 4 shows the resulting vector field u^* which corresponds to g^*. It shows clearly that the vertical transport of flow is increased.

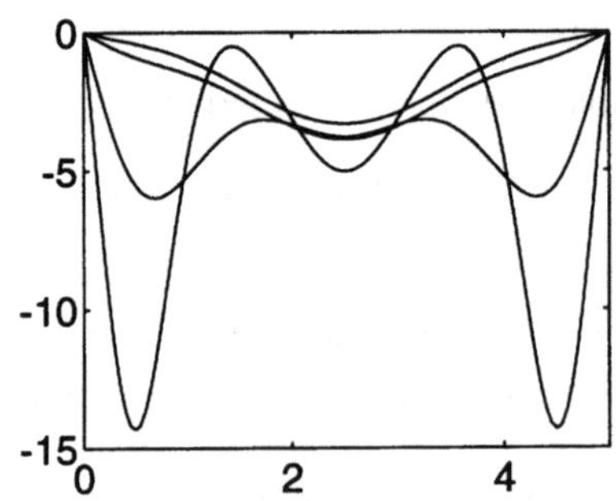

Figure 3. Optimal control iterates.

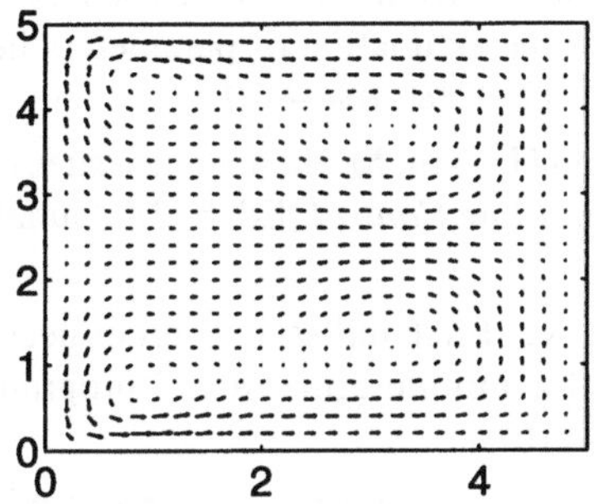

Figure 4. The controlled flow.

References

[1] F. Abergel and R. Temam; On some control problems in fluid mechanics, *Theor. Comp. Fluid Mech.* **1**, 1990, 303–325.

[2] D. Bertsekas; *Constraint Optimization an Lagrange Multiplier Methods*, Academic, Paris, 1982.

[3] P. Brown and Y. Saad; Hybrid Krylov methods for nonlinear systems of equations, *SIAM J. Sci. Statist. Comput.* **11**, 1990, 450–481.

[4] M. Desai and K. Ito; Optimal control of Navier-Stokes equations, *SIAM J. Cont.*, to appear.

[5] I. Ekeland and R. Temam; *Convex Analysis and Variational Problems*, North Holland, Amsterdam, 1976.

[6] R. Glowinski; *Numerical Methods for Nonlinear Variational Problems*, Springer, Berlin, 1984.

[7] V. Girault and P.-A. Raviart; *Finite Element Methods for Navier-Stokes Equations*, Springer, Berlin, 1984.

[8] M.R.Hestenes; Multiplier and gradient methods, *J. Opt. Theory Appl.* **4**, 1968, 303–320.

[9] K. Ito and K. Kunisch; The augmented Lagrangian method for equality and inequality constraints in Hilbert spaces, *Math. Prog.* **46** ,1990, 341–360.

[10] K. Ito and K. Kunisch; The augmented Lagrangian method for parameter estimation in elliptic systems, *SIAM J. Cont. & Optm.* **28**, 1990, 113–136.

[11] K. Ito and K. Kunisch; The augmented Lagrangian method with second order update and its application to parameter estimation in elliptic systems, to appear.

[12] K. Ito, M. Kroller, and K. Kunisch; A numerical study of an augmented Lagrangian method for the estimation of parameters in elliptic systems, *SIAM J. Sci. Stat. Comput.* **12**, 1991, 884–910.

[13] K. Ito, J. Scroggs, and H. Tran; Mathematical issues in optimal design of a vapor transport reactor, *Flow Control* (Ed. by M. Gunzburger), Springer, 1995, to appear.

[14] H. Maurer and J. Zowe; First and second-order necessary and sufficient optimality conditions for infinite-dimensional programming problems, *Math. Prog.* **16**, 1979, 98–110.

[15] M. Powell; A method for nonlinear constraints in minimization problems, in *Optimization* (Ed. by R. Fletcher), Academic, New York, 1969.

[16] V. Polyak and Tret'yakov; The method of penalty estimates for conditional extremum problems, *Z. Vychisl. Mat. i Mat. Fiz.* **13** 1973, 34–46.

[17] R. Temam; *Navier-Stokes Equations and Nonlinear Functional Analysis*, SIAM, Philadelphia, 1983.

[18] R. Temam; *Infinite Dimensional Dynamical Systems in Mechanics and Physics*, Springer, New York, 1988.

[19] G. Troianiello; *Elliptic Differential Equations and Obstacle Problems*, Plenum, New York, 1987.

Optimization Methods in the Process Control of Industrial Furnaces*

H. Jäger and E. Sachs
Universität Trier
FB IV – Mathematik
D-54286 Trier, Germany

H. Heidemüller, F. Maschler, and H. Klammer
GIWEP
Tristanstr. 12
D-45473 Mülheim (Ruhr), Germany

G. Woelk
Rhein.-Westf. Technische Hochschule Aachen
Lehrgebiet Industrieofenbau
Kopernikusstr. 16
D-52056 Aachen, Germany

The heating process of an industrial furnace is considered. The goal is to obtain a certain temperature for the ingots when leaving the furnace. The flow of the gas which heats the ingots is the control variable. There are requirements on the minimal temperature in various areas of the furnace which have to be satisfied. In an on-line calculation the controls are computed by the solution of systems of nonlinear equations to meet the requirements. As an alternative procedure we propose to model the physical problem as a minimization problem with constraints. This overcomes some of the difficulties encountered with the previous approach. The first numerical results indicate that this is an interesting method to be pursued further.

1. Introduction

In the production of high quality industrial ceramic products, see, e.g.,[1] and [9], it is necessary to heat the charge up to a certain temperature and to keep the temperature at a certain level. A similar process occurs in the heating of metals before they are processed, e.g., in steel rolling factories, [10].

We consider a furnace where the ingots enter the furnace from the left and leave to the right, see, e.g., [2]. The gas which heats the ingots flows in the opposite direction, so that the highest temperature of the ingots is on the right end of the furnace.

There are various engineering problems associated with the process of heating an ingot up to a certain temperature, see, e.g., [2] and [5]:

*This research was supported by the Bundesministerium für Forschung und Technologie through the research initiative "Anwendungsorientierte Verbundvorhaben auf dem Gebiete der Mathematik".

- Measurement of the ingot temperature is possible only at certain locations in the furnace, in particular when entering and leaving the furnace.
- Ingots of different size and material enter the furnace, requiring different temperature profiles.
- Unexpected delays or other irregularities may occur within the furnace.

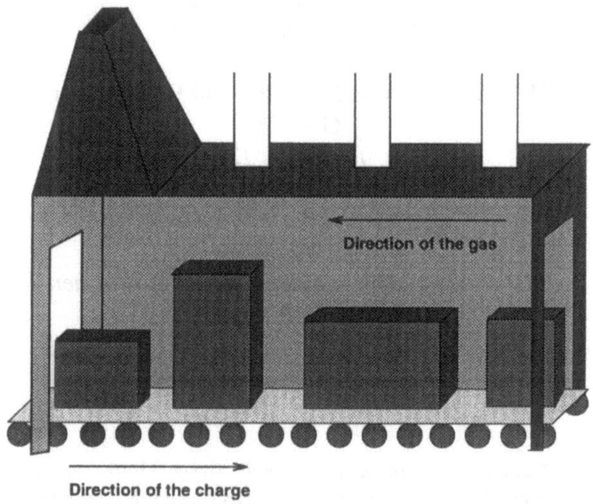

Figure 1. Schematic picture of furnace.

The temperature control of an industrial furnace is a highly complex system, see [8], which exhibits different aspects of applied mathematics. The temperature of the ingots is measured at certain places in the furnace. A model which takes into account the heat transfer by convection and radiation allows the computation of the temperature of the gas as a function of the gas flow rate. The control of the process is achieved by adjusting the gas flow rate. Each ingot requires a certain minimal temperature at every segment. However, the temperature inside the furnace can be controlled only per zone. Therefore, as a compromise, the temperature in each zone is adjusted according to the highest difference between the actual and required temperature of the segments in each zone. This calculation is based on data that change with time and therefore has to be repeated rather frequently.

In the next section we explain the modelling of the furnace. In mathematical terms, this leads to a description of the problem in the form of a system of equations to be solved at each time step. Section 3 describes an algorithm and contains some results in the form of graphs. This iterative procedure is efficient and commonly used. It can be seen that the formulation as a system of nonlinear equations can lead to a solution which is not desirable in practice. This is also indicated by an example in the form of a graph. In Section 4 we reformulate the problem as a nonlinear optimization. The new formulation uses different variables which lead to sparse matrices in the optimization code. The next section defines the algorithm which we use to solve the optimization problems. It is a Sequential Quadratic Programming algorithm (SQP method) which is considered the most successful numerical method in nonlinear programming. We use a standard software package to obtain some initial numerical results. These show that we avoid the undesirable results that can occur in the more common formulation. Results of several example problems are plotted.

2. Mathematical description of the model

To model the process, the furnace is partitioned into Z zones, in each of which the gas flow rate V is treated as constant. Each zone is subdivided into a certain number of segments. The temperatures of the gas T and the ingot y in each zone depend on a certain number of segments which make up each zone. This is illustrated in Figure 2. To reduce the use of subindices, we number the segments consecutively, $i = 1, ..., S$, independently of the zones in which they may lie. We let the index k_i indicate the zone in which the segment i lies. With these conventions for location, we will use the following notation

S		number of segments	
Z		number of zones	
k_i	$i = 1, ..., S$	index of the zone which contains segment i	(1)
$n(k)$	$k = 1, ..., Z$	number of segments in zone k	
$e(k)$		the last segment in zone k	

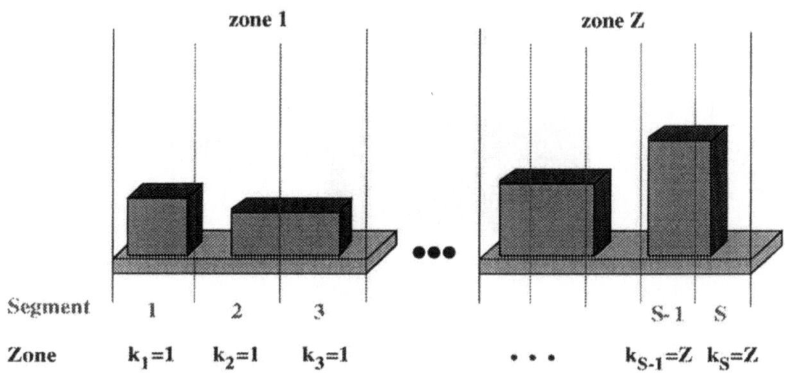

Figure 2. Diagram of zones and segments in a furnace.

For $i = 1, ..., S$ and $k = 1, ..., Z$ we define the following variables and constants which will be used throughout the paper to describe the equations.

T_i :	temperature of the gas in segment i $[K]$
y_i :	temperature of the charge in segment i $[K]$
V_k :	gas flow rate in zone k $\left[\frac{m^3}{s}\right]$
W_k :	cumulative gas flow rate in zone k $\left[\frac{m^3}{s}\right]$
H_u :	thermal value $\left[\frac{J}{m^3}\right]$
L_{min} :	minimal demand for air
c_{p^a} :	heat capacity of the air $\left[\frac{J}{m^3 K}\right]$
T_{air} :	preheated air $[K]$
v_F :	quantity of the gas
$c_{p^{ex}}$:	average heat capacity of the exhaust gas $\left[\frac{J}{m^3 K}\right]$
α_i :	heat transfer coefficient between gas and charge in segment i $\left[\frac{W}{m^2 K}\right]$
σ :	Stefan-Boltzmann constant $5.67\ 10^{-8}$ $\left[\frac{W}{m^2 K^4}\right]$
ϵ_i :	radiation exchange coefficient between gas and charge in segment i
A_i :	surface area of ingot in segment i $[m^2]$
A :	contact area between two segments $[m^2]$
ϵ :	constant radiation exchange coefficient between two segments

The heat exchange within a segment and with its neighboring segments is modelled as suggested in Figure 3.

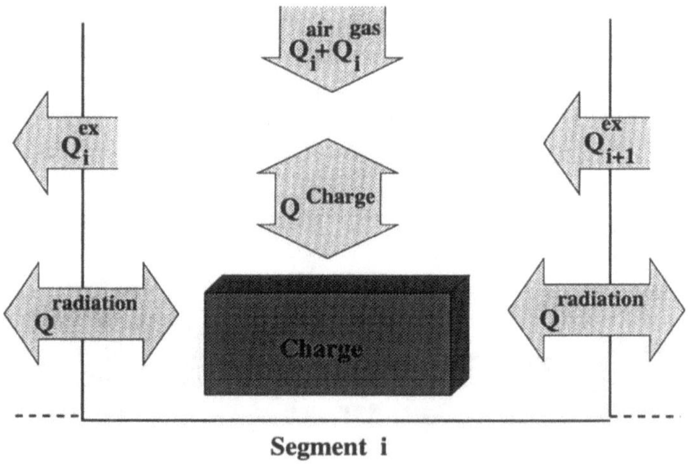

Figure 3. Diagram of heat exchange for a segment.

We now consider the heat fluxes which need to be balanced in the $i-th$ segment. The *gas enthalpy flux* is the product of the gas flow rate in the segment i and the thermal value H_u of the gas:

$$Q_i^{gas}(V) = \frac{V_{k_i}}{n(k_i)} H_u . \tag{2}$$

Similarly, the *air enthalpy flux* is given by:

$$Q_i^{air}(V) = \frac{V_{k_i}}{n(k_i)} L_{min} c_{p^a} (T_{air} - 298.15) , \tag{3}$$

where c_{p^a} is the heat capacity of the air and 298.15 is the reference temperature in Kelvin for the thermal value of the gas.

The *exhaust gas flux* is composed of two components, one entering the segment from the right and one leaving it to the left. The term entering the segment i is given by:

$$C_{i+1}^{ex}(V)(T_{i+1} - 298.15) ,$$

where

$$C_{i+1}^{ex}(V) = \sum_{j=i+1}^{S} \frac{V_{k_j}}{n(k_j)} v_F c_{p^{ex}} . \tag{4}$$

In the same way one can summarize the exhaust gas flux leaving segment i as:

$$C_i^{ex}(V)(T_i - 298.15) .$$

Heat transfer also takes place between the gas, the wall of the furnace, and the charge in the furnace. In the model used in practice, these effects are combined in terms which describe the heat flux between the charge and the gas and the heat flux between two segments. The *heat flux* between *charge and gas* consists of radiative and convective heat transfer

$$A_i \left(\varepsilon_i \sigma \left(T_i^4 - y_i^4 \right) + \alpha_i \left(T_i - y_i \right) \right) ,$$

where A_i is the surface area of the ingot in segment i, ε_i the radiation exchange coefficient between the ingots and the gas, σ the Stefan-Boltzmann constant and α_i the heat transfer coefficient between the ingots and the gas. The heat transfer by *radiation between segments* is given by

$$A\varepsilon\sigma \left(T_{i+1}^4 - T_i^4 \right) \text{ and } A\varepsilon\sigma \left(T_i^4 - T_{i-1}^4 \right) ,$$

where the first term concerns the segment to the right and the second term the segment to the left of the i-th segment. It is usual to include in the heat balance a term describing the *heat loss* Q^{loss}.

Over each segment, the heat terms should sum to zero. We consider this sum as a function of the temperature of the gas T_i in each segment and the gas flow rate

V_k in each zone. For segments $i = 2, \ldots, S-1$:

$$\begin{aligned} f_i(T,V) &= Q_i^{gas}(V) + Q_i^{air}(V) - Q_i^{loss} \\ &\quad + C_{i+1}^{ex}(V)(T_{i+1} - 298.15) - C_i^{ex}(V)(T_i - 298.15) \\ &\quad + A\,\varepsilon\sigma(T_{i+1}^4 - T_i^4) - A\,\varepsilon\sigma(T_i^4 - T_{i-1}^4) \\ &\quad - A_i(\varepsilon_i\sigma(T_i^4 - y_i^4) + \alpha_i(T_i - y_i)) \,. \end{aligned} \tag{5}$$

Similar expressions are obtained for the leftmost and rightmost segments:

$$\begin{aligned} f_1(T,V) &= Q_1^{gas}(V) + Q_1^{air}(V) - Q_1^{loss} \\ &\quad + C_2^{ex}(V)(T_2 - 298.15) - C_1^{ex}(V)(T_1 - 298.15) \\ &\quad + A\,\varepsilon\sigma(T_2^4 - T_1^4) - A_1(\varepsilon_1\sigma(T_1^4 - y_1^4) \\ &\quad + \alpha_1(T_1 - y_1)) \end{aligned} \tag{6}$$

and

$$\begin{aligned} f_S(T,V) &= Q_S^{gas}(V) + Q_S^{air}(V) - Q_S^{loss} - C_S^{ex}(V)(T_S - 298.15) \\ &\quad - A\,\varepsilon\sigma(T_S^4 - T_{S-1}^4) - A_S(\varepsilon_S\sigma(T_S^4 - y_S^4) + \alpha_S(T_S - y_S)) \,. \end{aligned} \tag{7}$$

Hence, given the gas flow rates in each zone, determining the temperature of the gas in each segment is equivalent to solving the following system of nonlinear equations:

For given $V \in I\!R^Z$ solve $f_i(T,V) = 0, \ \ i = 1, \ldots, S$ for $T \in I\!R^S$.

3. Algorithm for nonlinear system

Since the ingots have to achieve a certain temperature when they leave the furnace for further processing, the gas temperatures in the segments cannot be too low. Hence, in addition to solving the system of nonlinear equations, certain inequality constraints must be imposed. The problem in the industrial application is to find a proper gas flow rate such that the resulting temperatures satisfy certain minimal requirements: $T_i \geq T_i^{min} \ \ i = 1, \ldots, S$. It is further assumed that in each zone j there is at least one segment i_j where the minimal temperature is attained: $f_{i_j}(T_i^{min}, V) = 0 \ \ j = 1, \ldots, Z$. Then we obtain:

Nonlinear Equation

Find $V_1, \ldots, V_Z$ and $T_1, \ldots, T_S$ such that

$$\begin{aligned} f_i(T,V) &= 0 & i &= 1,\ldots,S, \\ f_{i_j}(T^{min},V) &= 0 & j &= 1,\ldots,Z, \\ T_i &\geq T_i^{min} & i &= 1,\ldots,S\,. \end{aligned}$$

The algorithm outlined below and used in practice, is based on the following arguments.

Initialize all gas flow rates V_j to 0. Then, from the system $f(T, V) = 0$, compute the resulting temperatures T. Therefore, at each iteration we have some T, V such that $f(T, V) = 0$. Since the gas flow rate is zero initially, the temperature will not always satisfy the lower bound constraint. For each zone $j = 1, ..., Z$ pick the segment i_j in the zone where the constraint $T_i \geq T_i^{min}$ is violated the most. For the next iteration, the gas flow rate in this zone is adjusted in such a way that the temperature in the segment exactly equals the minimal temperature, that is, solve

$$f_{i_j}(T^{min}, V) = 0, \quad j = 1, ..., Z$$

for the new iterate $V^+ \in I\!R^Z$. Then one can compute the new temperatures T^+ by solving $f(T, V^+) = 0$. In mathematical terms, the algorithm can be described as follows:

Algorithm

Step 0 Given $V \in I\!R^Z$.

Step 1 Solve $f_i(T, V) = 0, \ \ i = 1, ..., S$, for $T \in I\!R^S$.

Step 2 For each zone $j = 1, ..., Z$ find i_j such that

$$T_{i_j} - T_{i_j}^{min} = \min_{i \in \text{zone}_j} T_i - T_i^{min}.$$

Step 3 Solve $f_{i_j}(T^{min}, V^+) = 0, \ \ j = 1, ..., Z$, for $V^+ \in I\!R^Z$.

Step 4 Set $V = V^+$ and go to Step 1.

Note that while the system to be solved in Step 3 is linear, the system in Step 1 is nonlinear, and requires something like Newton's method. From a mathematical standpoint, it is not clear whether the iterates of this method converge. In practice this algorithm usually works quite well because good starting data are available.

However, cases have been reported where the choice of the index in Step 2 caused difficulties in Step 3. Consider a zone r with a moderate minimal temperature, but lying between two zones $r - 1$ and $r + 1$ with very high minimal temperatures. Forcing the temperature in zone r to equal the minimal temperature in at least one segment can lead to negative gas flow rates. This situation is highly undesirable in practice, because it indicates that energy is being wasted. If one sets the gas flow to zero where it should be negative in order to avoid this unpleasant result, then in the next iteration the same scenario could occur all over again. After that the code stops because the temperature has not been changed. This situation is illustrated in Figures 4 and 5 where the temperature of the gas and the gas flow rate are plotted as well as the minimum requirements for the temperature.

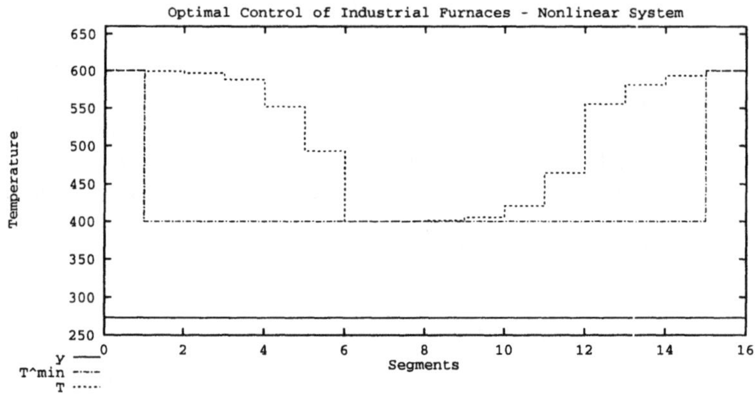

Figure 4. Graph of temperature; Example 1.

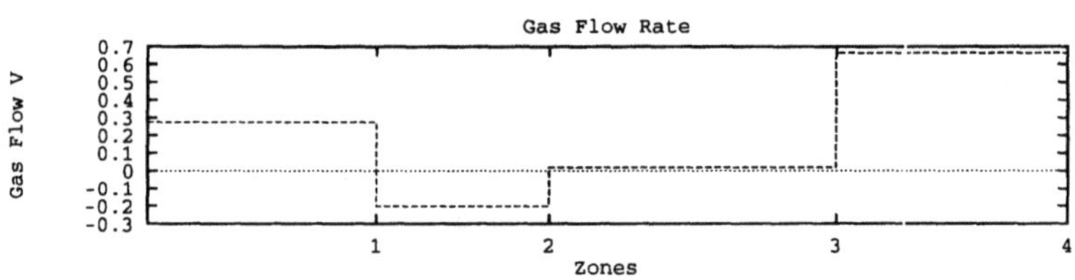

Figure 5. Graph of gas flow rate; Example 1.

4. Formulation as an optimization problem

In order to avoid the drawbacks of the algorithm mentioned above, we propose to model the problem as an optimization problem. The objective function is the sum of the gas flow rates, which corresponds to the cost of the process. As before, the variables are the gas flow rates and the gas temperatures. The equations are introduced as equality constraints. In addition to the inequality constraints given by the lower bounds on the temperature, we also require that the gas flow rates be nonnegative, in order to avoid effects like the ones mentioned above.

In order to introduce more sparsity into the problem we define new variables W, the cumulative gas flow rates:

Lemma 1 *For given $V_1, \ldots, V_Z$ define*

$$W_j = \sum_{k=j}^{Z} V_k, \quad j = 1, \ldots, Z.$$

Then we obtain the following functions $f_1, \ldots, f_S : \mathbb{R}^{S+Z} \to \mathbb{R}$. Let $i = 2, \ldots, S-1$ (Using the common convention on summation we set $W_{Z+1} = 0$):

$$\begin{aligned}
f_i(T, W) &= \frac{W_{k_i} - W_{k_i+1}}{n(k_i)}(H_u + L_{min}\, c_{p^a}\, (T_{air} - 298.15)) \\
&\quad + A\,\varepsilon\sigma(T_{i+1}^4 - 2T_i^4 + T_{i-1}^4) - A_i(\varepsilon_i\sigma(T_i^4 - y_i^4) + \alpha_i(T_i - y_i)) \\
&\quad - Q_i^{loss} + W_{k_i+1}\, v_F\, c_{p^{ex}}\, (T_{i+1} - T_i) \\
&\quad + \frac{W_{k_i} - W_{k_i+1}}{n(k_i)} v_F c_{p^{ex}}((e(k_i) - i)(T_{i+1} - T_i) - (T_i - 298.15)),
\end{aligned}$$

$$\begin{aligned}
f_0(T, W) &= \frac{W_0 - W_1}{n(0)}(H_u + L_{min}\, c_{p^a}\, (T_{air} - 298.15)) \\
&\quad + A\,\varepsilon\sigma(T_1^4 - T_0^4) - A_0(\varepsilon_0\sigma(T_0^4 - y_0^4) + \alpha_0(T_0 - y_0)) \\
&\quad - Q_0^{verl} + W_1\, v_F\, c_{p^{ex}}\, (T_1 - T_0) \\
&\quad + \frac{W_0 - W_1}{n(0)}\, v_F\, c_{p^{ex}}(e(0)(T_1 - T_0) - (T_0 - 298.15)),
\end{aligned}$$

$$\begin{aligned}
f_S(T, W) &= \frac{W_{k_S}}{n(k_S))}(H_u + L_{min}\, c_{p^a}\, (T_{air} - 298.15)) \\
&\quad - A\,\varepsilon\sigma(T_S^4 - T_{S-1}^4) - A_S(\varepsilon_S\sigma(T_S^4 - y_S^4) + \alpha_S(T_S - y_S)) \\
&\quad - Q_S^{loss} - \frac{W_{k_S}}{n(k_S)}\, v_F c_{p^{ex}}\, (T_S - 298.15)\,.
\end{aligned}$$

Proof Let $i = 2, \ldots, S-1$. Then

$$\begin{aligned}
f_i(T, V) &= Q_i^{gas}(V) + Q_i^{air}(V) - Q_i^{loss} \\
&\quad -(C_{i+1}^{ex}(V) - C_i^{ex}(V))(T_{i+1} - 298.15) + C_{i+1}^{ex}(V)(T_{i+1} - T_i) \\
&+ \quad A\varepsilon\sigma(T_{i+1}^4 - 2T_i^4 - T_{i-1}^4) - A_i(\varepsilon_i\sigma(T_i^4 - y_i^4) + \alpha_i(T_i - y_i))
\end{aligned}$$

Note that by definition of W $V_k = W_k - W_{k+1}$ and from (4)

$$C_{i+1}^{ex}(V) - C_i^{ex}(V) = -\frac{V_{k_i}}{n(k_i)}\, v_F c_{p^{ex}} = \frac{W_{k_i+1} - W_{k_i}}{n(k_i)}\, v_F c_{p^{ex}}$$

and with the definition of (1)

$$C_{i+1}^{ex}(V) = \sum_{j=i+1}^{S} \frac{V_{k_j}}{n(k_j)}\, v_F c_{p^{ex}} = \Big(\frac{e(k_i) - i}{n(k_i)}(W_{k_i} - W_{k_{i+1}}) + W_{k_i+1}\Big) v_F c_{p^{ex}}$$

Then we substitute these equations with (2) and (3) into (5) to obtain the identity in the Lemma. The rest can be verified similarly. □

This leads to a new formulation as an optimization problem:

Nonlinear Optimization Problem

Minimize W_1 subject to

$$\begin{array}{rcll} f_i(T, W) & = & 0 & i = 1, \dots, S \\ T_i - T_i^{min} & \geq & 0 & i = 1, \dots, S \\ W_i - W_{i+1} & \geq & 0 & i = 1, \dots, Z-1 \\ W_Z & \geq & 0 & \end{array}$$

5. Optimization method

In order to solve the constrained optimization problem efficiently, we propose the use of an SQP method as a numerical routine. For an introduction to these methods we refer to [3] and [4]. There are versions designed in particular for control problems with partial differential equations, such as a reduced SQP method in [6] or SQP combined with an interior point algorithm [7].

To describe the algorithm in a more compact form, we introduce a new variable x and the following matrices and vectors:

$$\begin{aligned} x &= (T_1, \dots, T_S \mid W_1, \dots, W_Z)^T \in I\!R^{S+Z} \\ c &= (0, \dots, 0 \mid 1, 0, \dots, 0)^T \in I\!R^{S+Z} \\ l &= (T_1^{min}, \dots, T_S^{min} \mid 0, \dots, 0)^T \in I\!R^{S+Z} \\ A &= \left(\begin{array}{c|c} I_S & 0 \\ \hline 0 & B \end{array}\right) \in I\!R^{(S+Z)\times(S+Z)} \quad \text{where} \\ B &= \begin{pmatrix} 1 & -1 & & \\ & \ddots & \ddots & \\ & & 1 & -1 \\ & & & 1 \end{pmatrix} \in I\!R^{Z\times Z} \end{aligned}$$

Then the optimization problem can be rewritten as

Nonlinear Optimization Problem

Minimize $c^T x$ subject to

$$\begin{aligned} f(x) &= 0 \\ Ax &\geq l \end{aligned}$$

For a given iterate x_k let the matrix $H_k \in \mathbb{R}^{(S+Z)\times(S+Z)}$ be an approximation for the Hessian of the Lagrangian

$$c^T x + \lambda^T f(x) + \mu^T (Ax - l)$$

where the Lagrange multipliers for the equality and inequality constraints are $\lambda \in \mathbb{R}^S$ and $\mu \in \mathbb{R}^{S+Z}$, respectively. A prototype of the SQP method then looks as follows: One needs to solve the following quadratic subproblem at each iteration k:

$$\min_{\delta x} \quad c^T \delta x + \tfrac{1}{2}\, \delta x^T H_k \delta x$$

$$\text{s.t.} \quad \begin{aligned} f'(x_k)\, \delta x &= -f(x_k) \\ A(x_k + \delta x) &\geq l\,. \end{aligned}$$

The next iterate is given by

$$x_{k+1} = x_k + \alpha \delta x_k$$

for some step size $\alpha > 0$ which is chosen on the basis of a proper decrease in a merit function. Locally, i.e., in a neighborhood of the solution x_*, the step size should be 1. In an implementation of this code one would add a suitable active set strategy. Examples are discussed in the references above.

In order to check the feasibility of the approach via optimization, we used a software package. The following results were obtained with code E04UCF from the NAG library. The constants used were:

H_u	=	$3.500\left[\frac{J}{m^3}\right]$	σ	=	5.67e-8 $\left[\frac{W}{m^2K^4}\right]$
L_{min}	=	10	ϵ_i	=	0.3
c_{p^a}	=	$1.300\left[\frac{J}{m^3K}\right]$	A_i	=	$5\,[m^2]$
T_{air}	=	$673\,[K]$	A	=	$35\,[m^2]$
v_F	=	10	ϵ	=	1
$c_{p^{ag}}$	=	$1.400\left[\frac{J}{m^3K}\right]$	S	=	16
α_i	=	$20\left[\frac{W}{m^2K}\right]$	Z	=	4

As an example, we used the same data as for the graph in Section 3. There the requirement was imposed that in each zone the temperature has to actually attain the minimal temperature in at least one segment. This led in a particular case to negative gas flow rates and caused the algorithm to stop. The same problem was solved with an SQP method and the graph of this solution is shown in Figures 6 and 7. The gas flow rates are all nonnegative and the temperatures lie above the required minimal temperatures. Figures 8 and 9 show the results when an increasing minimal temperature is imposed.

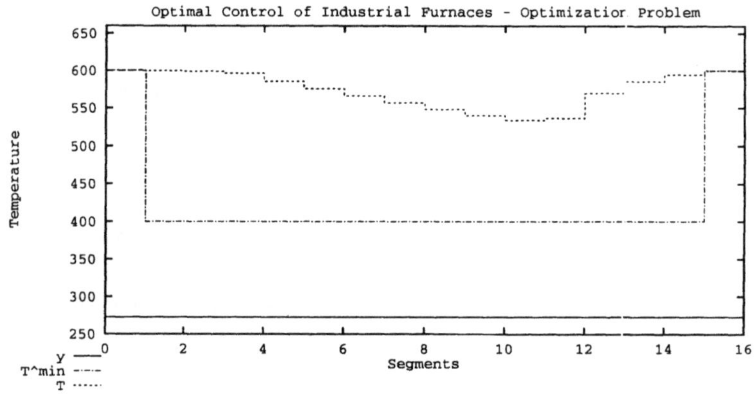

Figure 6. Graph of temperature; Example 1.

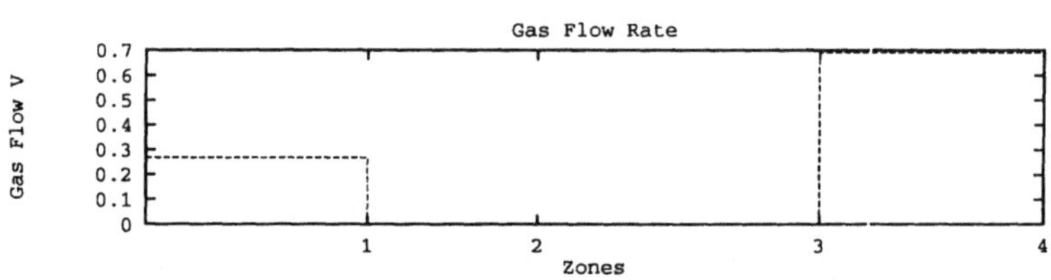

Figure 7. Graph of gas flow rate; Example 1.

The starting data in this case were far away from the solution, which caused the algorithm to use up to 80 iterations. This should improve when one can use as starting data the temperatures from the previous time step. The underlying structure of the optimization problem is quite sparse. Note that A is bidiagonal and that f' is a tridiagonal matrix. This has not been used in the code at this point and is a topic of further research and potential improvement.

6. Conclusion

We have shown that the modelling of the control of the heating gas in an industrial furnace can lead to various formulations. One way is the formulation as a nonlinear equation with inequality constraints on some of the variables. Since this system may be underdetermined, the solution algorithm starts with values equal to zero and determines a solution iteratively. This procedure is efficient and commonly used. We have seen, however, that this method can lead to undesirable results. Our proposed remedy reformulates the problem as a nonlinear optimization, to be solved by an SQP method. Cases which failed under the first formulation can now be solved in a proper way. Further research is required for an efficient implementation of the SQP method which takes the special structure of the problem into account. One could also add other constraints or change the objective function.

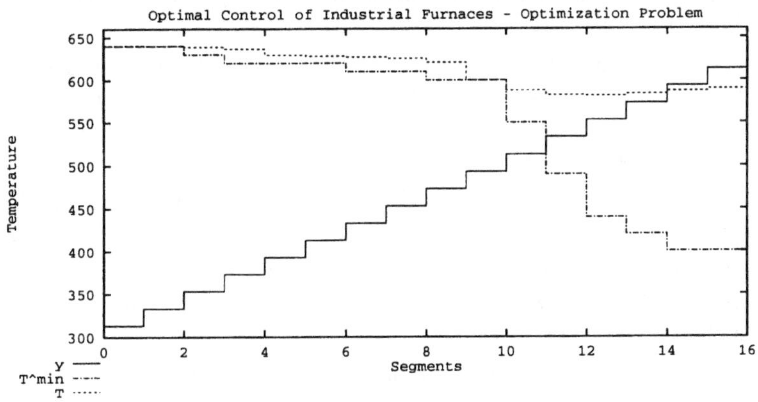

Figure 8. Graph of temperature; Example 2.

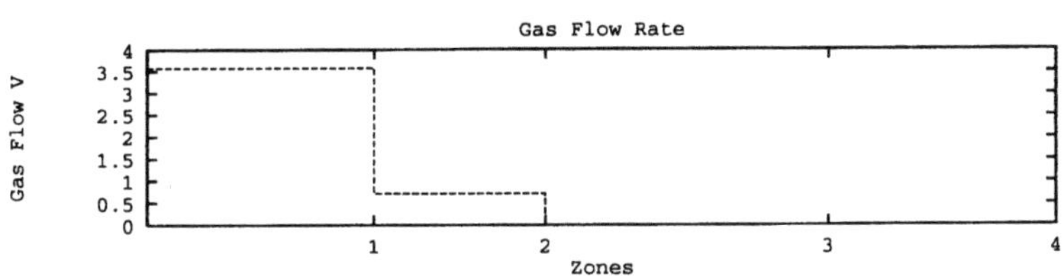

Figure 9. Graph of gas flow rate; Example 2.

References

[1] D. Barreteau, M. Hemati, J. Babary, and E. Weiland; On modelling and control of intermittent kilns in the ceramic industry. In *12th IMACS World Congress on Scientific Computation*, 1988, 339–341.

[2] J. Cüppers, H. Heidemüller, H. Klammer, E. Neuhäusler, and G. Woelk; Betriebserfahrungen mit verschiedenen prozeßrechnerunterstützten Erwärmungsstrategien an Hubbalkenöfen in der Buntmetall-Industrie, *Blech Rohre Profile* **11**, 1991, 1–11.

[3] R. Fletcher; *Practical Methods of Optimization*, Wiley, New York, 1987.

[4] P. E. Gill, W. Murray, and M. Wright; *Practical Optimization*, Academic, London, 1981.

[5] H. Klammer, G. Woelk, W. Brüsecke, and H. Heidemüller; Dynamisch-adaptive Prozeßführung eines 35-t/h-Drehherdofens mit Rechnervernetzung auf PC-Basis, *Stahl und Eisen* **110**, 1990, 103–109.

[6] F.-S. Kupfer and E. Sachs; Numerical solution of a nonlinear parabolic control problem by a reduced SQP method, *Comp. Optim. Appl.* **1**, 1992, 113–135.

[7] F. Leibfritz and E. Sachs; Numerical solution of parabolic state constrained control problems using SQP- and interior-point-methods, In *Large Scale Optimization: State of the Art* (Ed. by W. Hager, D. Hearn, and P. Pardalos), Kluwer, 1994, 251–264.

[8] W. Radermacher, G. Uetz, and G. Woelk; Prozeßrechnereinsatz an Erwärmungsöfen, *Technische Mitteilungen AEG-Telefunken* **70**, 1980, 10–15.

[9] E. Weiland and J. Babary; A solution to the IHCP applied to ceramic product firing, In *Proc. 5th Int. Conf. on Numerical Methods in Thermal Problems*, Chichester, 1988, 1358–1367.

[10] J. Yvon; Contrôle optimal d'un four industriel, Technical Report 22, INRIA, 1973.

Maximal Decay Rates and Asymptotic Behavior of Solutions in Nonlinear Elastic Structures

I. Lasiecka and W. Heyman
Department of Applied Mathematics
University of Virginia
Charlottesville VA 22903, USA

1. Introduction

In this paper we study the asymptotic behavior of systems arising in nonlinear elasticity. Of particular interest are problems related to quantitative description (in terms of the parameters in equations) of margin of stability achieved by implementation of a suitable nonlinear damping mechanism. To be more precise, we are interested in the following question: how and to which extent one can maximize the margin of stability by introducing a suitably large damping? Even in the case of a simple linear, constant coefficient problem with viscous damping, it is known that the maximum of decay's rate depends not only on the damping coefficient (representing the dynamic properties of the model) but also on the static properties of the model (for instance: the first eigenvalue of the Laplacian–the case of wave equation). In fact, in a recent paper [4] it was shown that the exact decay rates in *one-dimensional* wave equation with viscous damping are determined by the location of the eigenvalue corresponding to the wave operator. The above result, for a one-dimensional linear wave equation, is rather technical and it is based on Riesz property of associated eigenfunction. This, in turn, requires the so called "gap condition" which is known to fail for higher dimensional problems.

Thus, one of our goals is to establish a similar conclusion valid for two dimensional nonlinear plate models with nonlinear damping. A quantitative statement describing the balance between the "static" and "dynamic" properties of the model, in regard to maximalization of stability properties, is given in Theorem 3. The result of Theorem 3 is also essential in achieving our next goal which is the analysis of the asymptotic behavior of this model in the case of a load attached to the plate. We shall show (in Theorem 4) that all solutions are attracted by a global and compact attractor. This result extends the results obtained recently in [3], where an existence of a *weakly* compact attractor for a plate with *linear* damping was established. Indeed, our results deal with a strongly nonlinear damping and, moreover, provide the existence of strongly compact attractors. It should be noted, that a key role in our results is played by new "sharp regularity results" of Airy's stress function obtained in [5].

Finally, we note that the interest for studying this type of problems is motivated by problems arising in shape optimization/shape design where a transformation of the domain is reflected by the suitable variations in the coefficients and perturbations of the equation. Thus, an analysis of stability properties of the system with respect to changing coefficients of an underlined PDE will reveal the effects of changing the shape of the domain on the overall stability of the structure.

2. The model

We consider the model of a nonlinear plate with nonlinear damping. Here, the variable w represents a vertical displacement of a thin plate occupying a two dimensional domain Ω with a boundary Γ.

$$\rho h w_{tt} + D\Delta^2 w + \gamma(w_t) + \gamma_1^2 w = [w, \mathcal{F}(w)] + L(w) + F \quad \text{in } \Omega \times (0, \infty) \quad (1)$$

with boundary conditions:

$$w = \frac{\partial w}{\partial \nu} = 0 \quad \text{on } \Gamma \times (0, \infty). \tag{2}$$

$\mathcal{F}(w)$ is an Airy's stress function satisfying

$$\begin{aligned} \Delta^2 \mathcal{F}(w) &= -\frac{Eh}{2}[w, w] \qquad && \text{in } \Omega \\ \mathcal{F} = \frac{\partial}{\partial \nu}\mathcal{F} &= 0 && \text{on } \Gamma. \end{aligned} \tag{3}$$

Here, $D = \dfrac{Eh^3}{12(1-\mu^2)}$ is the modulus of flexual rigidity, $0 < \mu \leq \frac{1}{2}$ is Poisson's modulus, h represents thickness of the plate, E stands for Young's modulus and ρ is the density.

$$[\phi, \psi] \equiv \phi_{xx}\psi_{yy} + \phi_{yy}\psi_{xx} - 2\phi_{xy}\psi_{xy}.$$

The vector ν represents an outward normal. The damping of the plate is represented by a nonlinear function $\gamma(w_t)$ which satisfies the following hypotheses.

$H1$ $\quad \gamma : R^1 \to R^1$ is continuous and monotone decreasing.

The function $\gamma_1(x)$ is assumed continuous and such that $\gamma_1(x) \geq \gamma_1 \geq 0$, $x \in \bar{\Omega}$. $L(w)$ represents an arbitrary (unstructured) second order differential operator whose presence may result from a transformation of the domain. Function $F \in L_2(\Omega)$ represents a constant load attached to the plate.

Remark 1. (i) The results in this paper can be easily adapted to a more general structure of the fourth order elliptic operator in (1) under the sole condition of uniform ellipticity. Different types of boundary conditions can be considered as well. (ii) One could also consider a more general structure of load F, to allow for a nonlinear dependence with respect to the displacement w (see Remark 12).

The energy associated with unloaded plate model (1) is given by

$$E(t) = E_k(t) + E_p(t) \qquad \text{where } E_k(t) = \int_\Omega \rho h w_t^2, \tag{4}$$

$$E_p(t) \equiv \int_\Omega \left(D|\Delta w|^2 + \gamma_1^2(x) w^2 + \frac{2}{Eh}|\Delta \mathcal{F}(w)|^2 \right) d\Omega . \tag{5}$$

It is well known that $E_p(t)$ is topologically equivalent to $H^2(\Omega)$ topology. We shall denote by $H \equiv H^2(\Omega) \times L^2(\Omega)$. The result formulated below asserts the wellposedness of finite energy solutions.

Theorem 1 *For any intial data $w(t=0) = w_0 \in H^2(\Omega)$ and $w_t(t=0) = w_1 \in L^2(\Omega)$, there exists a* unique *solution of finite energy.*

Remark 2. Notice that the "uniqueness" statement in Theorem 1 is new even a simpler case when $\gamma = 0$, $L = 0$,$F = 0$. Indeed, in [9], [10], the problem of uniqueness of finite energy solution for a von Karman model has been addressed as an open problem.

Proof of Theorem 1. The key ingredient in the proof is so called "sharp" regularity of Airy's stress function which follows from the following result.

Proposition 2 *[5]. Let Δ^{-1} denote the solution operator corresponding to the bi-harmonic problem with zero boundary data. Then*

$$|\Delta^{-1}[z, v]|_{H^3(\Omega)} \le C|z|_{H^2(\Omega)} |v|_{H^2(\Omega)} \tag{6}$$

$$|[z, v]|_{L_2(\Omega)} \le C|z|_{H^2(\Omega)} |v|_{H^3(\Omega)} \tag{7}$$

Remark 3. Notice that the standard elliptic theory gives that $\forall \varepsilon > 0$

$$|\Delta^{-1}[z, v]|_{H^{3-\varepsilon}(\Omega)} \le C|z|_{H^2(\Omega)}|v|_{H^2(\Omega)} ,$$

which result *does* not imply (6,7). In fact, the proof of (6,7) is very technical and based on the theory of Hardy's spaces combined with methods of compensated compactness.

From (6–7), it follows that the operator $[\mathcal{F}(v), w]$ is locally Lipschitz: $H^2(\Omega) \to L^2(\Omega)$. Indeed,

$$[\mathcal{F}(w), w] - [\mathcal{F}(v), v] = [\mathcal{F}(w) - \mathcal{F}(v), w] + [\mathcal{F}(v), w - w] . \tag{8}$$

Since $\mathcal{F}(w) - \mathcal{F}(v) = -\frac{Eh}{2}\Delta^{-1}[w - v, w + v]$, from Proposition 2

$$|\mathcal{F}(w) - \mathcal{F}(v)|_{H^3(\Omega)} \le C|w - v|_{H^2(\Omega)} |w + v|_{H^2(\Omega)} \tag{9}$$

and by (7)

$$|[\mathcal{F}(w) - \mathcal{F}(v), w]|_{L^2(\Omega)} \le C|w - v|_{H^2(\Omega)} \left(|w|^2_{H^2(\Omega)} + |v|^2_{H^2(\Omega)} \right) . \tag{10}$$

As for the second term in (8), we recall that $\mathcal{F}(v) = -\frac{Eh}{2}\Delta^{-1}[v, v]$ and by Proposition 2

$$|\mathcal{F}(v)|_{W^2_\infty(\Omega)} \leq C|v|^2_{H^2(\Omega)} \,. \tag{11}$$

$$\begin{aligned} |[\mathcal{F}(v), w - v]|_{L_2(\Omega)} &\leq C|\mathcal{F}(v)|_{H^3(\Omega)}\, |w - v|_{H^2(\Omega)} \\ &\leq C|v|^2_{H^2(\Omega)}\, |w - v|_{H^2(\Omega)} \end{aligned}$$

which together with equations (8) and (10) proves a local Lipschitz property of $[\mathcal{F}(w), w]$. The next step is to prove local (in time) existence of weak solutions. To accomplish this, we construct a fixed point for the map $J : v \to w$, where $w = J(v)$ is defined as a solution to the following nonlinear problem

$$\begin{aligned} \rho h w_{tt} + D\Delta^2 w + \gamma(w_t) + \gamma_1^2 w &= [v, \mathcal{F}(v)] \\ &\quad + L(v) + F \quad \text{in } \Omega \times (0, \infty)\,, \\ w(0) = w_0 \in H^2(\Omega); \quad w_t(0) &= w_1 \in L^2(\Omega)\,, \\ w &= \frac{\partial w}{\partial \nu} \quad \text{on } \Gamma \times (0, \infty). \end{aligned} \tag{12}$$

By using the theory of monotone operators (see [2]), one shows that J is well defined on $H^2(\Omega)$. Moreover, estimates from the theory of nonlinear contractions together with local Lipschitz property of the right side of equation (12) allow us (in a similar manner to the arguments in [5]) to deduce, the existence of a fixed point in $H^2(\Omega)$ (for a small time T_0)). The obtained local result is then extended to the global one by the virtue of apriori bound. To obtain the latter, we multiply equation (1) by w_t and integrate by parts. This yields

$$\begin{aligned} E(t) + 2\int_0^t (\gamma(w_t), w(t))_\Omega\, dt &\leq E(0) \\ +2l \int_0^t |w(t)|^2_{H^2(\Omega)}\, |w_t(t)|_{L^2(\Omega)} &+ \int_0^t (F, w_t)_\Omega\, dt \end{aligned} \tag{13}$$

where $l = \sup_{|\Delta w|_{L^2(\Omega)}=1} |L(w)|_{L^2(\Omega)}$.

Using the structure of $E(t)$ and Gronwall's Inequality applied to (13), yields

$$E(t) \leq Ce^{\omega_0 t}(E(0), |F|_{L^2(\Omega)}); \quad \omega_0 > 0 \tag{14}$$

which is the desired apriori bound.

Remark 4. One can prove the result of Theorem 1 by replacing the assumption on monotonicity of γ by the requirement that γ is Lipschitz.

3. Formulation of the results

Our main goal is to study the asymptotic behavior of solutions to (1). We shall consider two cases: $F = 0$, (which corresponds to lack of load attached to the plate) and $F \neq 0$. In the first case, we shall show that all solutions decay to zero at the uniform rate. For the second case, we shall establish an existence of compact attractor, "attracting" when $t \to \infty$, all the solutions corresponding to the finite energy initial states. Our main results are formulated below.

Theorem 3 *Let $F \equiv 0$ in equation (1). We assume the following growth condition imposed on γ*

$$\gamma_0 s^2 \leq \gamma(s)s \leq M_\gamma s^2, \quad 0 < \gamma_0 \leq M_\gamma . \tag{15}$$

Moreover, we assume that the parameters γ_0 and γ_1 satisfy the following inequalities.

$$\begin{aligned} \gamma_1 &\geq \max\left\{ \frac{\sqrt{8}}{\sqrt{\rho h}} M_\gamma, \frac{4l}{\sqrt{D}} \right\} \\ \gamma_0 &\geq 4\sqrt{2}\frac{\sqrt{\rho h}}{D} l . \end{aligned} \tag{16}$$

Then $E(t) \leq Ce^{-\omega t} E(0)$ where $\omega = \dfrac{1}{3\rho h}\gamma_0$.

Remark 5. The result of Theorem 3 (proved in Section 4) allows us to obtain an arbitrary large exponential decay ($-\omega$) provided that both: the static (γ_1) and the dynamic (γ_0) coefficients are sufficiently large. This is to say that by playing with damping coefficients and the lower order term of static part of the equation only, we can achieve an arbitrary margin of stability. This fact is in agreement with the spectral behavior of a simple linear constant coefficients model which can be analyzed by elementary methods.

Remark 6. One could prove uniform (rather than the exponential) decay rates for solutions to (1) (with $F = 0$) *without* imposing any growth condition on the behavior of nonlinear damping $\gamma(\omega_t)$ at the origin (i.e. *without* assuming the first inequality in (15). This can be accomplished by using a technique introduced in [13] (and applied later, in the context of von Karman equations in [7]) and which is based on application of "comparison theory". This approach allows to deduce the appropriate decay rates by analyzing a suitable nonlinear ODE. For a sake of simplicity of exposition we did not consider here this more general case.

Remark 7. It would be interesting to see whether similar results hold with boundary dissipation only. Although, it is known that von Karman plate, with $L = 0$, $\gamma_1 = 0$, $\gamma = 0$, can be uniformly stabilized by means of boundary feedbacks only (see [9], [7]), the relations between the margin of stability and the parameters of equation are not clear at all. In particular, it is *not* known whether one can achieve an arbitrary large decay rates by "increasing" the boundary velocity feedback together with the lower order term represented by γ_1. Preliminary numerical experiments indicate a possibility of a negative result.

Remark 8. By using the same technique as in the proof of Theorem 3, one can show that the result of Theorem 3 remains valid for a full von Karman system where in-plane and transversal displacements are accounted for. This leads (after a suitable rescaling) to the model

$$\mathbf{u}_{tt} - \text{div}N(\mathbf{u}, w) + \gamma(\mathbf{u}_t) + \gamma_1 \mathbf{u} = 0 \quad \text{in } \Omega \times (0, \infty);$$

$$w_{tt} - h\Delta w_{tt} + D\Delta^2 w - \text{div}N(\mathbf{u}, w)\nabla w - \text{div}\gamma(\nabla w_t) + \gamma_1 w = 0$$

with, say, clamped boundary conditions.

$N(\mathbf{u}, \omega)$ is stress resultant given by $N(\mathbf{u}, \omega) \equiv C[\varepsilon(\mathbf{u}) + f(\nabla\omega)]$ where $\varepsilon(\mathbf{u})$ is the usual strain tensor, $f(s) \equiv \frac{1}{2}s \otimes s$, $\mathbf{u} = (u_1, u_2) : \Omega \to R^2$ and C is a symmetric and coercive tensor.

Our next result deals with the asymptotic behavior of solutions to a loaded plate.

Theorem 4 *Consider equation (1) with a parameter γ_1 satisfying: $\gamma_1 \geq \frac{4l}{\sqrt{D}}$. We assume that the nonlinear function γ satisfies*

$$\gamma_0 \leq \gamma'(s) \leq M_\gamma \quad \textit{where } \gamma_0 \textit{ is subject to (16).} \tag{17}$$

Then, all solutions $w(t)$ to equation (1) converge uniformly to a global attractor which is compact *in H.*

Remark 9. Note that if there is no lower order term $L(w)$ in equation (1), then we can take $\gamma_1 = 0$.

Remark 10. In the case of linear damping, one can show that the attractor of Theorem (4) is *finite dimensional.*

Remark 11. The problem of existence of compact attractors for a von Karman plate model with *linear* damping was considered in [C-1]. Indeed, Chueskov shows that in this linear case, there exists a *weakly* compact attractor. Our result, instead, deals not only with *nonlinear* damping, but also it shows that the attractor is compact in the strong topology of an underlined space H. The key ingredient in the proof of this result is regularity of Airy's stress function (see Proposition 2) together with multiplier's result in Proposition 5) below.

Remark 12. (i) Similar result can be obtained for a larger class of nonlinear function $F(w)$. Indeed, it suffices to assume that either

$$F(s)s \leq 0, \quad s \in R \text{ and } F \in C^1(R), \tag{18}$$

or

$$|F(s)| \leq C_1|s|^p + C_2; \quad \text{for } p < 1. \tag{19}$$

(ii) Assumption (17) can be relaxed by requiring these inequalities to be satisfied for *large* "s" only.

Remark 13. Similar result can be obtained for a von Karman model accounting for the rotational forces (see [11]).

Remark 14. If in addition we assume that either a) $L \in \mathcal{L}(H^{2Q}(\Omega); L_2(\Omega))$ for $Q < 1$ or we assume b) $L : L_2(\Omega) \to L_2(\Omega)$ is selfadjoint and $L \in \mathcal{L}(H^2(\Omega)$ $H^{-1}(\Omega))$ then the conclusion of part (i) holds true without assuming (16), i.e., $\forall \gamma_0 > 0$; (see [16]).

Remark 15. An example of a "lower order term" $L(w)$ arising in application, in the context of models of supersonic gas flow, is the term

$$L(w) = \delta \frac{\partial w}{\partial x},$$

where the parameter δ is determined by the velocity of the flow. Clearly, the operator L above satisfies condition (a) of Remark 14.

Another example, arising in the context of modeling an elastic sloping shell is with

$$L(w) = [w, f_1] + [\mathcal{F}(w), f_2] \equiv L_1(w) + L_2(w)$$

and

$$F = [f_1, f_2]$$

where

$$f_1 \in H^4(\Omega) \quad f_2 \in H^3(\Omega).$$

It can be easily verified that $L_1(w)$ complies with condition (b) of Remark 14 while $L_2(w)$ with (a) of Remark 14.

4. Proofs of Theorems 3 and 4

4.1. Proof of Theorem 3 (uniform decay rates)

In order to establish the decay rates for the energy function, we need to perform certain PDE calculations. These, in turn, require a sufficient regularity of the solutions which may not be available for the nonlinear problem (even if the initial data are smooth). To cope with this difficulty, a technique based on a special regularization of the original problem and passing through to the limit on desired final estimates was introduced in the context of boundary stabilization of von Karman system in [7]. The procedure is fairly technical and it can be adapted to the present problem. For a sake of clarity and simplicity of the exposition, these arguments will not be provided here. Thus, the computations performed below are, in a sense, formal. However, these can be rigorously justified by the procedure referred to above.

Multiplying equation (1) by w_t, integrating by parts over Ω and performing a number of calculations based on application of various forms of Green's Theorem gives

$$E_t(t) = -2\int_\Omega \gamma(w_t)w_t \, d\Omega + 2\int_\Omega L(w)w_t \, d\Omega\,. \tag{20}$$

We define

$$V(t) \equiv E(t) + c\gamma_0 \int_\Omega \rho h w_t w \, d\Omega , \tag{21}$$

where the constant c will be determined later.

$$\begin{aligned} V_t(t) &= E_t(t) + c\gamma_0 \int_\Omega \rho h w_{tt} w \, d\Omega + c\gamma_0 \int_\Omega \rho h w_t^2 \, d\Omega \\ &= -2\int_\Omega \gamma(w_t) w_t \, d\Omega + c\gamma_0 \int_\Omega \rho h w_t^2 \, d\Omega \\ &\quad + c\gamma_0 \int_\Omega \rho h w_{tt} w \, d\Omega + 2\int_\Omega L(w) w_t \, d\Omega . \end{aligned} \tag{22}$$

From (1)

$$\begin{aligned} \int_\Omega \rho h w_{tt} w \, d\Omega &= -\int_\Omega \left(D|\Delta w|^2 - \gamma_1^2 w^2 - |\Delta \mathcal{F}(w)|^2 \right) d\Omega \\ &\quad - \int_\Omega \gamma(w_t) w \, d\Omega + \int_\Omega L(w) w \, d\Omega \\ &= -E_p(t) + \int_\Omega (\gamma(w_t) w + L(w) w) \, d\Omega . \end{aligned} \tag{23}$$

On the other hand, for any ε, ε_1, $\varepsilon_2 > 0$ we have

$$\begin{aligned} \int_\Omega \gamma(w_t) w \, d\Omega &\leq M_\gamma |w_t|_\Omega |w|_\Omega \\ &\leq \frac{\varepsilon}{2} |w_t|_\Omega^2 + \frac{M_\gamma^2}{2\varepsilon} |w|_\Omega^2 \leq \frac{\varepsilon}{2} |w_t|_\Omega^2 + \frac{M_\gamma^2}{2\varepsilon\gamma_1^2} E_p . \end{aligned} \tag{24}$$

$$\int_\Omega L(w) w \, d\Omega \leq \varepsilon_1 |\Delta w|_\Omega^2 + \frac{l^2}{4\varepsilon_1 \gamma_1^2} E_p \leq \left(\frac{\varepsilon_1}{D} + \frac{l^2}{4\varepsilon_1\gamma_1^2} \right) E_p \tag{25}$$

$$\begin{aligned} \int_\Omega L(w) w_t \, d\Omega &\leq l |\Delta w|_\Omega |w_t|_\Omega \\ &\leq \varepsilon_2 |\Delta w|_\Omega^2 + \frac{l^2}{4\varepsilon_2} |w_t|_\Omega^2 \leq \frac{\varepsilon_2}{D} E_p + \frac{l^2}{4\varepsilon_2} |w_t|_\Omega^2 . \end{aligned} \tag{26}$$

Collecting (22) – (26) yields

$$\begin{aligned} V_t(t) &\leq -\gamma_0 \left(2 - c\rho h - c\frac{\varepsilon}{2} - \frac{l^2}{2\varepsilon_2\gamma_0} \right) |w_t|_\Omega^2 \\ &\quad - c\gamma_0 \left(1 - \frac{\varepsilon_1}{D} - \frac{l^2}{4\varepsilon_1\gamma_1^2} - \frac{M_\gamma^2}{2\varepsilon\gamma_1^2} - \frac{2\varepsilon_2}{D} \right) E_p(t) . \end{aligned}$$

Selecting the constants $c = \frac{1}{\rho h}$, $\varepsilon = \frac{\rho h}{2}$, $\varepsilon_2 = \frac{2l^2}{\gamma_0}$, $\varepsilon_1 = \frac{D}{8}$ and taking into consideration inequalities (16) we obtain

$$V_t(t) \leq -\frac{c\gamma_0}{2} E(t).$$

On the other hand, due to assumption (16) we also have that $|E(t) - V(t)| \leq \frac{1}{2}E(T)$, so that $E(t) \geq \frac{2}{3}V(t)$. ALso, $V_t(t) \leq -\frac{c\gamma_0^3}{3}V(t)$, which yields $V(t) \leq e^{-\frac{1}{3}c\gamma_0 t}V(0)$; hence, the desired result of Theorem 3.

4.2. Proof of Theorem 4 (compact attractors)

Step 1. Existence of the absorbing set. Here, the argument involves a construction of a Liapunov's function. Let $\tilde{E}(t) \equiv E_k(t) + \tilde{E}_p(t)$ where

$$\tilde{E}_p(t) \equiv -\int_\Omega Fw \, d\Omega + E_p(t). \tag{27}$$

We define

$$V(t) \equiv \tilde{E}(t) + c_0 \int_\Omega w_t w \, d\Omega$$

where c_0 is suitably small. Noting that $\tilde{E}_t = -2\int_\Omega (\gamma(w_t)w_t - L(w)w_t)\, d\Omega$ and performing computations similar to those in Section 4.1, we prove inequality

$$V_t(t) \leq -\omega_0 V(t) + M|F|^2_{L^2(\Omega)}$$

for suitable positive constants ω_0, M. Hence

$$V(t) \leq e^{-\omega_0 t}V(0) + \frac{M}{\omega_0}|F|^2_{L_2(\Omega)}. \tag{28}$$

Inequality in (28) together with the fact that $\tilde{E}_p(t)$ is radially unbounded (hence $\tilde{E}_p(t)$ is radially unbounded as well) imply the existence of a set $B \subset H$, with $B \in B(0, C|F|_{L^2(\Omega)})$ for a suitable constant $C > 0$ and such that for every initial data $(w_0, w_1) \in H$; there exist $T_0 > 0$ so that the solution $(w(t), w_t(t))$ remains in the set B for $T_0 > 0$. This is to say that $B \in H$ is absorbing for the nonlinear semigroup $T(t)$.

Step 2. We decompose the nonlinear semigroup $T(t)$ (associated with the solutions to (1)) into the following sum

$$T(t) = K(t) + S(t) \quad \text{where } z(t) \equiv K(t)(w_0, w_1) \text{ is given by} \tag{29}$$

$$\begin{aligned} &\rho h z_{tt} + D\Delta^2 z + \gamma(z_t) = [\mathcal{F}(w), z] - \gamma_1^2 z + L(z) + du + F \\ &z(0) = z_t(0) = 0 \\ &z = \frac{\partial}{\partial \nu} z = 0 \quad \text{on } \Gamma \end{aligned} \tag{30}$$

and $u(t) = S(t)(w_0, w_1)$ satisfies

$$\begin{aligned} &\rho h u_{tt} + D\Delta^2 u + \gamma(u_t + z_t) - \gamma(z_t) + \gamma_1^2 u + du - Lu = [\mathcal{F}(w), u] \\ &u(0) = w_0; \quad u_t(0) = w_1 \\ &u = \frac{\partial}{\partial \nu} u = 0 \quad \text{on } \Gamma . \end{aligned} \tag{31}$$

Here, the constant $d > 0$ will be selected later.

We shall prove later that $S(t)$ is uniformly decaying and $K(t)$ is compact in H. This, together with the result of Theorem 3.6 in [6], allows us to conclude the existence of compact attractors in H. The dimension (in the case of linear γ of this attractor can be estimated by using the technique presented in [11] together with Theorem 4.1 in [8].

Step 3. We shall establish the uniform decays for solutions $u(t)$. Noting that the function $\tilde{\gamma}(u) \equiv \gamma(u + z) - \gamma(z)$ is monotone and, by the result of Proposition 2 and of Step 1,

$$|[\mathcal{F}(\omega(t)), u(t)]|_{L^2(\Omega)} \leq C(B)|u(t)|_{H^2(\Omega)}, \quad t \geq T_0, \tag{32}$$

we are in a position to apply the arguments of Theorem 3 to infer uniform decay rates for $u(t)$. Indeed, with values γ_0 and γ_1 selected as in Theorem 3 and taking constant $d > 0$ *suitably large* (to compensate for the nonlinear term in (32)) we obtain that for $t > T_0$

$$\begin{aligned} &|u(t)|_{H^2(\Omega)} + |u_t(t)|_{L^2(\Omega)} \\ &\quad \leq C(B)e^{-\omega t}\left(|w_0|_{H^2(\Omega)} + |w_1|_{L^2(\Omega)}\right) \quad \text{for } t > 0 \end{aligned} \tag{33}$$

Step 4. We shall see that $K(t)$ is compact $H \to H$ for all $t \geq 0$. A key role, here in the proof is played by the result of Proposition 2 together with the following

Proposition 5 *For all* $0 \leq \alpha \leq \frac{1}{2}$ *we have*

$$|\,[\mathcal{F}(w), z]\,|_{H^\alpha(\Omega)} \leq C|w|^2_{H^2(\Omega)}\,|z|_{H^{2+\alpha}(\Omega)}\,.$$

Remark 16. The result of Proposition 5 is based on "sharp" regularity of Airy stress function (see Proposition 2)

Proposition 6 *The nonlinear function γ is bounded: $H^\alpha(\Omega) \to H^\alpha(\Omega)$; for any $\alpha \leq 1$.*

Proof The proof follows from hypothesis (17) imposed on function γ together with nonlinear interpolation theorem in [15].

In order to prove the sought compactness of $K(t)$, we shall describe a solution to (30) via the variation of constants formula. To accomplish this, we introduce the following operator: $A : H \to H$ given by

$$\begin{aligned} A(u, v) &\equiv \left(v, \frac{D}{\rho H}\Delta^2 u + \gamma_1^2 u + Lu\right) \\ \mathcal{D}(A) &= \left\{(u, v) \in [H_0^2(\Omega)]^2;\ u \in H^4(\Omega)\right\} \end{aligned}$$

It is well known that A generates on H a C_0 semigroup denoted by e^{At}. The solution to (20) can be written as

$$\begin{pmatrix} z(t) \\ z_t(t) \end{pmatrix} = \int_0^t e^{A(t-\tau)} \begin{pmatrix} 0 \\ \gamma(z_t) + [\mathcal{F}(w), z] - du(\tau) + F \end{pmatrix} d\tau . \tag{34}$$

By the results of Propositions 5 and 6 together with the identification

$$\mathcal{D}(A^{\alpha/4}) \approx H^{2+\alpha}(\Omega) \times H^{\alpha}(\Omega) \tag{35}$$

for $0 \le \alpha \le 1/2$ we obtain

$$\begin{aligned} &\left| \int_0^t e^{A(t-\tau} \begin{pmatrix} 0 \\ \gamma(z_t(\tau)| + [\mathcal{F}(w), z(\tau)] \end{pmatrix} d\tau \right|_{\mathcal{D}(A^{\alpha/4})} \\ &\le C_t(B) \left(\int_0^t |z_t(\tau)|^2_{H^2(\Omega)} + |z(\tau)|^2_{H^{2+\alpha}(\Omega)}\, d\tau \right)^{1/2} \end{aligned} \tag{36}$$

Moreover, by direct computations and using the fact that e^{At} is a C_0 semigroup on $\mathcal{D}(A^{\alpha/4})$ we obtain

$$\begin{aligned} &\left| \int_0^t e^{A(t-\tau)} \begin{pmatrix} 0 \\ -du(\tau) + \mathcal{F} \end{pmatrix} d\tau \right|_{\mathcal{D}(A^{\alpha/2})} \\ &\le C_t |u|_{C(o,t;H^{\alpha}(\Omega)} + \left| \int_0^t Ae^{A(t-\tau)} F\, d\tau \right|_{L^2(\Omega)} . \end{aligned} \tag{37}$$

On the other hand, by semigroup methods

$$\left| \int_0^t Ae^{A(t-\tau)} F\, d\tau \right|_{L^2(\Omega)} \le C_t |(e^{At} - I)F|_{L^2(\Omega)} \le C_t |F|_{L^2(\Omega)} . \tag{38}$$

Combining (34) – (38) yields

$$\begin{aligned} |z_t(t)|^2_{H^{\alpha}(\Omega)} + |z(t)|^2_{H^{2+\alpha}(\Omega)} &\le C_t \int_0^t \Big(|z_t(\tau)|^2_{H^{\alpha}(\Omega)} \\ &+ |z(\tau)|_{H^{2+\alpha}(\Omega)} \Big) d\tau + |F|_{L^2(\Omega)} + |u|_{C(0,t;H^{\alpha}(\Omega))} . \end{aligned} \tag{39}$$

Recalling (34) applied to the last term in (38) and Gronwell's inequality we obtain

$$|z_t(t)|_{H^{\alpha}(\Omega)} + |z(t)|_{H^{2+\alpha}(\Omega)} \le C_t \left(|F|_{L^2(\Omega)} + |w_0|_{H^2(\Omega)} + |w_1|_{H^2(\Omega)} \right)$$

which in view of compact Sobolev's imbeddings (see [1]), proves the compactness of $K(t)$.

References

[1] R. Adams; *Sobolev Spaces*, Academic, 1975.

[2] V. Barbu; *Nonlinear Semigroups and Differential Equations in Banach Spaces*, Noordhof, 1976.

[3] I. Chueshov; Strong solutions and the attractors of the von Karman equations, *Math. USSR Sbornik* **29**, 1981, 25–36.

[4] S. Cox and E. Zuazua; The rate at which energy decays in a damped string, to appear.

[5] A. Favini, M. Horn, I. Lasiecka, and D. Tataru; Global existence, uniqueness and regularity of solutions to a von Karman system with nonlinear boundary dissipation, IMA Preprint Series #1168. Also in *Differential and Integral Equations*, to appear.

[6] J. Hale; *Asymptotic Behavior of Dissipative Sets*, AMS, Providence, 1988.

[7] M. Horn and I. Lasiecka; Uniform decay of weak solutions to a von Karman plate with nonlinear boundary conditions, *J. Diff. Int. Eqns.* **7**, 1994, 885–908.

[8] O. Ladyzenskaya; On the finite dimensionality of bounded invariant sets for the Navier Stokes system and other dissipative systems, *J. Soviet Math.* **28**, 1985, 714–726.

[9] J. Lagnese; *Boundary Stabilization of Thin Plates*, SIAM, Philadelphia, 1989.

[10] J. Lions; *Quelques Methods de Resolution des Problemes aux Limites Nonlinearies*, Dunod, 1969.

[11] I. Lasiecka; Finite dimensional attractors for von Karman Equations with nonlinear boundary damping, *J. Diff. Eq.*, to appear.

[12] I. Lasiecka; Local and global compact attractors arising in nonlinear elasticity. Use of noncompact nonlinearity and nonlinear dissipation, *J. Math. Anal. Appl.*, to appear.

[13] I. Lasiecka and D. Tataru; Uniform boundary stabilization of semilinear wave equations with nonlinear boundary damping, *Diff. and Integral Eq.* **6**, 1993, 507–533.

[14] V. Mazya and M. Saposnikova; *Multipliers in Spaces of Differentiable Function*, Pittman, 1985.

[15] L. Tartar; Interpolation nonlinéaire et régularité, *J. Funct. Anal.* **9**, 1972, 469–489.

[16] I. Lasiecka and W. Heyman; Aymptotic behavior of solutions in nonlinear dynamic elasticity, *Disc. Dynam. Syst.*, to appear.

Preparation of Advanced CFD Codes for use in Sensitivity Analyses and Multidisiplinary Design Optimization

Perry A. Newman*
NASA Langley Research Center
Hampton VA 23681, USA

Several topics concerning computational methodologies relevant to preparation and efficient use of advanced CFD codes in sensitivity analyses (SA) and multidisciplinary analysis and design optimization (MDO) are briefly discussed. This work was initiated several years ago by NASA Langley and has continued to involve government and university personnel. The several topics to be discussed and illustrated with sample results are as follow:

Efficient calculation of both geometric and nongeometric sensitivity derivatives (SD) from a 2–D thin layer Navier-Stokes code and a 3–D supersonic (marching) Euler code using incremental iterative methods.

Automatic differentiation (AD) of a 3–D thin layer Navier-Stokes code to obtain:

- geometric and nongeometric SD typical of those needed for MDO
- CFD algorithm and turbulence modeling SD, illustrating feasibility for SA of complex math models.

Initial application of AD to incremental iterative forms for both first and second derivatives for a 2–D thin layer Navier-Stokes code.

Development of a method for Simultaneous Aerodynamic Analysis and Design Optimization, SAADO.

Some of these techniques have already been suggested for doing parts of the problem more efficiently; the contribution here is demonstrating how they fit together in a mutually beneficial way.

1. Introduction

For a single discipline design, adjoint-variable based methods appear to be computationally much more efficient than the optimizer-based sensitivity analysis (SA) methods since a single cost function can be defined that is composed of only the single discipline analysis solution output functions. However, in multidisciplinary design optimization (MDO) problems the objective and constraint functions are now generally composed of output functions from several disciplines and at multiple design points. Each single-discipline analysis code is then to supply not only the output functions required for the constrained optimization process and other discipline analysis inputs, but also the derivatives of all these output functions with respect to input variables. These variables include not only the multidisciplinary design (MdD) variables, but also output functions from other disciplines which implicitly depend on the MdD variables.

The use of current advanced computational fluid dynamics analysis codes in MDO studies and applications via SA requires efficient and accurate calculation

*See acknowlegments for others involved in the work reported on in this paper.

of individual discipline sensitivity derivatives (SD). These SD are not customarily obtained as part of the nonlinear iterative computational fluid dynamics (CFD) solution process; an overview of a number of recent research activities concentrating on the calculation of SD from advanced CFD codes is given in [1]. The present paper summarizes only our work on issues related to preparation of advanced CFD codes for SA and MDO applications. Most of the published work by other researchers are cited within the references listed herein. The direction taken in this work was outlined in [2]; results obtained from this work are given in the appropriate following sections.

The several topics concerning computational methodologies relevant to preparation and efficient use of advanced CFD codes in SA and MDO are briefly discussed in the following four sections. These sections are: incremental iterative method (IIM) for SD; automatic differentiation (AD) of Navier-Stokes (N-S) codes for SD; application of AD for IIM SD calculation; and simultaneous aerodynamic analysis and design optimization (SAADO).

2. Incremental iterative method for SD

The IIM is among the most accurate and efficient yet proposed or demonstrated for obtaining SD from nonlinear CFD codes [1]–[3]. Derivative results from a two-dimensional (2–D) thin-layer Navier-Stokes code for both nongeometric (flow) and geometric (shape) design variables are shown in [2] and [3], whereas those from 3–D Euler codes are shown in [4]–[6]. In addition to accurate consistent discrete SD obtained with computational efficiency (for both CPU time and memory requirements), the IIM also allows using approximate solution operators of choice for further efficiency, parallelization, or robustness, for example. In all of the above cited works, the discretized flow residuals were differentiated "by hand" (also called the quasi-analytic (QA) method) and assembled to obtain SD by an IIM. Even though the method has been outlined in [1]–[4], the equations will be repeated here since they are needed to develop the second-order SD IIM forms.

2.1. Equation summary for functions and (first-order) SD

A summary of the basic flow equations and IIM forms for SD is as follows (using the notation of [1] where more details are available):

Nonlinear, steady-state flow equations:

$$\mathbf{R}(\mathbf{Q}(\mathbf{D}), \mathbf{X}(\mathbf{D}), \mathbf{D}) = 0\,, \tag{1}$$

where $\mathbf{Q}$ is the vector of field (state) variables, $\mathbf{X}$ is the vector of grid coordinates, and $\mathbf{D}$ is the vector of design (input) variables.

Aerodynamic output functions:

$$\mathbf{F} = \mathbf{F}(\mathbf{Q}(\mathbf{D}), \mathbf{X}(\mathbf{D}), \mathbf{D})\,. \tag{2}$$

Associated direct-differentiation equations:

$$\mathbf{R}' = \frac{\partial \mathbf{R}}{\partial \mathbf{Q}}\mathbf{Q}' + \frac{\partial \mathbf{R}}{\partial \mathbf{X}}\mathbf{X}' + \frac{\partial \mathbf{R}}{\partial \mathbf{D}} = 0 \tag{3}$$

$$\mathbf{F}' = \frac{\partial \mathbf{F}}{\partial \mathbf{Q}}\mathbf{Q}' + \frac{\partial \mathbf{F}}{\partial \mathbf{X}}\mathbf{X}' + \frac{\partial \mathbf{F}}{\partial \mathbf{D}}, \tag{4}$$

where $\mathbf{Q}' = \frac{d\mathbf{Q}}{d\mathbf{D}}$, $\mathbf{X}' = \frac{d\mathbf{X}}{d\mathbf{D}}$, $\mathbf{R}' = \frac{d\mathbf{R}}{d\mathbf{D}}$, and $\mathbf{F}' = \frac{d\mathbf{F}}{d\mathbf{D}}$ are SD with respect to the design variables.

Associated adjoint-variable equations:

$$\left(\frac{\partial \mathbf{R}}{\partial \mathbf{Q}}\right)^T A + \left(\frac{\partial \mathbf{F}}{\partial \mathbf{Q}}\right)^T = 0 \tag{5}$$

$$\mathbf{F}' = \left(\frac{\partial \mathbf{F}}{\partial \mathbf{X}} + A^T \frac{\partial \mathbf{R}}{\partial \mathbf{X}}\right)\mathbf{X}' + \frac{\partial \mathbf{F}}{\partial \mathbf{D}} + A^T \frac{\partial \mathbf{R}}{\partial \mathbf{D}}, \tag{6}$$

where A is the adjoint variable matrix associated with $\mathbf{F}$, and T denotes matrix transpose.

Incremental iterative forms:

$$\begin{aligned} -\frac{\partial \widetilde{\mathbf{R}^n}}{\partial \mathbf{Q}}\Delta \mathbf{Q} &= \mathbf{R}^n \\ \mathbf{Q}^{n+1} &= \mathbf{Q}^n + \Delta \mathbf{Q} \qquad n = 1, 2, 3, \ldots, \end{aligned} \tag{1*}$$

$$\begin{aligned} -\frac{\widetilde{\partial \mathbf{R}}}{\partial \mathbf{Q}}\Delta \mathbf{Q}' &= \mathbf{R}'^m \\ \mathbf{Q}'^{m+1} &= \mathbf{Q}'^m + \Delta \mathbf{Q}' \qquad m = 1, 2, 3, \ldots, \end{aligned} \tag{3*}$$

$$\begin{aligned} -\left(\frac{\widetilde{\partial \mathbf{R}}}{\partial \mathbf{Q}}\right)^T \Delta A &= \left(\frac{\partial \mathbf{R}}{\partial \mathbf{Q}}\right)^T A^n + \left(\frac{\partial \mathbf{F}}{\partial \mathbf{Q}}\right)^T \\ A^{n+1} &= A^n + \Delta A \qquad n = 1, 2, 3, \ldots, \end{aligned} \tag{5*}$$

where n and m are iteration indices and $\frac{\widetilde{\partial \mathbf{R}}}{\partial \mathbf{Q}}$ denotes "approximate operator of convenience."

The nonlinear flow (1) is solved for $\mathbf{Q}$ given the grid $\mathbf{X}$ and design variables $\mathbf{D}$. The required iterative solution algorithm can be written symbolically as (1*), an incremental iterative form for $\mathbf{Q}$. Equations (3) or (5) are solved for $\mathbf{Q}'$ or A, respectively. These are called the standard forms, whereas, (3*) or (5*) are the corresponding incremental iterative forms.

2.2. Sample results

The sample results shown here are examples of both 2–D Navier-Stokes [3] and 3–D supersonic Euler [4, 6] aerodynamic SD with respect to geometric and flow (nongeometric) variables. These SD have been obtained using the QA IIM and are compared with those obtained by finite differencing, either central (CD) or forward (FD). An application for coupled static aeroelastic IIM SD, using 3–D transonic small-disturbance and equivalent plate approximations [7], is mentioned. Sample results for aerodynamic shape optimization (improvement) using 3–D supersonic Euler IIM SD [6] are also given.

2.2.1. 2–D viscous IIM SD

Numerical results, which compare fifteen (first-order) SD, were obtained from variations of the same thin-layer Navier-Stokes code studied in [3]. For details about the code, one is referred to [3] and the references cited therein. The flow conditions are subsonic low Reynolds number laminar flow and transonic high Reynolds number turbulent flow over a NACA 1406 airfoil. The solutions are computed on a C-mesh with 257×65 grid points. These fifteen (first-order) SD are for the three output functions: (coefficients of) lift, drag, and pitching moment (C_L, C_D, C_M respectively), with respect to five design variables. These D_i were the two nongeometric variables: Mach number, M, and angle-of-attack, α; and the three geometric variables: airfoil thickness-to-chord ratio, T, maximum camber parameter, C, and chordwise location of maximum camber, L.

The laminar flow results comparing SD for accuracy are shown in Table 1. The Mach number M is 0.6, angle of attack α is 1.0°, and the Reynolds number is 5.0×10^3. These comparisons are for very well converged solutions (from all codes) and are presented as SD ratios, with SD_{CD} always the numerator. As shown in [3] and seen here, these QAII results for laminar, subsonic flow agreed very well with the CD results.

SD Ratio	Design Var., D	$\frac{dC_L}{dD}$	$\frac{dC_D}{dD}$	$\frac{dC_M}{dD}$
$\frac{SD_{CD}}{SD_{QAII}}$	T	1.000	1.000	1.000
	C	1.000	1.000	1.000
	L	1.000	1.000	1.000
	α	1.000	1.000	1.000
	M	1.000	1.000	1.000

Table 1. Comparison of SD ratios for three output functions with respect to three geometric and two nongeometric design variables. (NACA 1406 airfoil, subsonic laminar low Reynolds number flow.)

The turbulent flow results comparing SD for accuracy are shown in Table 2a; here, the turbulence is simulated using the well-known algebraic model of Baldwin and Lomax. The Mach number, M, is 0.8, angle of attack, α, is 1.0°, and the

Reynolds number is 5.0×10^6. As was shown in [3] and seen here, the CD and QAII do not agree. This was expected, since the turbulence model had not been "hand differentiated"; the disagreement, however, was much larger than expected for four of the ratios (19%–116%) shown in Table 2a. This case will be discussed in subsequent sections.

Comparison of computational memory and solution time (CPU seconds) requirements for the CD and QAII methods are shown in Table 2b. Two results are given for the Cray Y-MP computer. The first uses default compile-option execution; the second, the aggressive-option execution. It is seen that only about 0.5% memory increase is required for the aggressive option. It appears that there is only a modest increase in memory for the QAII work over the CD (original) code. This occurs since the original code provides arrays for both conserved and primitive variables; the additional Q associated arrays for the QAII version code are stored in what corresponded to the conserved variable arrays of the original code.

SD Ratio	Design Var., D	$\frac{dC_L}{dD}$	$\frac{dC_D}{dD}$	$\frac{dC_M}{dD}$
$\frac{SD_{CD}}{SD_{QAII}}$	T	2.157	1.028	1.262
	C	1.021	1.021	1.025
	L	0.810	1.029	0.947
	α	1.068	1.027	1.287
	M	1.012	1.003	1.027

Turbulence model *not* differentiated.

a. Comparison of SD ratios.
Turbulence model *not* differentiated.

Solution Method	Memory (Mbytes)	CPU Time* (sec)
CD	5.31	7,100
CD**	5.34	2,520
QAII	7.16	693
QAII**	7.28	413

*Cray Y-MP **Vectorization improvements

b. Computational memory requirements and timing comparisons.

Table 2. Results for calculation of three output function SD with respect to three geometric and two nongeometric design variables. (NACA 1406 airfoil, transonic turbulent high Reynolds number flow.)

For both the CD and QAII codes, it can be seen that use of Cray Y-MP automatic loop unrolling and aggressive compile options, called vectorization improvements in Table 2b, results in significant CPU time reductions. Run times without these improvements are quoted since such times are given in [3]. All results here are given for (or estimated to) 1,280 iteration runs where solutions are very well converged. The hand-differentiated QAII method is seen to be more efficient than the CD method for calculating SD.

2.2.2. 3–D inviscid IIM SD

Initial results from a 3–D marching Euler code (MARSEN) for nongeometric design variables have appeared in [4]. The present paper presents sample results for efficient calculation of geometric (shape) SD for the same 3–D marching Euler code; more complete results are given in [6]. These SD have been generated for a filleted wing-body configuration approximating a Mach 2.4 high-speed civil transport (HSCT).

Comparisons of SD results for the six-component force and moment coefficients with respect to nongeometric variables (taken from the expanded version of [4]) are shown in Table 3a. These SD have been computed on a grid of about 67 thousand points (37 streamwise × 121 circumferential × 15 normal). The nongeometric variables (flight conditions) are M (Mach number) = 2.4, α (angle-of-attack) = 0°, and β (yaw angle) = 0°. Sensitivity derivative ratios of six output functions (C_x, C_y, C_z, C_{M_x}, C_{M_y}, C_{M_z}) with respect to M, α, and β are given. These calculated SD ratios, FD (at perturbation size, $\Delta D = 1.0 \times 10^{-5}$) to QA derivatives are seen to be unity, to four significant figures. However, the computational cost of the finite-difference method is approximately seven times greater, as can be seen from Table 3b which gives computational timing comparisons; both the baseline time and convergence levels used for all solutions are given in the footnote there.

Most modern advanced CFD analysis codes use body fitted computational grids that are required as input. That is, the grid generation for a specified configuration is not part of the flow code, but rather, a separate entity. This has several consequences regarding flow sensitivity analysis with respect to geometric or shape design variables. First, these variables affect the CFD code only through their effect on the generated body-fitted grid, i.e., these variables set boundary conditions in the grid generation code. Second, CFD grid generation appears to be a separate discipline, which now must be coupled to the flow code not only through the body-fitted grid, but also through its SD with respect to the geometric variables specifying the configuration or those deemed to be design variables. Several approaches can be taken to obtain these latter SD of the grid (or its movement) with respect to design variables; these approaches and related matters are also discussed in [1].

The geometry processing and grid generation codes used here ([8], [9]) take “wave-drag deck” descriptions of configuration parts as their input. The various component surfaces are first intersected and filleted [8] into a continuous surface, and then suitable computational grids are generated [9]. These codes had been linked together, front-ended with a 42–variable geometric parameterization, and automatically differentiated by Eric Unger [10]. His parameterization of the HSCT 24E geometry was divided into three types: 7 planform variables, 15 section thickness variables (5 each at root, break, and tip sections), and 20 camber surface variables. A different camber parameterization and more extensive results than discussed below are given in [6].

$\frac{SD_{FD}}{SD_{QA}}$	Design Variables, D		
	Mach	*Alpha*	*Beta*
$\frac{\partial C_x}{\partial D}$	0.9999	1.0000	a
$\frac{\partial C_y}{\partial D}$	a	a	0.9999
$\frac{\partial C_z}{\partial D}$	0.9999	1.0000	a
$\frac{\partial C_{M_x}}{\partial D}$	a	a	0.9999
$\frac{\partial C_{M_y}}{\partial D}$	0.9999	1.0000	a
$\frac{\partial C_{M_z}}{\partial D}$	a	a	0.9999

[a] Ratio for extremely small quantities is meaningless, i.e., $O(\epsilon)$ or $O(\epsilon)/O(\epsilon)$.

a. Comparison of SD ratios.

Solutions	Number of Solutions	Ratio
Baseline	1	1.000[b]
Forward Finite	3	3.426
Quasi-Analytical	3	0.487

[b] Baseline solution runtime for (R^n_{rms}/R^1_{rms}) reduction to $\epsilon = 10^{-8}$ on Cray-2 is 827 sec.

b. Computational timing comparisons .

Table 3. Results for calculation of six output function SD with respect to three nongeometric design variables. (HSCT 24E, supersonic inviscid flow.)

Sample results comparing SD obtained by CD and the IIM are given for several geometric variables of each type. However, because these results are computed on a half-space grid (only 49 points in the circumferential direction, with a symmetry plane at $y = 0$), some forces, moments, and SD are not balanced by their images and therefore do not vanish. These nonvanishing components do not affect the geometric SD comparisons shown in Tables 4 through 6. These tables are in the same format and contain information analogous to that in Table 3 for the nongeometric SD. Sample SD comparisons are shown for planform variables in Tables 4, several section thickness variables in Table 5, and several camber surface variables in Table 6. In all cases the SD agree to about four significant figures and are obtained much faster by the IIM.

$\frac{SD_{CD}}{SD_{QA}}$	Design Variables, D					
	Root Chord	Break Chord	Tip Chord	Inboard Span	Outboard Span	X-Break Location
$\frac{\partial C_x}{\partial D}$	1.00000	1.00000	1.00000	0.99999	1.00000	1.00000
$\frac{\partial C_y}{\partial D}$	1.00000	0.99999	1.00000	0.99999	1.00000	1.00000
$\frac{\partial C_z}{\partial D}$	1.00000	0.99999	1.00000	1.00000	1.00000	0.99999
$\frac{\partial C_{M_x}}{\partial D}$	1.00000	0.99999	1.00000	1.00000	1.00000	0.99999
$\frac{\partial C_{M_y}}{\partial D}$	1.00000	0.99999	1.00000	1.00000	1.00000	1.00000
$\frac{\partial C_{M_z}}{\partial D}$	1.00000	1.00000	0.99999	1.00000	1.00000	1.00000

a. Comparison of SD ratios.

Solutions	Number of Solutions	Ratio
Baseline	1	1.000[a]
Central Finite	12	12.001
Quasi-Analytical	6	0.811

[a] Baseline solution runtime for (R^n_{rms}/R^1_{rms}) reduction to $\epsilon = 10^{-11}$ on Cray-2 is 451 sec.

b. Computational timing comparisons.

Table 4. Results for calculation of six output function SD with respect to six geometric planform design variables. (HSCT 24E, supersonic inviscid flow.)

$\frac{SD_{CD}}{SD_{QA}}$	Design Variables, D		
	t/c Root	t/c Break	t/c Tip
$\frac{\partial C_x}{\partial D}$	1.0000	1.0000	1.0000
$\frac{\partial C_y}{\partial D}$	0.9999	1.0000	0.9999
$\frac{\partial C_z}{\partial D}$	1.0000	1.0000	1.0000
$\frac{\partial C_{M_x}}{\partial D}$	1.0000	1.0000	1.0000
$\frac{\partial C_{M_y}}{\partial D}$	1.0000	1.0000	1.0000
$\frac{\partial C_{M_z}}{\partial D}$	1.0000	1.0000	1.0000

Table 5. Comparison of SD ratios for six output functions with respect to three section thickness design variables. (HSCT 24E, supersonic inviscid flow.)

$\frac{SD_{CD}}{SD_{QA}}$	Design Variables, D		
	Camber C_1	Camber C_2	Camber C_3
$\frac{\partial C_x}{\partial D}$	1.0008	a	1.0016
$\frac{\partial C_y}{\partial D}$	1.0000	1.0001	1.0000
$\frac{\partial C_z}{\partial D}$	0.9999	1.0000	1.0001
$\frac{\partial C_{M_x}}{\partial D}$	1.0000	1.0000	1.0000
$\frac{\partial C_{M_y}}{\partial D}$	a	1.0000	1.0002
$\frac{\partial C_{M_z}}{\partial D}$	1.0000	1.0000	0.9999

[a] Ratio for extremely small quantities
is meaningless, i.e., $O(\epsilon)$ or $O(\epsilon)/O(\epsilon)$.

Table 6. Comparison of SD ratios for six output functions with respect to three section camber design variables. (HSCT 24E, supersonic inviscid flow.)

A recent pilot study [7] applied the IIM to a coupled (integrated) 3–D static aeroelastic problem where the fluid flow was modeled using a transonic (nonlinear) small-disturbance potential equation and the wing structure was approximated using an equivalent plate deflection model. Table 7 depicts the simultaneous iterative solution algorithm for the incremental (Δ) updates of: the flow potential (ϕ) and its SD with respect to aero-design variables (X_{DA}); the structural deflection (δ) and its SD with respect to structural design variables (X_{DS}); and the system coupling derivatives of flow potential with respect to structure deflection and structural deflection with respect to pressure loading. Global system SD are calculated when the iteration converges. Sample results for an unswept, twisted, rectangular-planform wing of aspect ratio 3 at supercritical flow conditions are given in [7].

Do (iterative solution loop)
Sweep flowfield by cross planes Solve flowfield equations for $\Delta\phi_{ijk}$ Solve aerodynamic sensitivity eqs. for $\Delta\left(\frac{\partial\phi}{\partial X_{\text{DA}}}\right)$ Solve structural equations for $\Delta\delta_{ijk}$ Solve structural sensitivity eqs. for $\Delta\left(\frac{\partial\delta}{\partial X_{\text{DS}}}\right)$ Solve coupling derivative eqs. for $\Delta\left(\frac{\partial\phi}{\partial\delta_{ii,jj,kk}}\right)_{ijk}$ Solve coupling derivative eqs. for $\Delta\left(\frac{\partial\delta}{\partial\Delta C_{p\,ii,jj,kk}}\right)_{ijk}$ Update aerodynamic and structural boundary conditions Iterate until convergence
Continue
Solve system sensitivity eqs. for $\frac{d\Delta C_p}{dX_D}$, $\frac{d\Delta C_L}{dX_D}$, $\frac{d\delta}{dX_D}$, etc.

Table 7. Procedure for application of IIM to integrated static aeroelastic problem.

2.2.3. 3–D inviscid optimization using IIM SD

Initial applications of the 3–D marching Euler code (MARSEN) with efficient geometric SD calculations via IIM are shown for aerodynamic optimization studies where the CFD and grid-generation codes are considered as separate disciplines. In these studies, the Automated Design Synthesis (ADS) program [11] is used as the "black-box" optimization code. Two sample case results for an HSCT 24E configuration shape optimization are discussed: a section thickness variable study and a planform variable study. In these initial results, only a limited number of design variables are active in each study and the objective function is maximized lift over drag (i.e., min($-C_z/C_x$), subject to a constraint on the wing bending moment (i.e., C_{M_x}) and side constraints (bounds) on design variables. The goal here is to ascertain that the changes predicted, when *only* the supersonic cruise point is considered, are reasonable.

The planform variable study was limited to two variables that control the outboard wing planform when the inboard wing is held fixed. These two are the wing-tip leading-edge axial location and the tip chord. For supersonic flow considerations alone, this outboard panel should be swept much more than in the HSCT 24E; Figure 1 shows that the optimization procedure produced this effect. Table 8 gives the initial and final values of objective, constraint, and design variables for this case. It can be seen that there was a design improvement after 4 optimization cycles. This improvement required 5 function evaluations (grid and CFD), 4 gradient evaluations (2 design variables), and about 15 minutes of Cray-2 CPU time.

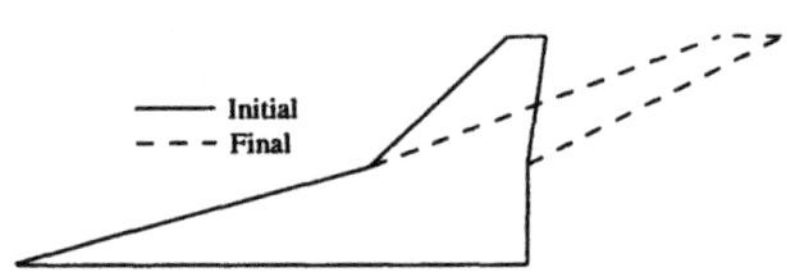

Figure 1. Sample planform change in design optimization study using only two outboard wing planform variables. (HSCT 24E, supersonic inviscid flow.)

The section thickness variable study involved six variables; the thickness-to-chord ratio, t/c, and the chordwise location of maximum thickness, X_m, at the wing root, break, and tip sections. For supersonic flow considerations alone, one would expect the wing to become thinner; results in Table 9 confirm this expectation. Again it can be seen that there was an improvement in the objective function and, after 3 optimization cycles, 3 of the design variables were constrained by their respective input lower bounds. This improvement required 4 function evaluations (grid and CFD), 3 gradient evaluations (6 design variables), and about 21 minutes of Cray-2 CPU time.

	Initial	Final	Comments
Objective, C_z/C_x	16.95	20.01	+18.066%
Constraint, C_{M_x}	5.62E-3	2.68E-3	Not Violated
Tip Chord	9.303	13.025	Large Change
X-Loc., Tip LE	13.84	20.068	Upper Bound

Table 8. Optimization study using two planform variables, better design after 4 optimization iterations. (HSCT 24E, supersonic inviscid flow.)

	Initial	Final	Comments
Objective, C_z/C_x	8.767	9.745	+11.15%
Constraint, C_{M_x}	9.74E-4	9.86E-4	Active
t/c Root	2.971	2.6739	Lower Bound
t/c Break	2.500	2.25	Lower Bound
t/c Tip	2.500	2.25	Lower Bound
X_m Root	6.000	3.6	Large Change
X_m Break	5.000	4.995	Small Change
X_m Tip	5.000	5.000	No Change

Table 9. Optimization study using six section thickness variables, better design after 3 optimization iterations. (HSCT 24E, supersonic inviscid flow.)

In [6], more complete shape optimization study results involving many more geometric design variables are given. These include studies involving camber surface and flap deflection variables in addition to others for wing thickness and planform shape.

3. Automatic differentiation of N-S codes for SD

The concept and various techniques for implementation of AD have been known for some time [12]–[14]; AD is a chain-rule-based technique for evaluating the derivatives of functions, defined by computer programs, with respect to their input variables. A recent implementation called ADIFOR (AD Fortran) has been developed and is discussed in [15] and [16]. Applications of ADIFOR, prompted by the NASA Langley interest in MDO, have appeared as or in [6], [10], and [16]–[23]. Since the primary purpose in this paper is to indicate how advanced CFD codes might be prepared for use in SA and MDO, the reader is directed to the quoted references for details about AD and ADIFOR. One aspect of AD, however, is relevant to further discussion—the straightforward application of ADIFOR to iterative solution algorithms, such as those encountered in advanced CFD codes.

3.1. AD of iterative algorithms

Application of the precompiler tool ADIFOR to an iterative solution algorithm produces an iterative algorithm to compute derivatives. However, as is pointed out in [2], [16], [18], [19], and [24], this latter iterative algorithm obtained from a straightforward AD application may be neither computationally efficient nor robust; and, it is not an IIM form even if the original solution algorithm was. The straightforward applications of ADIFOR to advanced CFD codes ([16], [18]–[21], and [23]) produced iterative algorithms for derivative calculations in which the entire solution algorithm is differentiated ([16] and [18]). That is (using the notation of [1]), the iterative solution algorithm can be depicted as

$$\mathbf{Q}^{n+1} = \mathbf{Q}^n - P^n \mathbf{R}^n \qquad n = 1, 2, 3, \dots , \tag{7}$$

where the operator P^n can be viewed as a preconditioner. A straightforward application of ADIFOR to (7) produces

$$\mathbf{Q}'^{n+1} = \mathbf{Q}'^n - P^n \mathbf{R}'^n - P'^n \mathbf{R}^n \qquad n = 1, 2, 3, \dots , \tag{8}$$

It can be seen that the iterative solution algorithm as depicted by (7) is that given by the IIM form (1*) if $P^n = \widetilde{(\frac{\partial \mathbf{R}^n}{\partial \mathbf{Q}})}^{-1}$. However, the resulting iterative derivative solution algorithm of (8) is not the IIM form (3*) until the residual $\mathbf{R}^n$ *and* $P'^n \mathbf{R}^n$ vanish, as discussed in [18] and [24]. The previously cited successful calculations of SD by straightforward application of ADIFOR to generate iterative CFD algorithms ([16], [18]–[21], and [23]) indicate that the contribution from the $P'^n \mathbf{R}^n$ term in (8) must be vanishing as the iteration proceeds. Unfortunately though, differentiation through the complete iterative CFD solution algorithm and repeated calculation of its derivatives, denoted as P'^n, although unnecessary, is *not* avoided in the straightforward application of the presently available ADIFOR.

3.2. Sample results

The examples shown in the following subsections are from straightforward application of AD. The point to be noted is that consistent discrete SD are obtained; the turbulence models have been differentiated. Both 2–D and 3–D sample results are given.

3.2.1. 2–D viscous SD via AD

These SD results are for the same turbulent flow case as previously discussed and for which the QA results were compared with the CD results in Table 2a. It was noted there that the turbulence terms had not been differentiated by hand and rather large differences were noted in several of the SD when compared to those obtained by CD. The straightforward AD results for SD, denoted $\mathrm{SD_{AD}}$ are compared with the same CD results in Table 10a. With the turbulence terms differentiated by AD, it can be seen that the SD results agree to three significant figures. Table 10b shows, however, the penalty paid for this accuracy in terms of both computer memory and runtime (CPU) requirements. The memory required for AD is about six times that of the original code (CD), just as is expected for five design variables; the runtime is also larger. Note that vectorization improvements were previously discussed. Both memory and runtime aspects are addressed by forming an AD in IIM; this is discussed in the next section.

SD Ratio	Design Var., D	$\frac{dC_L}{dD}$	$\frac{dC_D}{dD}$	$\frac{dC_M}{dD}$
	T	0.999	1.000	1.000
	C	1.000	1.000	1.000
$\frac{SD_{CD}}{SD_{AD}}$	L	1.000	1.000	1.000
	Alpha	1.001	1.000	1.000
	Mach	0.999	0.999	0.999

a. Comparison of SD ratios .

Solution Method	Memory (MBytes)	CPU Time* (sec)
CD	5.31	7,100
CD**	5.34	2,520
QAII	7.16	693
QAII**	7.28	413
AD	30.92	9,390
AD**	31.09	5,740

*Cray Y-MP **Vectorization improvements

b. Computational memory requirements and timing comparisons.

Table 10. Results for calculation of three output function SD with respect to three geometric and two nongeometric design variables. (NACA 1406 airfoil, transonic turbulent high Reynolds number flow.)

3.2.2. 3–D viscous SD via AD—nongeometric

The nongeometric SD examples shown here have been taken from [18] and [19] and might find use in either SA or MDO. A brief summary is also given in [21]. These SD, with respect to the (free-stream) flow variables or modeling parameters within the flow code, are obtained on a fixed computational grid so that SD of the grid are not required.

All of the results in this subsection are computed from the three dimensional thin-layer Navier-Stokes code TLNS3D [25]. This code employs a multigrid (MG) acceleration technique to an explicit multistage Runge-Kutta time-stepping scheme with central spatial differencing to efficiently obtain steady-state high-Reynolds-number turbulent flow solutions. The TLNS3D code is a highly vectorized code, and this aspect contributed greatly to its overall computational efficiency. However, ADIFOR inserts loops of length equal to the number of design variables in order to compute all intermediate "gradient-objects"; this results in loss of the original code's long-loop vectorization efficiency unless there are many design variables. Some relief can be obtained by post-ADIFOR hand processing of the derivative code.

Relative accuracy results for nongeometric SD with respect to flow variables [18] at transonic turbulent flow conditions, $M = 0.84$, $\alpha = 3.06$, and $Re = 11.7 \times 10^6$, are shown in Table 11a for an ONERA M-6 wing and computed on a coarse ($97 \times 25 \times 17$) 3–D grid by current CFD analysis standards. It should also be noted that these SD ratios are computed with the Baldwin-Lomax (B-L) turbulence model [26] and are the inverse of those SD ratios given in earlier tables. The agreement between the SD from the CD and AD methods is seen to be three or four significant digits for these very well converged results.

An indication of the computational time and memory requirements for the derivative code in its current form compared with those for the original code can be seen in Table 11b. These are the runtime statistics for the results shown in Table 11a. The AD-version code requires about 3 times the memory and 2.5 times more runtime than that needed using the original code and FD for the problem considered. However, these are simply initial results and little has yet been done to refine the iterative derivative paradigm with regard to just what convergence levels are required.

An indication of the derivative accuracy as a function of the derivative residual (R) convergence level for the AD TLNS3D code is given in Table 11c. For agreement to two significant digits, the AD runtime is shown to be essentially equal to the FD runtime. The accuracy results given in Table 11a are the last column in Table 11c.

Examples of nongeometric SD with respect to modeling parameters are taken from [19] There, it was proposed that an AD-based SA could be useful for investigating the adequacy of approximation models within a system. That is, an SA can be used to adjust the approximation model within the system in order to improve its physical simulation (i.e., match with experimental data) or obtain desired numerical solution properties such as stability or convergence acceleration. In this regard, exact nongeometric SD of the wing lift, drag, and pitching moment coefficients with respect to CFD algorithm and turbulence modeling parameters were obtained from TLNS3D via ADIFOR. Initial SD results with respect to modeling parameters were given for the flow conditions, B-L turbulence model, and grid of the previous example.

$\frac{SD_{AD}}{SD_{FD}}$	C_L	C_D	C_M
M	1.0000	1.0000	1.0000
α	1.0000	1.0000	1.0000
Re	0.9991	1.0000	0.9961

a. Comparison of SD ratios.

Solution Procedure	Number Solutions	Time (sec[a])	Storage (MW)
FD	4 Nonlinear	3960	2.47
CD	6 Nonlinear	5940	2.47
AD	1 NL, 3 Linear	10290	7.63

[a] Cray Y-MP.

b. Computational memory requirements and timing comparisons.

AD Convergence[a]	10^{+1}	10^{0}	10^{-1}	$10^{-1.5}$
AD/FD Agreement[b]	2	3	4	4+
AD time/FD time	1.01	1.45	2.07	2.59

[a] Norm of $\mathbf{R}' = d\mathbf{R}/dx$.

[b] Number of significant digits.

c. Computational accuracy and timing comparisons versus SD convergence level.

Table 11. Results for calculation of three output function SD with respect to three nongeometric design variables. (ONERA M-6 wing, transonic turbulent high Reynolds number flow.)

The CFD algorithm parameters chosen as independent variables were essentially the coefficients of the second- and fourth-order artificial dissipation terms. These variables [25] are $\kappa^{(2)}$ and $\kappa^{(4)}$; these are proportional to VIS2 and VIS4, respectively, in the TLNS3D code input. The iterative paradigm for using the AD-generated code to calculate the SD is as follows: (1) run the original code to a reasonably good convergence, and (2) start the SD calculation from a restart file of that solution. The iterative convergence histories for the residuals (R) for both TLNS3D and the AD code, as well as that for the derivative (R') from the AD code, are shown in Figure 2a. The three distinct peaks at the start are caused by the grid sequencing that was used to initiate the finest grid solution. The relative accuracy of SD via AD and FD is shown in Table 12a as the ratios SD_{AD}/SD_{FD}. Approximately three-significant-digit agreement exists at the convergence levels obtained. All of these SD are $O(10^{-3})$ or (10^{-4}). Since the FD step size Δ is 10^{-6} times the baseline values of the independent variables, the third digit in the Table 12a entries corresponds to the 11th digit of the dependent variables which are subtracted to form the numerator of the SD_{FD}!

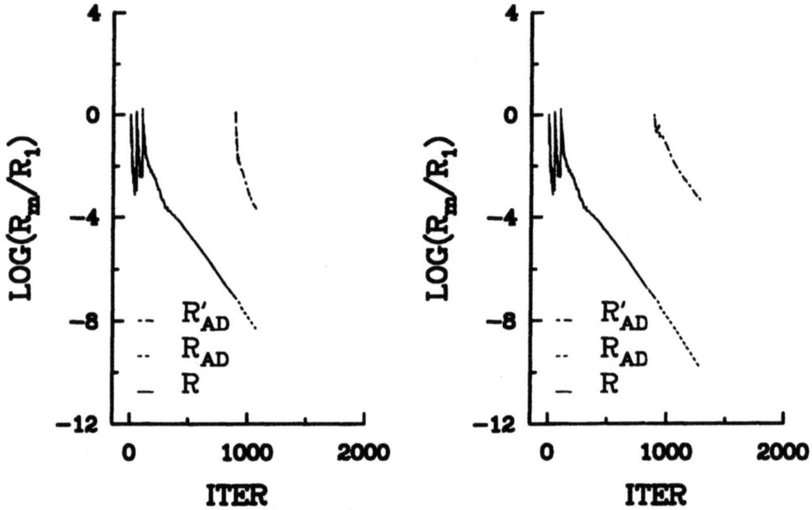

a. Algorithm parameters. b. Viscosity parameters.

Figure 2. Iterative convergence histories of TLNS3D solution and SD residuals on 97x25x17 grid. (ONERA M-6 wing, transonic turbulent high Reynolds number flow.)

For ADIFOR-generated SD with respect to turbulence model parameters, the Baldwin-Lomax algebraic model was used initially. The two parameters chosen are the Clauser constant $K(= 0.0168$, in [27]) of the outer-region eddy viscosity coefficient and the constant $A^+(= A$, in [27]) of the Van Driest correction to the Prandtl mixing length of the inner-region viscosity coefficient. The same TLNS3D restart file that was used for the previous example was used here. Figure 2b shows the iterative convergence histories for the residual R from the original TLNS3D and the AD code, as well as the convergence history for R' from the AD code. The relative accuracy of SD via AD and FD is shown in Table 12b as the ratios SD_{AD}/SD_{FD}. Again, nearly three-significant-digit agreement exists at the convergence levels obtained.

Results on a $97 \times 25 \times 17$ grid demonstrate that the ADIFOR-generated derivatives are accurate. That is, the "mechanical aspect" of ADIFOR differentiation, which *does not* depend upon grid size, has been verified for both the CFD algorithm and turbulence modeling parameters considered. The basic flow solution itself, however, does exhibit truncation-error and perhaps also resolution effects. Both of these effects, of course, can affect the SD values, whether computed by the AD or the FD methods. Since initial runs of the original TLNS3D code with the Johnson-King (J-K) turbulence model ([28], [29]) for both two-level and three-level MG on a $97 \times 25 \times 17$ grid could not be converged before the residual became extremely "noisy", a finer grid was required to obtain solutions needed to verify SD for the J-K turbulence model.

$\frac{SD_{AD}}{SD_{FD}}$	C_L	C_D	C_M
VIS2	0.9972	1.0008	0.9954
VIS4	0.9943	1.0015	1.0016

a. Two nongeometric algorithm parameters.

$\frac{SD_{AD}}{SD_{FD}}$	C_L	C_D	C_M
K	0.9880	0.9882	0.9879
A^+	0.9997	0.9950	0.9997

b. Two nongeometric viscosity parameters.

Table 12. Comparison of SD ratios for three output functions with respect to two pairs of parameters. (ONERA M-6 wing, transonic turbulent high Reynolds number flow.)

$\frac{SD_{AD}}{SD_{FD}}$	C_L	C_D	C_M
K	0.9999	1.0000	0.9998
A^+	0.9997	1.0001	0.9999

a. B-L model.

$\frac{SD_{AD}}{SD_{FD}}$	C_L	C_D	C_M
K	0.9940	0.9826	0.9935
A^+	0.9694	0.9544	1.0191

b. J-K model.

Table 13. Comparison of SD ratios for three output functions with respect to two nongeometric viscosity parameters. (ONERA M-6 wing, transonic turbulent high Reynolds number flow.)

The problem shown previously for the B-L turbulence modeling SD on a $97 \times 25 \times 17$ grid was run on a finer ($193 \times 49 \times 33$), more highly stretched grid. As shown in Table 13a, the agreement among all the SD ratios is significantly improved over those in Table 12a. The TLNS3D code with the J-K turbulence model frequently appears to have significantly different asymptotic convergence characteristics than that with the B-L model, as illustrated by the comparison of the normalized residuals in Figure 3. At this low angle-of-attack condition, the J-K residual converges several orders of magnitude and then "hangs up" in a low amplitude noise (which may be oscillatory); the B-L residual can be driven down to machine zero as seen in Figure 3. Noise also exists in each J-K solution for the dependent variables required in the SD_{FD} computations. This finding has dire implications for obtaining SD via FD, as illustrated in [19]. To obtain a meaningful correlation between the SD obtained by AD and FD for this case, averaging the dependent variable results over a number of iterations was required for the FD. The lift, drag, and pitching moment coefficients were averaged and the SD were then constructed with these averaged output coefficients for various perturbations in the independent variables. These averaged results, denoted as $SD_{\overline{FD}}$, are shown in the SD ratios of Table 13b. These ratios are all significantly improved over those FD results found by other means.

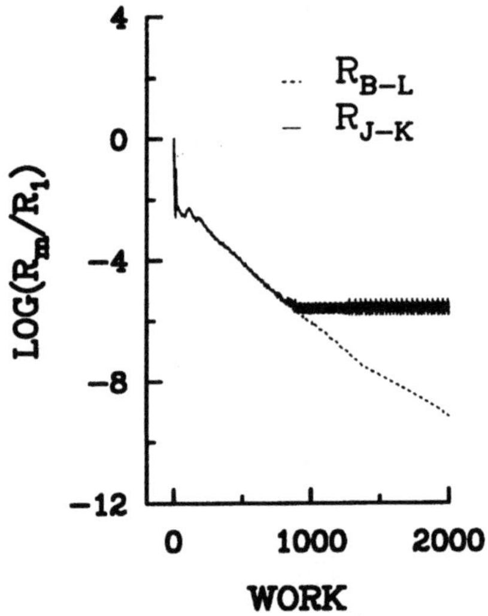

Figure 3. Comparison of iterative convergence history for TLNS3D solution residuals on 193x49x33 grid for B-L and J-K turbulence models. (ONERA M-6 wing, transonic turbulent high Reynolds number flow.)

3.2.3. 3-D viscous SD via AD—geometric

The geometric SD examples shown here were selected from those presented in [20] and briefly summarized in [21]. Such SD would, for example, be used in aerodynamic shape optimization or MDO studies of aerospace vehicles. As previously was discussed, most modern advanced CFD codes use body-fitted computational grids, which are generated in a separate code and treated essentially as a different discipline. Since the geometric design variables determine the body shape or changes to it, then these variables set the boundary values in the grid-generation problem or process. Thus, SD of the aerodynamic output functions with respect to these geometric design variables must be chain-ruled back to the grid generation code or grid movement algorithm. That is, for grids that are not solution adaptive,

$$\frac{dC_L}{dD} = \frac{dC_L}{dX} \times \frac{dX}{dD}, \tag{9}$$

where dX/dD is the computational grid SD with respect to the geometric design variable D. One must now provide the flow solver code not only the grid, X, but also its sensitivity. This is depicted in Figure 4. In the previous section, the nongeometric SD were discussed; they are denoted on this figure as dAero with respect to Stream, Physics, Algorithm. The computational grid SD with respect to geometry is denoted as dMesh/dShape and the aerodynamic SD with respect to geometry as dAero/dShape. Customarily, the wing geometry design variables are divided into those governing the planform, section thickness, and section camber shapes.

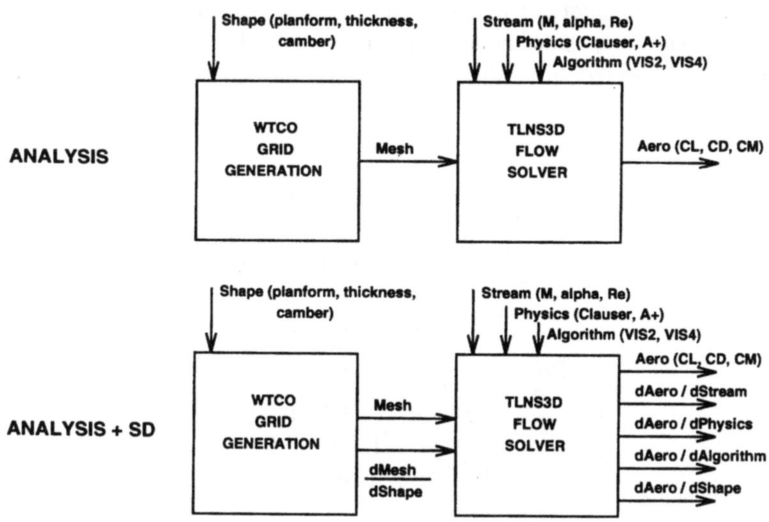

Figure 4. Data flow for grid generation — flow solver code interaction in CFD analysis and sensitivity analysis.

The TLNS3D flow solver was briefly described in the previous section, and it is used in conjunction with the WTCO grid-generation code as indicated in Figure 4. The WTCO code is a batch-mode, algebraic, transfinite-interpolation wing grid-generation program. The code includes the capability to generate grids around wings of at least two airfoil sections, described by ordinates or the NACA four-digit series airfoils. The user specifies spacing constants at several points on the grid boundaries, types of boundary surface, grid interior interpolation, as well as the amount of smoothing required. This code was chosen because it has been used reliably with the TLNS3D flow solver on many previous occasions; it is noniterative and does not include any coding extraneous to the grid generation (such as graphics). In addition, the source code necessary for ADIFOR application was readily available. Application of ADIFOR to WTCO produces a Fortran code that computes not only the grid, X, but also the grid sensitivities, dX/dD. These are passed to the ADIFOR-ed TLNS3D flow code as a "seed matrix," S (see [18]).

Accuracy comparisons of aerodynamic SD with respect to three planform variables are shown as ratios SD_{AD}/SD_{FD} in Tables 14. The planform variables are wing tip chord (C_{tip}), which controls wing taper; wing-tip leading-edge x coordinate ($x_{tip,LE}$), which controls the wing sweep angle; and wing-tip leading-edge z coordinate ($z_{tip,LE}$) which controls the wing span. These results are for the ONERA M-6 wing at transonic flow conditions ($M = 0.84$ and $\alpha = 3.06°$) and computed on $97 \times 33 \times 17$ grids. Results in Table 14a are for inviscid (Euler) flow whereas those in Table 14b are for high Reynolds number (11.7×10^6) viscous flow using the B-L turbulence model. The agreement is much better for the inviscid results; there, the grid is not stretched nearly as hard as that for the viscous case. The agreement is not good for two SD in this latter viscous case and is still being investigated on a finer grid.

$\frac{SD_{AD}}{SD_{FD}}$	C_L	C_D	C_M
C_{tip}	1.0000	1.0074	1.0000
$x_{tip,LE}$	1.0000	1.0000	1.0000
$z_{tip,LE}$	1.0000	1.0000	1.0000

a. Transonic inviscid flow.

$\frac{SD_{AD}}{SD_{FD}}$	C_L	C_D	C_M
C_{tip}	0.9939	0.5938	0.9629
$x_{tip,LE}$	1.0062	1.0007	0.9981
$z_{tip,LE}$	0.9745	0.7690	1.0781

b. Transonic turbulent high Reynolds number flow.

Table 14. Comparison of SD ratios for three output functions with respect to three geometric planform design variables. (ONERA M-6 wing.)

Both section camber (including twist) and thickness design variable SD are compared for accuracy in Tables 15. The ratios SD_{AD}/SD_{FD} for four design variables as determined on a $97 \times 25 \times 17$ grid are given in Table 15a whereas those for a $193 \times 49 \times 33$ grid are given in Table 15b. These variables are all for the tip section: twist angle (α_{tip}), maximum camber ($CMAX_{tip}$), chordwise location of $CMAX$ ($x_{CMAX,tip}$), and thickness-to-chord ratio (τ_{tip}). Here the planform is that of the ONERA M-6 wing but the airfoil sections are the NACA 2412. The conditions are high Reynolds number (11.7×10^6) transonic ($M = 0.84, \alpha = 0.0°$) flow. Both grids are highly stretched, and the finer grid SD agreement is better (max difference about 4% rather than 10%) than that for the coarser grid. Again, these were expected to be even better and are still being investigated.

$\frac{SD_{AD}}{SD_{FD}}$	C_L	C_D	C_M
α_{tip}	1.0011	1.0007	1.0016
$CMAX_{tip}$	0.9971	0.9987	0.9973
$x_{CMAX,tip}$	0.9950	1.0022	0.9970
τ_{tip}	1.0643	1.0041	1.1013

a. 97 × 25 × 17 grid.

$\frac{SD_{AD}}{SD_{FD}}$	C_L	C_D	C_M
α_{tip}	1.0006	1.0003	1.0008
$CMAX_{tip}$	0.9962	0.9985	0.9965
$x_{CMAX,tip}$	0.9943	1.0017	0.9967
τ_{tip}	1.0234	1.0040	1.0389

b. 193 × 49 × 33 grid.

Table 15. Comparison of SD ratios for three output functions with respect to four geometric camber and thickness design variables. (ONERA M-6 plan — NACA 2412 foil, transonic turbulent high Reynolds number flow.)

The runtime and memory requirements for AD generated SD can be summarized as follow: the AD runtime is about equal to a single evaluation of SD by CD; the AD memory scales by the number of design variables. Some improvement in both of these aspects should be realized from an AD IIM implementation.

Again, the accuracy and ease with which very complex code can be differentiated using ADIFOR can neither be overlooked nor ignored. These few examples are the only reported SD results for high Reynolds number turbulent flow. Obtaining them more efficiently is the goal of continuing work and is addressed in the next section.

4. Application of AD for IIM SD calculation

The current ADIFOR technology is used to obtain SD from IIM forms as first suggested in [2]. All of the derivative terms which may be needed to construct the IIM form, (3*) are found on the right-hand sides ("residual") of the equations, (1*) and (3*). Note that the left-hand side operators are essentially those of the (efficient) original CFD solver, (1*). Once these residuals have been identified and isolated into separate (from all left-hand side operators) subroutines, then ADIFOR can be applied only where needed. However, the derivative flow-dependence path must be constructed through the code from input design variables, D, to output functions, F, in order to construct the SD F'.

4.1. First derivative ADII

The procedure for application of ADIFOR "by hand" to get first derivatives, as outlined above, is depicted in Table 16. Once the residual subroutine(s) have been isolated, separate applications of ADIFOR are made to produce the derivatives $(\frac{\partial \mathbf{R}}{\partial \mathbf{Q}}, \frac{\partial \mathbf{R}}{\partial \mathbf{X}}, \frac{\partial \mathbf{R}}{\partial \mathbf{D}})$ and $\mathbf{X}'$. These are the derivative terms needed for the right-hand side of (3*). The corresponding seed matrices are $\mathbf{Q}'^m$, $\mathbf{X}'$, and I, the identity matrix as indicated in Table 16. Note that straightforward application of ADIFOR would, among other things, put the design variable loop (over j) inside the incremental iteration (II) loop (over m); however, with IIM the loops are to be reversed (or only partially reversed for vectorization efficiency regarding Cray automatic loop-unrolling). Only the Jacobian–"variable" seed matrix product $(\frac{\partial \mathbf{R}}{\partial \mathbf{Q}} \times \mathbf{Q}'^m_j)$ needs to be calculated in the innermost loop; the other derivatives in G_j do not depend on m. This innermost loop is then the solution of (3*) for the design variable j. This procedure was done for the SD Navier-Stokes code of [3] in order to obtain (first-order) SD for the three output functions (coefficients of) lift, drag, and pitching moment (C_L, C_D, C_M, respectively) with respect to five design variables.

1. Define residual routine(s)
2. Obtain ADIFOR-ed derivative code: $\frac{\partial \mathbf{R}}{\partial \mathbf{Q}}$; $\frac{\partial \mathbf{R}}{\partial \mathbf{X}}$; $\frac{\partial \mathbf{R}}{\partial \mathbf{D}}$; and $\mathbf{X}'$
3. Set seed matrices for Jacobians above; $\mathbf{Q}'^m$; $\mathbf{X}'$; I
4. Construct iteration: Do (design variables loop over j) $G_j = \frac{\partial \mathbf{R}}{\partial \mathbf{X}} \cdot \mathbf{X}'_j + \frac{\partial \mathbf{R}}{\partial \mathbf{D}_j} \cdot I$ Do (II loop over m) $\mathbf{R}'^m_j = \frac{\partial \mathbf{R}}{\partial \mathbf{Q}} \cdot \mathbf{Q}'^m_j + G_j$ $\mathbf{Q}'^{m+1}_j = \mathbf{Q}'^m_j + \Delta \mathbf{Q}'_j$ Continue Continue

Table 16. Procedure for application of ADIFOR by-hand to IIM.

Initial SD results via ADII, for the same 2–D turbulent flow case as previously discussed in comparing both the QAII and straightforward AD results with CD results, are given in Table 17. Relative accuracy results are shown in Table 17a as the SD ratios $\mathrm{SD_{CD}}/\mathrm{SD}_{ADII}$; the agreement is good, to almost three significant digits. Of more interest, however, are the ADII results for memory and runtime requirements given in Table 17b. These SD are seen to be very competitive and consistent with those from the CD method whereas those from the seemingly efficient QAII were not consistent due to the undifferentiated turbulence terms. Many more results for first derivative (order) SD are presented in [22]. The ADII form with its reduced memory requirements should be even more amenable to coarse-grain parallel calculation of SD than the straightforward AD code which was discussed in [23].

SD Ratio	Design Var., D	$\frac{dC_L}{dD}$	$\frac{dC_D}{dD}$	$\frac{dC_M}{dD}$
$\frac{SD_{CD}}{SD_{ADII}}$	T	1.005	1.002	1.004
	C	0.999	0.997	0.999
	L	0.998	1.002	1.000
	α	1.000	1.000	1.000
	M	1.002	0.998	1.002

a. Comparison of SD ratios.

Solution Method	Memory (MBytes)	CPU Time* (sec)
CD	5.31	7,100
CD**	5.34	2,520
QAII	7.16	693
QAII**	7.28	413
AD	30.92	9,390
AD**	31.09	5,740
ADII	14.35	6,190
ADII**	14.48	1,268

*Cray Y-MP **Vectorization improvements

b. Computational memory requirements and timing comparisons.

Table 17. Results for calculation of three output function SD with respect to three geometric and two nongeometric design variables. (NACA 1406 airfoil, transonic turbulent high Reynolds number flow.)

4.2. Second derivative ADII

Second-order SD are obtained here by differentiating first-order discretized sensitivity equations again. These are of interest for several reasons. For example, aerodynamic stability derivatives are required by the controls discipline as an input; therefore, inclusion of controls in a gradient based design optimization means that the sensitivities of stability derivatives are needed. These are second-order SD, such as $\frac{\partial C_{L\alpha}}{\partial D} = \frac{\partial^2 C_L}{\partial D \partial \alpha}$. Secondly, in constructing function approximations for nonlinear flow behavior, the expansions using first-order derivative information are only of limited usefulness. That is, for I design variables,

$$\mathbf{F}(\mathbf{D}+\Delta\mathbf{D}) \approx \mathbf{F}(\mathbf{D}) + \sum_{i=1}^{I} \frac{\partial \mathbf{F}}{\partial D_i}\Delta D_i \tag{10}$$

is a linear approximation. If the derivatives are available, then

$$\mathbf{F}(\mathbf{D}+\Delta\mathbf{D}) \approx \mathbf{F}(\mathbf{D}) + \sum_{i=1}^{I} \frac{\partial \mathbf{F}}{\partial D_i}\Delta D_i + \frac{1}{2}\sum_{i=1}^{I}\sum_{j=1}^{I} \frac{\partial^2 \mathbf{F}}{\partial D_i \partial D_j}\Delta D_i \Delta D_j \tag{11}$$

may exhibit much of the nonlinear behavior of $\mathbf{F}$. Thirdly, and more important, second-order optimization techniques can be employed. Second-order SD have been discussed in [30] for additional sensitivity analysis and design optimization applications in other disciplines. The objective here is not to review previous non–CFD related work in this area but rather to derive one of the discrete IIM forms for second-order aerodynamic SD. Four combinations of direct-differentiation (DD) and adjoint-variable (AV) methods are possible for second order SD as discussed in [30]. The equation set derived here is the DD.DD combination, and equations for other combinations are given in [22]; there are advantages for the DD.AV combination [30].

Equation (3) is a matrix equation and the vector corresponding to the design variable D_j can be written as

$$\frac{d\mathbf{R}}{dD_j} = \frac{\partial \mathbf{R}}{\partial \mathbf{Q}}\frac{\partial \mathbf{Q}}{\partial D_j} + \frac{\partial \mathbf{R}}{\partial \mathbf{X}}\frac{\partial \mathbf{X}}{\partial D_j} + \frac{\partial \mathbf{R}}{\partial D_j} = 0 \tag{12}$$

or symbolically as

$$\mathbf{R}'_j = \frac{\partial \mathbf{R}}{\partial \mathbf{Q}}\mathbf{Q}'_j + \frac{\partial \mathbf{R}}{\partial \mathbf{X}}\mathbf{X}'_j + \frac{\partial \mathbf{R}}{\partial \mathbf{D}_j} = 0 \tag{12a}$$

or

$$\mathbf{R}'_j = \mathbf{R}'_j\Big(\mathbf{Q}(\mathbf{D}), \mathbf{X}(\mathbf{D}), \mathbf{Q}'_j(\mathbf{D}), \mathbf{X}'_j(\mathbf{D}), \mathbf{D}\Big) = 0\,. \tag{12b}$$

Differentiation of (12b) with respect to D_i gives

$$\frac{d\mathbf{R}'_j}{dD_i} = \frac{\partial \mathbf{R}'_j}{\partial \mathbf{Q}}\frac{\partial \mathbf{Q}}{\partial D_i} + \frac{\partial \mathbf{R}'_j}{\partial \mathbf{X}}\frac{\partial \mathbf{X}}{\partial D_i} + \frac{\partial \mathbf{R}'_j}{\partial D_i} + \frac{\partial \mathbf{R}'_j}{\partial \mathbf{Q}'_j}\frac{\partial \mathbf{Q}'_j}{\partial D_i} + \frac{\partial \mathbf{R}'_j}{\partial \mathbf{X}'_j}\frac{\partial \mathbf{X}'_j}{\partial D_i} = 0 \tag{13}$$

But from (12a) it is seen that

$$\frac{\partial \mathbf{R}'_j}{\partial \mathbf{Q}'_j} = \frac{\partial \mathbf{R}}{\partial \mathbf{Q}} \quad \text{and} \quad \frac{\partial \mathbf{R}'_j}{\partial \mathbf{X}'_j} = \frac{\partial \mathbf{R}}{\partial \mathbf{X}}\,. \tag{14}$$

Recognizing other notations (such as $\mathbf{Q}'_j = \frac{\partial \mathbf{Q}}{\partial D_j}$) from (12) and using (14), (13) can be written as

$$\frac{\partial \mathbf{R}}{\partial \mathbf{Q}}\frac{\partial^2 \mathbf{Q}}{\partial D_j \partial D_i} + \frac{\partial \mathbf{R}}{\partial \mathbf{X}}\frac{\partial^2 \mathbf{X}}{\partial D_j \partial D_i} + \frac{\partial \mathbf{R}'_j}{\partial \mathbf{Q}}\mathbf{Q}'_i + \frac{\partial \mathbf{R}'_j}{\partial \mathbf{X}}\mathbf{X}'_i + \frac{\partial \mathbf{R}'_j}{\partial D_i} = 0\,. \tag{15}$$

Again, from (12a), the last three terms of (15) can be written as

$$\frac{\partial \mathbf{R}'_j}{\partial \mathbf{Q}}\mathbf{Q}'_i = \left(\frac{\partial^2 \mathbf{R}}{\partial \mathbf{Q}^2}\mathbf{Q}'_j + \frac{\partial^2 \mathbf{R}}{\partial \mathbf{Q}\partial \mathbf{X}}\mathbf{X}'_j + \frac{\partial^2 \mathbf{R}}{\partial \mathbf{Q}\partial D_j}\right)\mathbf{Q}'_i \tag{16}$$

$$\frac{\partial \mathbf{R}'_j}{\partial \mathbf{X}}\mathbf{X}'_i = \left(\frac{\partial^2 \mathbf{R}}{\partial \mathbf{Q}\partial \mathbf{X}}\mathbf{Q}'_j + \frac{\partial^2 \mathbf{R}}{\partial \mathbf{X}^2}\mathbf{X}'_j + \frac{\partial^2 \mathbf{R}}{\partial \mathbf{X}\partial D_j}\right)\mathbf{X}'_i \tag{16a}$$

$$\frac{\partial \mathbf{R}'_j}{\partial D_i} = \frac{\partial^2 \mathbf{R}}{\partial \mathbf{Q}\partial D_i}\mathbf{Q}'_j + \frac{\partial^2 \mathbf{R}}{\partial \mathbf{X}\partial D_i}\mathbf{X}'_j + \frac{\partial^2 \mathbf{R}}{\partial D_i \partial D_j}\,. \tag{16b}$$

Equation (15) is the standard-form linear equation which would be solved to obtain the second-order SD $\frac{\partial^2 \mathbf{Q}}{\partial D_j \partial D_i}$ once the state variables, $\mathbf{Q}$, and first-order SD, $\mathbf{Q}'$, have been determined from either (1) and (3) or (1*) and (3*), respectively, assuming that $\mathbf{X}$ and $\mathbf{X}'$ are known from the grid and grid derivative code solutions. That is, it can be seen from (16) that, with $\mathbf{Q}$ and $\mathbf{Q}'$ known, the last three terms in (15) are known. The second term is also known since it is assumed that $\frac{\partial^2 \mathbf{X}}{\partial D_j \partial D_i}$ is determined from a second-derivative grid code solution. Therefore, if one iteratively solves for $\frac{\partial^2 \mathbf{Q}}{\partial D_j \partial D_i}$, all of these four terms are fixed (do not depend on $\frac{\partial^2 \mathbf{Q}}{\partial D_j \partial D_i}$). The IIM form of (15) is then

$$\begin{aligned} -\widetilde{\frac{\partial \mathbf{R}}{\partial \mathbf{Q}}}\Delta\frac{\partial^2 \mathbf{Q}}{\partial D_j \partial D_i} &= \left(\frac{d\mathbf{R}'_j}{dD_i}\right)^n \\ \left(\frac{\partial^2 \mathbf{Q}}{\partial D_j \partial D_i}\right)^{n+1} &= \left(\frac{\partial^2 \mathbf{Q}}{\partial D_j \partial D_i}\right)^n + \Delta\frac{\partial^2 \mathbf{Q}}{\partial D_j \partial D_i} \qquad n = 1, 2, 3, \ldots . \end{aligned} \tag{15*}$$

This IIM form for the second-order SD, $\frac{\partial^2 \mathbf{Q}}{\partial D_j \partial D_i}$, can be solved with the same approximate operator (left-hand side) as that used to obtain the solution, $\mathbf{Q}$, and the first-order SD, $\mathbf{Q}'$, as can be seen by comparing (1*), (3*), and (15*); only the right-hand sides are different. Thus, all of the advantages associated with solving (1*) and (3*), as opposed to solving (1) and (3) (discussed in [2] and [3]), also apply to solving (15*) instead of (15).

Initially the ADIFOR tool was applied to the hand-differentiated (first-order) SD code of [3] for the nongeometric variables Mach number, M_∞ and angle of attack, α, using the IIM forms, (1*), (3*) and (15*). That is, the DD.DD form is QAII.ADII. Initial results for the first Q component, (i.e., $\frac{\partial^2 Q_1}{\partial D_1^2} = \frac{\partial^2 \rho}{\partial M_\infty^2}$), were obtained using first-order CFD algorithm approximations for $\frac{\partial \mathbf{R}}{\partial \mathbf{Q}}$, etc., on *both* right-

and left-hand sides of the equations. These initial results for the IIM calculation of a second-order SD are given in Figure 5, as contour plots; results obtained by CD and DD.DD are shown in Figure 5(a) and (b), respectively. For the DD.DD result, the first differentiation is the hand differentiation QAII, which is followed by the ADII. The DD.DD solutions and those solutions used to construct the CD results were converged to 10^{-11} residual ratios (R^N/R^1), i.e., to machine zero. Much noise is apparent in the CD results (Figure 5(a)) and using it to determine the accuracy of the QAII. ADII could certainly be suspect. Reference [22] contains a number of second-order SD results and an extensive detailed discussion of them.

5. Simultaneous aerodynamic analysis and design optimization

Simultaneous aerodynamic analysis and design optimization (SAADO) is a technique which incorporates design improvement within the aerodynamic analysis in order to gain computational efficiency. This algorithm enables the simultaneous reduction of the objective function and the correction of the constraint violations; in this strategy, the constraints include the residuals of the aerodynamic equations. This design optimization technique requires that the steady-state aerodynamic equations be satisfied only at the final optimum solution. This requirement obviates the need to obtain the complete steady-state flow solutions accurately at and within every optimization iteration, which is characteristic of conventional design optimization procedures and accounts for the bulk of the central processing unit (CPU) time required in such procedures.

The SAADO algorithm and its variants have been successfully implemented on two quasi-one-dimensional nozzle-flow problems ([31] and [32]), including a sample problem where a discontinuity exists (i.e., a normal shock) in the flow field. A substantial increase in the number of design variables and constraints occurs when the field variables are treated as part of the set of design variables and when the aerodynamic flow equations are considered to be part of the set of constraints; this is nullified by the introduction of an adjoint-variable vector. Furthermore, the appropriate introduction of direct Newton iterations (including under-relaxation in the design optimization process) helps to control the value of the residual of the aerodynamic equations. Depending upon the number of flow-solver iterations used and their placement in the design optimization scheme, a trade-off can be achieved between the values of the residual of the aerodynamic equations (at the end of the design process) and the objective function and the CPU time. In both sample problems, a considerable savings in terms of equivalent Newton iterations (and, hence, overall CPU time) is achieved with the SAADO scheme when compared with the conventional black-box design optimization procedure.

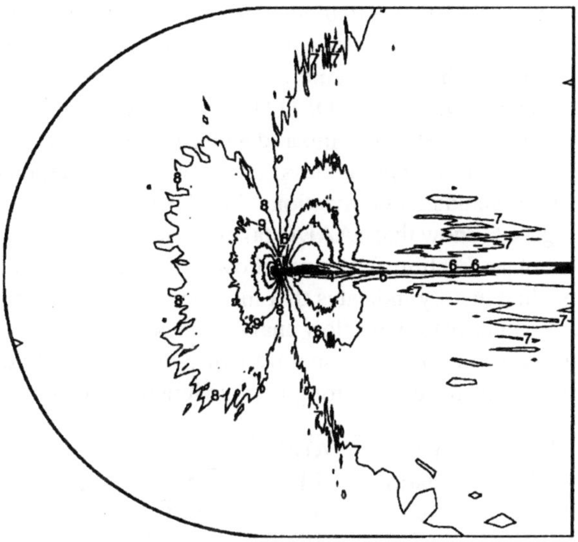

a. Via CD.

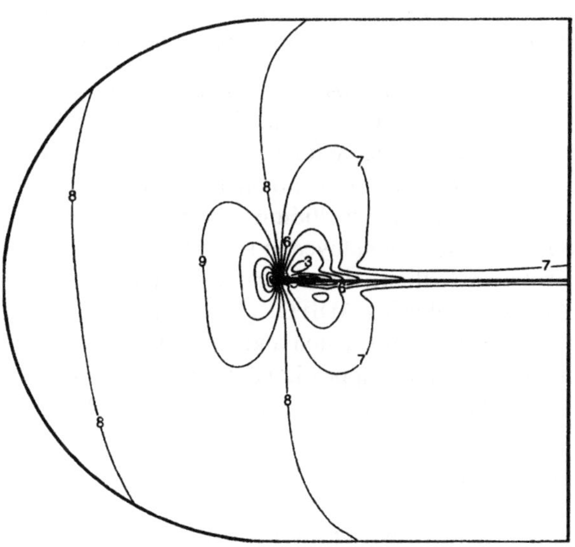

b. Via QAII.ADII.

Figure 5. Contour plots of the initial results for the calculation of the second-order SD $\partial^2\rho/\partial M_\infty^2$. (NACA 1406 airfoil; subsonic, laminar, low Rey-nolds number flow.)

The feasibility of SAADO is demonstrated for both supersonic and transonic flow described by a quasi 1–D Euler equation. Optimization results for a supersonic nozzle design are shown in Table 18. Standard optimization obtained from the Automated Design Synthesis (ADS) [11] software using both (one-sided) FD and QA differentiation for SD are compared with SAADO; fewer equivalent Newton Raphson (NR) iteration steps and less computer time are required for SAADO than for either of the standard optimizations. For the transonic nozzle, residual (R) history plots, Figure 6, show that SAADO takes about the same CPU time to reach the optimal solution as that time for a single flow analysis. The oscillating nature of the SAADO residual history indicates that incorporation of design changes alters the normal convergence pattern for flow analysis; these oscillations cease when the design changes cease. In both of these nozzle designs the objective was to match a prescribed velocity distribution throughout the length of the nozzle.

Optimizer	ADS	ADS	
SD Calculation	FD	QA	SAADO
Objective	8.8E-4	8.9E-4	8.9E-5
Equiv. NR iter.	174	102	80
Cray-2 (sec)	2.19	1.29	0.69

Table 18. Feasibility for simultaneous aerodynamic analysis and design optimization. (Quasi 1–D nozzle, supersonic inviscid flow.)

The primary objective of the SAADO formulation is a significant reduction in the CPU time; the time needed due to repeated analyses required by standard optimization is impractical. The potential for SAADO is clearly demonstrated in these sample 1–D nozzle design optimization problems. In principle, the SAADO algorithm is extendable to large-scale multidimensional flow problems. However, in these cases, the IIM should be employed to solve the linear equations that are involved in this new optimization scheme, and the required complicated Jacobian matrices should be obtained using ADIFOR. Unfortunately, the use of an adjoint-variable vector requires the construction and efficient evaluation of the Jacobian transpose–adjoint vector product appearing in (5) and (5*). For AD this is done in the "reverse mode", which is not possible with the current ADIFOR.

6. Conclusions and challanges

Conclusions and challenges relevant to the computational methodologies discussed for the preparation and efficient use of advanced CFD codes in sensitivity analysis (SA) and multidisciplinary design optimization (MDO) are grouped according to the paper sections: Incremental Iterative Method (IIM) for Sensitivity Derivatives (SD); Automatic Differentiation (AD) of Navier-Stokes (N-S) Codes for SD; Application of AD for IIM SD Calculation; and Simultaneous Aerodynamic Analysis and Design Optimization (SAADO).

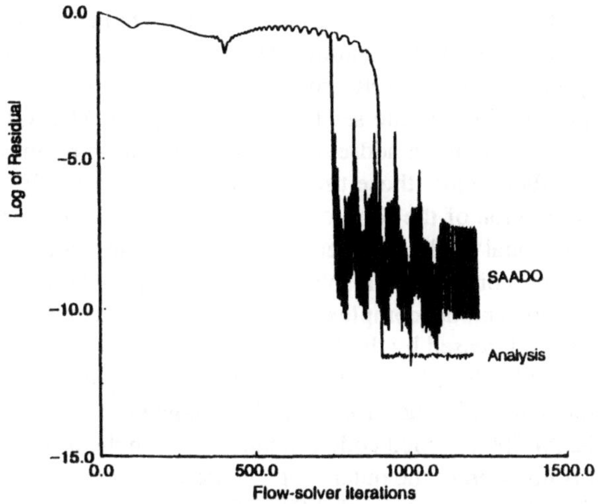

Figure 6. Comparison of iterative convergence history for analysis and SAADO residuals. (Quasi 1–D nozzle, transonic inviscid flow.)

The IIM is among the most accurate and efficient yet proposed or demonstrated for obtaining SD from nonlinear CFD codes. Derivative results from a 2–D thin-layer Navier-Stokes code and a 3–D Euler code for both nongeometric (flow) and geometric (shape) design variables have been shown. In addition to accurate consistent discrete SD obtained with computational efficiency (for both CPU time and memory requirements), the IIM also allows using approximate solution operators of choice for further efficiency, parallelization, or robustness, for example. Furthermore the method permits the use of symbolic or automatic differentiation to obtain the needed Jacobian or seed matrices. However, for most existing codes that employ an IIM for the basic solution algorithm, it can be a challenge to isolate the residual calculation in order to construct the efficient IIM for SD calculation. The 3–D supersonic Euler code with IIM SD calculation has been used in optimization studies.

Computational differentiation of advanced CFD codes employing ADIFOR in order to obtain SD of output flow properties with respect to nongeometric and geometric input variables, as well as algorithm and viscosity parameters, has been quantitatively demonstrated. This is a significant and encouraging result for several reasons: the SD convergence can be monitored during code execution; small derivative values and those from "noisy" solutions can be obtained; SD verification can be accomplished on coarse grids; the TLNS3D code is an efficient, complex, state-of-the-art 3–D CFD code and its SD have been obtained; the computational efficiency of TLNS3D is based upon the rather delicate multigrid acceleration algorithm, and the successful application of ADIFOR yielded similar convergence rates for the SD; ADIFOR also successfully differentiated the Baldwin-Lomax and

Johnson-King turbulence models; accurate SD were obtained using AD in a relatively short lead time, essentially treating ADIFOR, TLNS3D, and WTCO as black boxes; and coupling of the ADIFOR-processed grid generation and flow codes was readily accomplished through the seed matrix concept of ADIFOR. Challenges, however, remain and are being addressed. Computational run times to obtain SD are, at best, only about equal those times for CD; this is partially due to the unnecessary differentiation of the solution process. The black-box AD memory requirement, proportional to the number of design variables, precludes the use of many design variables on the high resolution meshes required for detail CFD or the large meshes needed to model complex configurations. Both of these aspects, however, are favorably addressed by the IIM. Currently post-processing of ADIFOR-produced code must be done by hand in order to recover vector efficiency. It may be simpler to maintain efficiency for the ADIFOR-generated versions of parallel CFD codes. The ADIFOR added code can interfere with the inner loop vectorization whereas it is not seen in the outer loop parallelization.

The procedure for application of ADIFOR to obtain first- and second-order SD in an IIM form has been outlined. Initial ADII results calculated from a 2–D Navier-Stokes code for first-order SD at transonic turbulent flow and second-order SD at subsonic laminar flow have been given. The relative accuracy between the first-order SD via CD and those via ADII is shown to be almost three significant digits. However, the ADII results for computer memory and runtime requirements are very competitive and consistent with those from the CD method, whereas those from the the seemingly efficient QAII were not consistent due to the undifferentiated turbulence terms. The ADII form with its reduced memory requirements should be even more amenable to coarse-grain parallel calculation of SD than the straightforward AD code. The 2–D Navier-Stokes ADII code is being run in airfoil optimization studies, using the first-order SD. Calculation of second-order SD using ADIFOR applied to the hand-differentiated first-order QAII code has just been done. The ADIFOR tool is required for the differentiation of complex turbulent terms and construction of the many second-order derivative pieces.

The SAADO algorithm has been successfully implemented on quasi 1–D nozzle flow problems where its potential for a significant reduction in computational time, as compared to that for conventional optimization, was demonstrated. Implementation of the SAADO scheme in an existing aerodynamic design optimization capability for 2–D, transonic, high Reynolds number, turbulent flow is being tested. The turbulence terms are processed by ADIFOR, but the adjoint form is not used. Use of AD to obtain the adjoint-required form is being investigated. More rigorous mathematical analysis of SAADO convergence and extension to a 3–D implementation are also planned.

In summary then, several computational techniques have been applied to existing advanced CFD codes in order to extract accurate, consistent, discrete SD information. The efficiency with which such SD can be calculated is both code and machine dependent. In regard to new CFD analysis codes, for example, hand-generated solution derivatives or self-contained modular construction of the residuals for later differentiation should be provided. The challenge for new CFD algorithm and analysis code development is to recognize at the outset that efficient SD calculation is required for gradient-based design applications.

Acknowledgments

Thanks are due and extended to Dr. John Burns, Director of the Center for Optimal Design and Control at Virginia Polytechnic Institute and State University, for the invitation and support to present this paper. It represents the joint work of a number of people, initiated several years ago by the (former) Interdisciplinary Research Office, the Computational Aerodynamics Branch, and the U.S. Army Aeroflightdynamics Directorate Joint Research Program Office of NASA Langley. It has continued to involve government and university personnel, under the auspices of the Computational Sciences Branch of NASA Langley. The several groups and associated personnel which have been involved are shown in the following table; the NASA funding instrument is indicated but should not be taken to mean that no other sources of funding were involved.

NASA Langley Research Center
Jean Barthelemy
Larry Green
Kara Haigler
Laura Hall (CSC)
Henry Jones (U.S. Army)
Perry Newman
Eric Unger (NRC)

Argonne National Laboratory
(NASA PO L25935D)
Chris Bischof
George Corliss (Marquette University)
Andreas Griewank (Dresden University)
Peyvand Khademi
Tim Knauff
Andy Mauer

Old Dominion University
(NASA Grant NAG-1-1265)
Gene Hou
Mohan Korivi
Satish Mani
Venkat Maroju
Pravnav Patwa
Laura Sherman
Art Taylor III

Rice University
(NASA Coop. Agreement NCW-0027)
Alan Carle

Texas A&M University
(NASA Grant NAG-1-793)
Alan Arslan
Leland Carlson
Hesham Elbanna (Cairo University)

References

[1] A. Taylor, P. Newman, G. Hou, and H. Jones; Recent advances in steady compressible aerodynamic sensitivity analysis, in *Flow Control* (Ed. by M. Gunzburger), Springer, 1995, 341–356. (Presented at the IMA Workshop on Flow

Control, Institute for Mathematics and Its Applications, University of Minnesota, MN, Nov. 1992.)

[2] P. Newman, G. Hou, H. Jones, A. Taylor, and V. Korivi; Observations on computational methodologies for use in large-scale gradient-based multidisciplinary design, *Proc. Fourth AIAA/USAF/NASA/OAI Symposium on Multidisciplinary Analysis and Optimization*, AIAA, Cleveland, 1992, 531–542. (AIAA Paper 92–4753–CP, 1992.)

[3] V. Korivi, A. Taylor, P. Newman, G. Hou, and H. Jones; An approximately-factored incremental strategy for calculating consistent discrete aerodynamic sensitivity derivatives, *Proc. Fourth AIAA/USAF/NASA/OAI Symposium on Multidisciplinary Analysis and Optimization*, AIAA, Cleveland, 1992, pp. 465–478 (AIAA Paper 92–4746–CP, 1992.)

[4] V. Korivi, A. Taylor, G. Hou, P. Newman, and H. Jones; Sensitivity derivatives for three-dimensional supersonic Euler code using incremental iterative strategy, *A Collection of Technical Papers, Part 2, 11th AIAA Computational Fluid Dynamics Conference*, AIAA, Orlando, 1993, 1053--1054. (Expanded version: *AIAA J.* **32**, 1994, 1319–1321.)

[5] G. Burgreen and O. Baysal; Three-dimensional aerodynamic shape optimization of wings using sensitivity analysis, AIAA Paper 94–0094, 1994.

[6] V. Korivi, P. Newman, and A. Taylor; Aerodynamic optimization studies using a 3–D supersonic Euler code with efficient calculation of sensitivity derivatives,*A Collection of Technical Papers, 5th AIAA/NASA/USAF/ISSMO Symposium on Multidisciplinary Analysis and Optimization*, AIAA, Panama City, 1994, 170–194. (AIAA Paper 94–4270, 1994.)

[7] A. Arslan and L. Carlson; Integrated determination of sensitivity derivatives for an aeroelastic transonic wing,*A Collection of Technical Papers, 5th AIAA/NASA/USAF/ISSMO Symposium on Multidisciplinary Analysis and Optimization*, AIAA, Panama City, 1994, 1286–1300. (AIAA Paper 94–4400, 1994.)

[8] R. Barger and M. Adams; Automatic computation of wing-fuselage intersection lines and fillet inserts with fixed-area constraint, NASA TM 4406, 1993.

[9] R. Barger, M. Adams and R. Ramakrishnan; Automatic computation of Euler marching grids and subsonic grids for wing-fuselage configurations, NASA TM 4573, 1994.

[10] E. Unger and L. Hall; The use of automatic differentiation in an aircraft design problem, *5th AIAA/NASA/USAF/ISSMO Symposium on Multidisciplinary Analysis and Optimization*, AIAA, Panama City, 1994, 64–72. (AIAA Paper 94–4260, 1994.)

[11] G. Vanderplaats; ADS — a Fortran program for automated design synthesis, NASA CR–17785, NASA Ames Research Center, 1985.

[12] L. Rall; *Automatic Differentiation: Techniques and Applications*, Springer, Berlin, 1981.

[13] A. Griewank; On automatic differentiation, in *Mathematical Programming: Recent Developments and Applications*, (Ed. by M. Iri and K. Tanabe), Kluwer, Boston, 1989, 83–108.

[14] A. Griewank and G. Corliss (Eds.); *Automatic Differentiation of Algorithms: Theory, Implementation, and Application*, SIAM, Philadelphia, 1991.

[15] C. Bischof, A. Carle, G. Corliss, A. Griewank, and P. Hovland; ADIFOR: generating derivative codes from Fortran programs, *Scient. Prog.* **1**, 1992, 1–29.

[16] C. Bischof and A. Griewank; ADIFOR: a Fortran system for portable automatic differentiation, *Proc. Fourth AIAA/USAF/NASA/OAI Symposium on Multidisciplinary Optimization*, AIAA, Cleveland, 1992, 433–441. (AIAA 92–4744–CP, 1992.)

[17] J. Barthelemy and L. Hall; Automatic differentiation as a tool in engineering design, *Proc. Fourth AIAA/USAF/NASA/OAI Symposium on Multidisciplinary Optimization*, AIAA, Cleveland, 1992, 424–432. (AIAA 92–4743–CP, 1992.)

[18] C. Bischof, G. Corliss, L. Green, A. Griewank, K. Haigler, and P. Newman; Automatic differentiation of advanced codes for multidisciplinary design, *Comput. Syst. Engnrg.* **3**, 1993. (Presented at the Symposium on High-Performance Computing for Flight Vehicles, Dec. 7–9, 1992, Arlington, VA.)

[19] L. Green, P. Newman, and K. Haigler; Sensitivity derivatives for advanced CFD algorithm and viscous modelling parameters via automatic differentiation, AIAA 93–3321, 1993.

[20] L. Green, C. Bischof, A. Carle, A. Griewank, K. Haigler, and P. Newman; Automatic differentiation of advanced CFD codes with respect to wing geometry parameters for MDO, *Abstracts from Second U.S. National Congress on Computational Mechanics*, Washington, 1993, 136.

[21] A. Carle, L. Green, C. Bischof, and P. Newman; Applications of automatic differentiation in CFD, AIAA 94–2197, 1994.

[22] L. Sherman, A. Taylor, L. Green, P. Newman, G. Hou, and M. Korivi; First- and second-order aerodynamic sensitivity derivatives via automatic differentiation with incremental iterative methods, *A Collection of Technical Papers, 5th AIAA/USAF/NASA/ISSMO Symposium on Multidisciplinary Analysis and Optimization*, AIAA, Panama City, 1994, 87–120. (AIAA 94–4262, 1994.)

[23] C. Bischof, L. Green, K. Haigler, and T. Knauff; Parallel calculation of sensitivity derivatives for aircraft design using automatic differentiation, *A Collection of Technical Papers, 5th AIAA/USAF/NASA/ISSMO Symposium on Multidisciplinary Analysis and Optimization*, AIAA, Panama City, 1994, 73–86. (AIAA 94–4261, 1994.)

[24] A. Griewank, C. Bischof, G. Corliss, A. Carle, and K. Williamson; Derivative convergence of iterative equation solvers, Technical Report MCS–P333–1192, Mathematics and Computer Science Division, Argonne National Laboratory, 1992.

[25] V. Vatsa and B. Wedan; Development of a multigrid code for 3–D Navier-Stokes equations and its applications to a grid-refinement study, *Comp. Fluids* **18**, 1990, 391–403.

[26] B. Baldwin and H. Lomax; Thin layer approximation and algebraic model for separated turbulent flow, AIAA-78–257, 1978.

[27] C. Hirsch; Numerical computation of internal and external flows, *Computational Methods for Inviscid and Viscous Flows, Vol. 2*, Wiley, New York, NY 1990.

[28] D. Johnson and L. King; A mathematically simple turbulence closure model for attached and separated turbulent boundary layers, *AIAA J.* **23**, 1985, 1684–1692.

[29] R. Abid, V. Vatsa, D. Johnson, and B. Wedan; Prediction of separated transonic wing flows with nonequilibrium algebraic turbulence model, *AIAA J.* **28**, 1990, 1426–1431.

[30] G. Hou and J. Sheen; Numerical methods for second-order shape sensitivity analysis with applications to heat conduction problems, *Inter. J. Numer. Meth. Engrg.* **36**, 1993, 417–435.

[31] G. Hou, A. Taylor, S. Mani, and P. Newman; Simultaneous aerodynamic analysis and design optimization, *Abstracts from Second U.S. National Congress on Computational Mechanics*, Washington, 1993, 130.

[32] S. Mani; *Simultaneous Aerodynamic Analysis and Design Optimization*, M.S. Thesis, Old Dominion University, Norfolk, 1993.

An Introduction to the Formation Theory of Active Materials

David L. Russell
Department of Mathematics
Virginia Tech
Blacksburg VA 24061, USA

1. Introduction

We will use the term *formation theory* to describe a certain class of optimal design and control problems for elastic systems. The subject matter concerns the deliberate alteration of the equilibrium configuration of an elastic structure by means of attached or embedded actuators responding to an external control signal. To the extent that the controlled configuration is maintained constant over an extended time interval, or varies slowly enough so that inertial effects can be ignored, we have what we designate as *static* formation theory. It will be seen as we proceed that, from the mathematical point of view, the subject has much in common with *parameter identification* problems in a similar context, the optimal design part of the work primarily involving the specification of parameter functions in a particular class of partial differential equations.

The name *formation theory* is suggested by a comparison of the viewpoint appropriate to this theory with that commonly evident in classical elasticity studies. In elasticity the term *deformation* usually refers to a displacement away from some nominal equilibrium configuration of the system. In the *control theory* of elastic systems such deformations are usually viewed as undesirable and controls are designed so as to suppress them. The emphasis is on *restoration* of the equilibrium state.

Formation theory takes as its point of departure the realization that in the age of self-actuated, or "smart", materials it may be both feasible and desirable to achieve a variety of different geometric configurations for the same elastic structure at different times in response to varying operational requirements, these to be realized through specific *actuator distribution patterns* within the structure, ordinarily independent of individual configurations but, perhaps, designed in the context of a particular set of configurations to be achieved, with certain frequencies, in the long term operation of the system, and equally specific *actuator control strategies* to be employed to realize a particular configuration at a given moment. Since the envisioned mode of operation involves repeated change from one configuration to another, one might reasonably use the term *reformation theory*, but we have decided at the outset not to revive ancient controversies.

We should note that most instances of structural deformation in the context of current and historical technology involve the process of *articulation*; different components of the structure undergo movement relative to each other via rotation about axes corresponding to hinges, axles, universal joints, etc., or are caused to translate relative to each other by means of rack and pinion systems, sliding tube structures, worm gear units, etc. There is a little doubt that articulation will continue indefinitely as the dominant modality by which structural deformation is accomplished. But articulation brings with it attendant vulnerabilities to damage; "sand" in the gears, frictional damage to joints, etc. It also introduces significant com plications in regard to aerodynamic or hydrodynamic efficiency, since the points of articulation tend to compromise achievement of smooth "streamlined" exterior surfaces. It is significant that natural selection has not endowed any creature with "pure" articulation; motion of body parts relative to each other is accomplished either through structural deformation, as in the contraction of muscles, e.g., or through hybrid systems involving articulated skeletal structures embedded in tissue systems capable of continuous, smooth deformation to accommodate articulate motions and providing a continuous protective covering for the articulated structure. In beginning the passage from the purely articulated structures, such as locomotives, chain drives, typewriters, etc., etc., characteristic of the early industrial age, to smoothly deformed structures only now feasible in the age of advanced materials, we are simply returning to a basic pattern which has long been the plan of choice in the natural world.

The requirements of formation theory include *model reformulation* to include various types of attached or embedded actuators, determination of optimal *actuator spatial distributions* - the *design* problem, and determination of optimal signals to induce actuation, the *control* problem. The last two tasks also involve the formulation and use of *optimality criteria* involving best *fit to target* notions and *actuator cost* considerations. None of these tasks can be carried out wholly in the abstract; they must take into account the application in which the active structure involved is to be employed. Some of these applications are shown in Figures 1–6 at the end of this expository article. Figure 1 shows a *non-articulated valve*, essentially an elastic plate capable of reconfiguration into a curved shell; it is envisioned that the flow of a gas or fluid is closed off when the plate is in its flat, unactuated, configuration while such flow does take place, past the edges of the valve, when the plate is actuated into the curved configuration. It is perhaps significant that, as this article was being written, media announcements were being made concerning what we may regard as a halfway step: magnetically activated valve systems requiring no cams. In Figure 2 we show what we call a *systolic pump*; a tube is constructed so as to be capable of expansion or constriction in its variouus segments so that a fluid can be forced through the tube without a conventional pump being involved. Figure 3 shows a possible space application; troublesome folding (i.e., articulated) antenna panels being replaced by a structure which is coiled when under actuation, presumably in effect during transportation into space, unrolling into a flat, rigid panel after deployment. Figure 4 can be described as showing a *conveyor belt*, but with no moving parts; a "wavelet", produced by localized actuation, moves along a stationary belt, causing a body on the belt to move from one end of the belt to the other (this is clearly very similar to the systolic pump of Figure 2). A *non-articulated sorting device* is shown in Figure 5; this consists of a ramp-like structure which can be actuated to match the level of an output slot appropriate to the current product moving along the ramp; this can be contrasted with devices on most current copier machine sorters, e.g., which involve gross motion of the sorting tray relative to a fixed output slot. Finally, Figure 6 indicates just one of many possibilities for silent propulsion, involving the use of paddle structures enabling a craft to "swim" through the water in much the same way as many aquatic creatures do, without any sort of noise-producing propeller device.

To begin the mathematical discussion let us consider a general, though over-simplified, formulation in the context of linear elasticity. As shown in Figure 7, we suppose that we have a body consisting of a *matrix* material along with embedded or attached *actuator* elements. The attachment/embedding process results in both systems being constrained to a *common geometry*, which we suppose parametrized by a vector (or vector function) w in an appropriate inner product space, the inner product being conveniently represented by $< u, v > \equiv v^*u$. Different w correspond to different configurations of the material. In the static problem, which is what we consider here, the combined matrix/actuator structure has an associated *potential energy* form which, in the simplest situation, can be written in the form

$$V = \frac{1}{2}\Big(w^*C^*ECw + (Dw + Bu)^*K(Dw + Bu)\Big).$$

Here $Cw \rightarrow ECw$ is the stress/strain relationship of the matrix material while $Dw + Bu \rightarrow K(Dw + Bu)$, involving the external control signal u, which in application alters the equilibrium configuration of the actuator structure, is the corresponding stress/strain relationship for the actuator structure. Minimization of V leads to the *formal equilibrium equation*:

$$(C^*EC + D^*KD)w + D^*KBu = 0\,.$$

The *stiffness operators* E and K of the respective matrix and actuator structures,

$$E = E(\Phi, \alpha, \Psi, \beta, \ldots), \qquad K = K(\Phi, \alpha, \Psi, \beta, \ldots)\,,$$

depend on, among other things, the *spatial densities* Φ, Ψ, ... of the various types of actuators present, and corresponding *orientation parameters* α, β, ... which describe the manner in which the actuators are installed in the matrix. The operators E and K are, in general, nonnegative, with K generally increasing as Φ, Ψ, ... increase while the operator E generally decreases due to replacement of matrix material by actuator material. Recapitulating the equilibrium equation, we have

$$\Big(C^*E(\Phi, \alpha, \Psi, \beta, \ldots)C + D^*K(\Phi, \alpha, \Psi, \beta, \ldots)D\Big)w = \\ -D^*K(\Phi, \alpha, \Psi, \beta, \ldots)Bu\,.$$

The presence of the operator $D^*K(\Phi, \alpha, \Psi, \beta, \ldots)D$, generally increasing with $\Phi, \Psi \ldots$ constitutes what we call the *actuator stiffening effect* whereby the material is made more rigid as actuator material, ordinarily having a higher modulus of elasticity than the matrix material, is added to the structure.

For fixed control inputs constituting the components of the vector u, the problem of using $\Phi, \alpha, \Psi, \beta, \ldots$ to match the system configuration w to the target configuration z is mathematically very similar to the parameter/coefficient *identification problem*, long familiar in the application of partial differential equations to hydrology and petrology, location of faults in elastic structures, etc.. On the other hand, for fixed $\Phi, \alpha, \Psi, \beta, \ldots$ determination of u so as best to match w to z is closer in spirit to the standard optimal control problem. Actually either of these tasks could be viewed in either light.

2. Formation theory of the Timoshenko beam

To better illustrate the concepts which we have, so far, introduced somewhat abstractly, we now turn to a specific example. The interested reader is referred to [1], [2], [3] for development of the Timoshenko model of the elastic beam, which incorporates both bending and shear effects. Here it will be enough to note that the Timoshenko potential energy takes, in the static case, the form

$$\mathcal{E} = \frac{1}{2}\int_0^L \left[EI(x)\big(\psi'(x)\big)^2 + K(x)\big(\psi(x) - w'(x)\big)^2 \right] dx\,,$$

where EI is the standard *bending modulus*, K is the *shear modulus*, w is the *lateral displacement*, and ψ is the *element rotation angle*, as illustrated in Figure 8 at the end of this article. Corresponding to these different displacement modes, we consider two types of actuators, which we call *bending actuators* and shear actuators. These are also shown in Figure 8. In the case of the bending actuators, which may be thought of as piezoelectric strips, or filaments, aligned with the elastic axis of the beam, contracting on one side of that axis as they expand on the other side, we here assume that the density dist ribution in the lateral direction is fixed (optimization questions in regard to this distribution are discussed in [4]). Letting that distribution be $F(y)$, we suppose the overall planar distribution to be $F(y)\Phi(x)$. In the case of the shear actuators, which are mounted in pairs at angles α and $\pi - \alpha$ relative to the elastic axis, one expanding as the other contracts to produce a shearing force in the beam, we describe the spatial actuator distribution in the x direction by means of the function $\Theta(x)$. It can then be shown that the potential energy of the actuator structure, not taking replacement effects into account here, takes the form

$$\hat{\mathcal{E}} = \tfrac{1}{2}\int_0^L \Phi(x)(\psi'(x), u(x)) \begin{pmatrix} ei(x) & -fi(x) \\ -fi(x) & gi(x) \end{pmatrix} \begin{pmatrix} \psi'(x) \\ u(x) \end{pmatrix} dx$$

$$+\int_0^L k(x)\Theta(x)\Big(\frac{\sin 2\alpha}{2}(\psi(x) - w'(x)) - v(x)\Big)^2\, dx\,.$$

Here $u(x)$ and $v(x)$ are, respectively, the bending actuator and shear actuator control signals and $k(x)$ is the stiffness parameter for the shear actuator material. The quadratic form appearing in the first integral is positive definite with $gi(x)$, $fi(x)$ and $ei(x)$ denoting the zero, first and second order moments of $e(x)F(y)$, where $e(x)$ is the stiffness parameter for the bending actuator material and $F(y)$ is the lateral actuator distribution density referred to earlier. The total potential energy is the sum of $\mathcal{E}$ and $\hat{\mathcal{E}}$. The equilibrium configuration, which depends on Φ, Θ, u and v in addition to other quantities regarded as fixed here, is obtained by minimization of this composite potential energy form.

Minimization of the total potential energy with respect to small variations in ψ and w, respectively, yields, along with corresponding boundary conditions, the equilibrium equations

$$\begin{aligned}
&\frac{d}{dx}\left(\big(EI(x) + \Phi(x)ei(x)\big)\frac{d\psi}{dx}\right)\\
&- \left(K(x) + k(x)\Theta(x)\left(\frac{\sin 2\alpha}{2}\right)^2\right)(\psi(x) - w'(x))\\
&- \frac{d}{dx}\Big(\Phi(x)fi(x)u(x)\Big) + k(x)\Theta(x)\left(\frac{\sin 2\alpha}{2}\right)v(x) = 0
\end{aligned}$$

$$\begin{aligned}
&\frac{d}{dx}\left(\left(K(x) + k(x)\Theta(x)\left(\frac{\sin 2\alpha}{2}\right)^2\right)\frac{dw}{dx}\right)\\
&- \frac{d}{dx}\left(\left(k(x)\Theta(x)\left(\frac{\sin 2\alpha}{x}\right)^2\right)\psi(x)\right)\\
&+ \frac{d}{dx}\left(k(x)\Theta(x)\left(\frac{\sin 2\alpha}{2}\right)v(x)\right) = 0\,.
\end{aligned}$$

A formation problem for the Timoshenko beam is posed by choosing a *target configuration* $z(x)$ and then selecting Φ, Θ, u, v so as to try to match the beam configuration $w(x)$ to this target configuration, subject to the constraints imposed by the above equilibrium equations and corresponding boundary conditions. Additional cost terms can be used to realize additional objectives. For example, if it is desired to approximate a given target configuration $z(x)$ for the lateral displacement while not inducing excessive shear in the structure, one is led to a class of problems of which the following is a prototype:

$$\min_{\Phi,\Theta,u,v} J(w, \psi, \Phi, \Theta, u, v)$$

where

$$
\begin{aligned}
&J(w, \psi, \Phi, \Theta, , u, v) \\
&\quad = \tfrac{1}{2}\int_0^L \Big(\gamma_1(x))(w(x) - z(x))^2 + \gamma_2(x)(\psi(x) - w'(x))^2\Big)\, dx
\end{aligned}
$$

+cost terms corresponding to unconstrained boundary components

(e.g., in the cantilever configuration: $\epsilon(w(L) - z(L))^2 + \epsilon_2(w'(L) - z'(L))^2$)

+control/actuator cost terms, as described below.

Initially, one might think of using a control/actuator cost of the form

$$\frac{1}{2}\int_0^L \left(\delta_1 u(x)^2 + \delta_2 v(x)^2 + \delta_3 \Phi(x)^2 + \delta_4 \Theta(x)^2\right)\, dx\,.$$

This is usable but rather artificial because the beam configuration is primarily influenced by the products $\Phi(x)u(x)$ and $\Theta(x)v(x)$ rather than by $\Phi(x)$, $\Theta(x)$, $u(x)$, $v(x)$ in linear combination. It turns out that a rather natural cost expression is obtained through use of what we call the *power/density criterion*. If one thinks, for example, of $u(x)$ and $v(x)$ in terms of voltage, since expended power is proportional to the product of voltage and current, and since the current is proportional, in general, to the product of voltage and conductivity and, since the latter may be assumed proportional to the quantity of actuator material used, expended power can be expressed in terms of an integral involving $u(x)^2\Phi(x)$ and $v(x)^2\Theta(x)$. The one we have most commonly used takes the form

$$\frac{1}{2}\int_0^L \left(\delta_1 \Phi(x)^2 + \delta_2 \Theta(x)^2 + \delta_3 u(x)^4 \Phi(x)^2 + \delta_4 v(x)^4 \Theta(x)^2\right)\, dx\,.$$

It then turns out that, for small actuator densities $\Phi(x)$ and $\Theta(x)$, the optimization problem can be effectively solved in terms of the products $f(x) = \Phi(x)u(x)$ and $g(x) = \Theta(x)v(x)$, leaving one with the elementary problems of minimizing $\delta_1\Phi(x)^2 + \delta_3 u(x)^4\Phi(x)^2$ subject to the constraint $f(x) = \Phi(x)u(x)$ and $\delta_2\Theta(x)^2$ $+\delta_4 v(x)^4\Theta(x)^2$ subject to the constraint $g(x) = \Theta(x)v(x)$. Considering only the first of these, one arrives at the problem

$$\min_u \left\{\delta_1\left(\frac{f(x)}{u(x)}\right)^2 + \delta_3\big(f(x)u(x)\big)^2\right\},$$

yielding $u(x)$ as a "bang-bang" control

$$u(x) = \pm\sqrt[4]{\delta_1/\delta_3}\,.$$

In fact, once $f(x)$ has been determined, the non-negativity of $\Phi(x)$ dictates that we should have

$$\Phi(x) = \frac{|f(x)|}{\sqrt[4]{\delta_1/\delta_3}}$$

while the sign of the optimal control $u(x)$ is specified as

$$u(x) = \text{sgn}(f(x))\sqrt[4]{\delta_1/\delta_3}\,.$$

The actuator density $\Theta(x)$ and the control $v(x)$ may be similarly expressed in terms of $g(x)$ once that function has been determined.

In travelling wave applications, such as the stationary conveyor belt described earlier, the desired configuration is created at different spatial locations at different times. Again taking the nonnegativity of the actuator density into account, exchanging the roles of the actuator and the control in the cost function, i.e.,

$$\frac{1}{2}\int_0^L \left(\delta_1 u(x)^2 + \delta_2 v(x)^2 + \delta_3 \Phi(x)^4 u(x)^2 + \delta_4 \Theta(x)^4 v(x)^2\right)\,dx\,,$$

leads to

$$\Phi(x) = \sqrt[4]{\delta_1/\delta_3}, \qquad u(x) = \frac{f(x)}{\sqrt[4]{\delta_1/\delta_3}}\,.$$

However convenient this may be, it is not easy to assign a natural significance to this modified cost functional.

3. A computational example; figures

Let us consider a formation problem of the sort that one might encounter in connection with the "conveyor belt" described in Section 1, but with a static target configuration rather than a moving one. In Figure 9 we show the target function as a piecewise linear function $z(x)$. We use a cost functional which includes the term

$$\frac{1}{2}\int_0^L \left(w(x) - z(x)\right)^2\,dx$$

and a control/actuator cost of the power/density type discussed in Section 2. Assuming that actuator volume is small it is possible to obtain the functions $f(x)$ and $g(x)$ introduced in Section 2 by solving standard linear-quadratic optimization problems, as in, e.g., [5], [6]. The smooth curve in Figure 9 corresponds to the achieved configuration assumed by a Timoshenko beam with the use of both bending and shear actuators. Figure 10 shows the computed bending actuator density and the bending control while Figure 11 gives the same data for the shear actuators. The bending and shear actuator densities are shown superimposed in Figure 12.

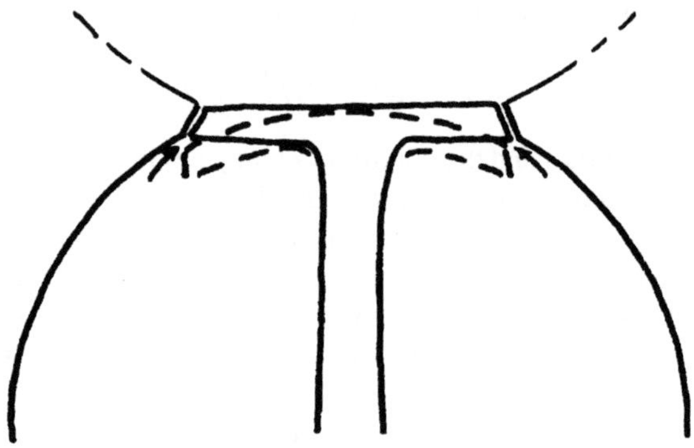

Figure 1. Non-articulated valve.

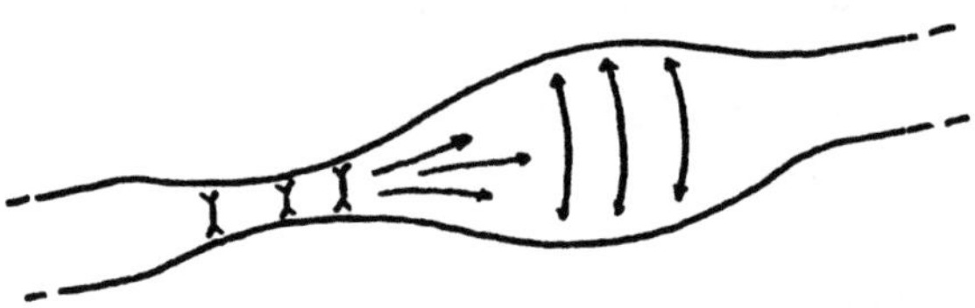

Figure 2. Systolic pump.

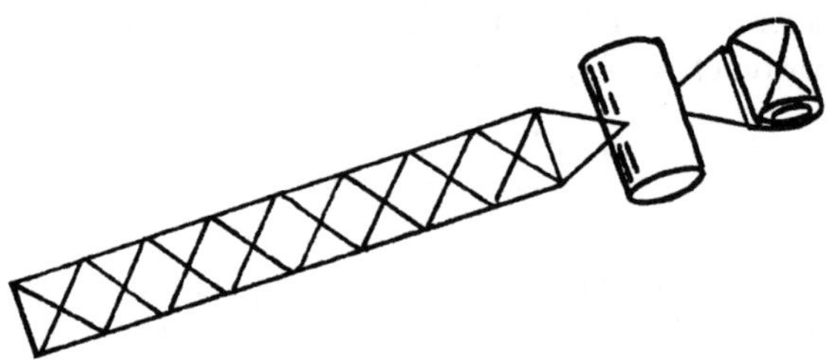

Figure 3. Retractable antenna.

Figure 4. Conveyor belt with no moving parts.

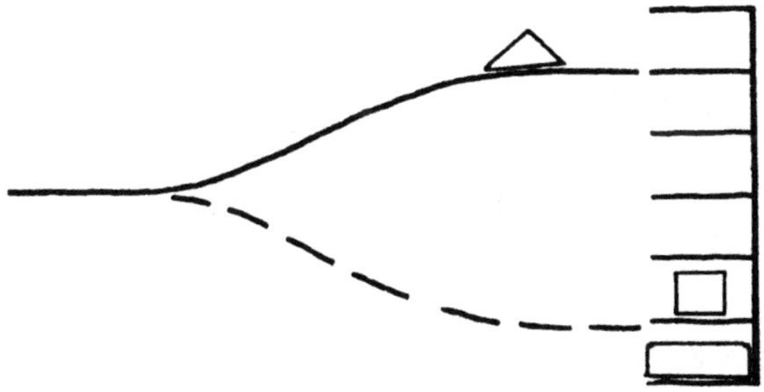

Figure 5. Non-articulated sorting device.

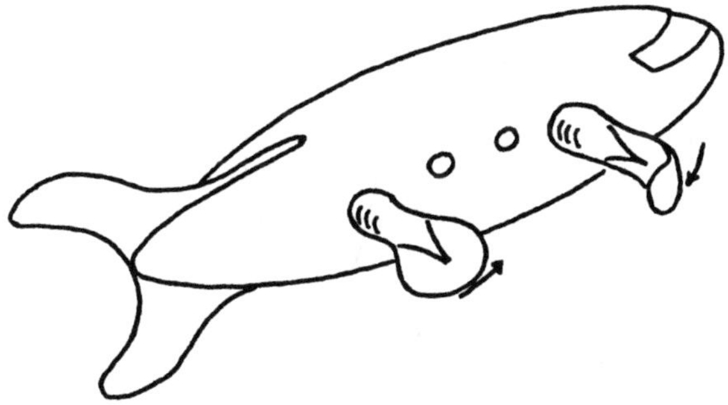

Figure 6. Silent propulsion.

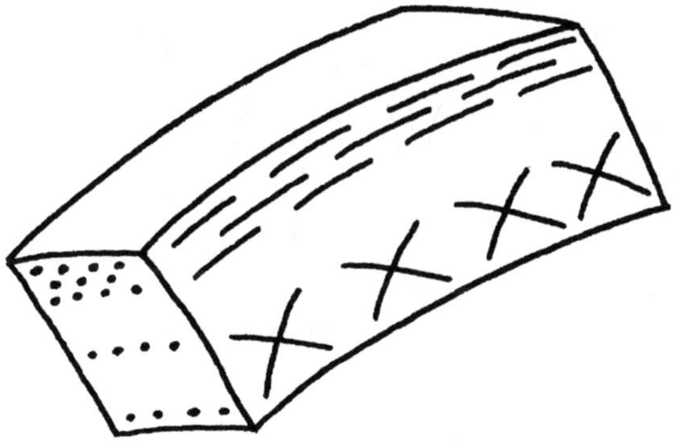

Figure 7. Matrix/actuator structure.

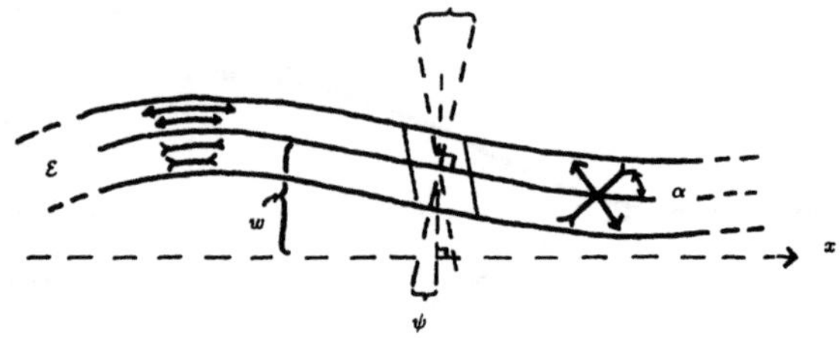

Figure 8. Timoshenko beam; bending and shear actuators.

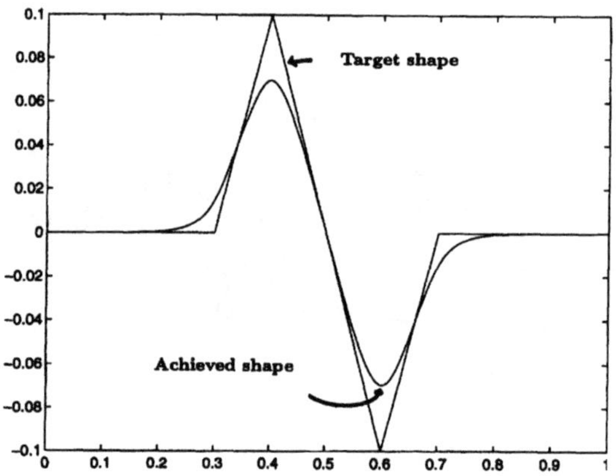

Figure 9. Target and achieved configuration; conveyor belt example.

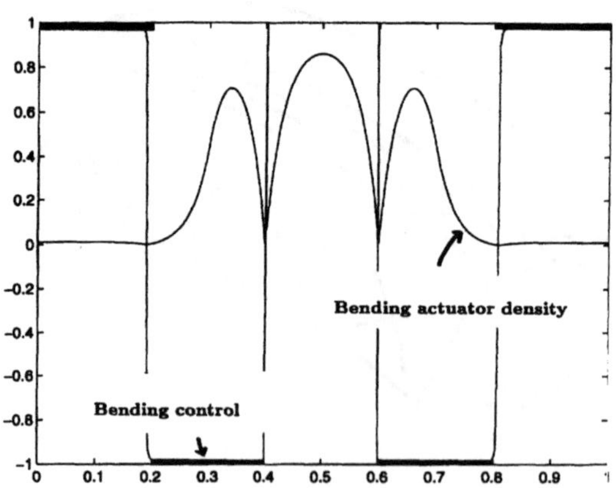

Figure 10. Computed optimal bending actuator density and control.

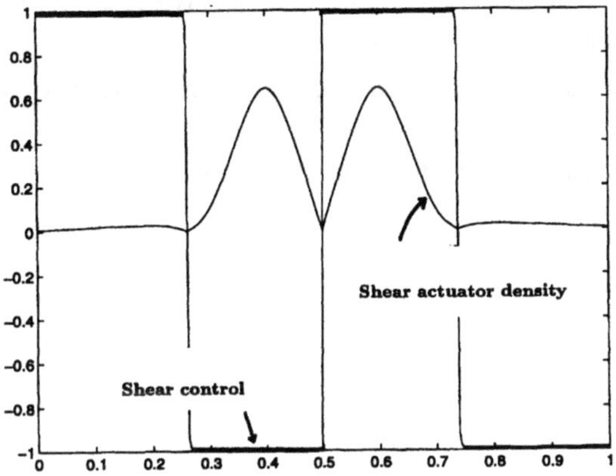

Figure 11. Computed optimal shear actuator density and control.

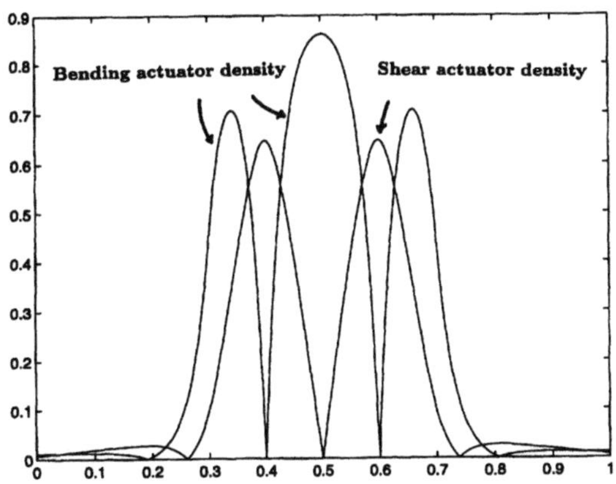

Figure 12. Superimposed bending and shear actuator densities.

References

[1] S. Timoshenko; *Vibration Problems in Engineering*, 2nd Ed., D. Van Nostrand, New York, 1937.

[2] Y. Fung; *Foundations of Solid Mechanics*, Prentice-Hall, Englewood Cliffs, 1965.

[3] D. Russell; On mathematical models for the elastic beam with frequency-proportional damping, in *Control and Estimation in Distributed Parameter Systems* (Ed. by H.T. Banks). SIAM, Philadelphia, 1993.

[4] D. Gao and D. Russell; An extended Timoshenko beam theory with applications to advanced materials, to appear.

[5] B. Anderson and J. Moore; *Linear Optimal Control*, Prentice-Hall, Englewood Cliffs, 1971.

[6] D. Russell; *Mathematics of Finite Dimensional Control Systems*, Marcel Dekker, New York, 1979.

Progress in Systems and Control Theory

Series Editor

Christopher I. Byrnes
Department of Systems Science and Mathematics
Washington University
Campus P.O. 1040
One Brookings Drive
St. Louis, MO 63130-4899

Progress in Systems and Control Theory is designed for the publication of workshops and conference proceedings, sponsored by various research centers in all areas of systems and control theory, and lecture notes arising from ongoing research in theory and applications control.

We encourage preparation of manuscripts in such forms as LATEX or AMS TEX for delivery in camera-ready copy which leads to rapid publication, or in electronic form for interfacing with laser printers.

Proposals should be sent directly to the editor or to: Birkhäuser Boston, 675 Massachusetts Avenue, Cambridge, MA 02139, U.S.A.

PSCT1 Computation and Control: Proceedings of the Bozeman Conference, Bozeman, Montana, 1988
K. Bowers and J. Lund

PSCT2 Perspectives in Control Theory: Proceedings of the Sielpia Conference, Sielpia, Poland, 1988
B. Jakubczyk, K. Malanowski, and W. Respondek

PSCT3 Realization and Modelling in System Theory: Proceedings of the International Symposium MTNS-89, Volume 1
M. A. Kaashoek, J. H. van Schuppen, and A. C. M. Ran

PSCT4 Robust Control of Linear Systems and Nonlinear Control: Proceedings of the International Symposium MTNS, Volume II
M. A. Kaashoek, J. H. van Schuppen, and A. C. M. Ran

PSCT5 Signal Processing, Scattering and Operator Theory, and Numerical Methods: Proceedings of the International Symposium MTNS-89, Volume III
M. A. Kaashoek, J. H. van Schuppen, and A. C. M. Ran

PSCT6 Control of Uncertain Systems: Proceedings of an International Workshop, Bremen, West Germany, June 1989
D. Hinrichsen and B. Mårtensson

PSCT7 New Trends in Systems Theory: Proceedings of the Università di Genova - The Ohio State University Joint Conference, July 9-11, 1990
G. Conte, A. M. Perdon, and B. Wyman

PSCT8 Analysis of Controlled Dynamical Systems: Proceedings of a Conference in Lyon, France, July 1990
B. Bonnard, B. Bride, J. P. Gauthier and I. Kupka

PSCT9 Nonlinear Synthesis: Proceedings of a IIASA Workshop held in Sopron, Hungary, June 1989
Christopher I. Byrnes and Alexander Kurzhansky

PSCT10 Modeling, Estimation and Control of Systems with Uncertainty: Proceedings of a Conference held in Sopron, Hungary, September 1990
Giovanni B. Di Masi, Andrea Gombani and Alexander B. Kurzhansky

PSCT 11 Computation and Control II: Proceedings of the Second Conference at Bozeman, Montana, August 1990
Kenneth Bowers and John Lund

PSCT 12 Systems, Models, and Feedback: Theory and Applications. Proceedings of a U.S.-Italy Workshop in honor of Professor Antonio Ruberti, held in Capri, Italy, 15-17, June 1992
A. Isidori and T.J. Tarn

PSCT 13 Discrete Event Systems: Modelling and Control. Proceedings of a Joint Workshop held in Prague, Poland, August 1992
S. Balemi, P. Kozak, and R. Smedinga

PSCT 14 Essays on Control: Perspectives in the Theory and its Applications
H. L. Trentelman and J. C. Willems

PSCT 15 Computation and Control III: Proceedings of the Third Bozeman Conference, Bozeman, Montana, August 1992
K. L. Bowers and J. Lund

PSCT 16 Set-Valued Analysis and Differential Inclusions, A Collection of Papers from a Workshop held in Pamporovo, Bulgaria, September 1990
A. B. Kurzhanski and Vladimir M. Veliov

PSCT 17 Advances in Nonlinear Dynamics and Control: A Report from Russia
A. B. Kurzhanski

PSCT 18 Modeling Techniques for Uncertain Systems, Proceedings of a Conference held in Sopron, Hungary, July 1992
A. B. Kurzhanski and Vladimir M. Veliov

PSCT 19 Optimal Design and Control, Proceedings of the Workshop on Optimal Design and Control, Blacksburg, Virginia, April 8-9, 1994
J. Borggaard, J. Burkardt, M. Gunzburger, and J. Peterson